Applied Statistics for Social and Management Sciences

Abdul Quader Miah

Applied Statistics for Social and Management Sciences

 Springer

Abdul Quader Miah
Asian Institute of Technology
Bangkok
Thailand

ISBN 978-981-10-0399-8   ISBN 978-981-10-0401-8   (eBook)
DOI 10.1007/978-981-10-0401-8

Library of Congress Control Number: 2015960235

Printed on acid-free paper

This Springer imprint is published by Springer Nature
The registered company is Springer Science+Business Media Singapore Pte Ltd.

When you can measure what you are speaking about, and express it in numbers, you know something about it.

When you cannot measure it, when you cannot express it in numbers, your knowledge is of a meager and unsatisfactory kind. It may be the beginning of knowledge, but you have scarcely in your thoughts advanced to the stage of science.

Lord Kelvin

Preface

The beauty of statistics lies in the fact that almost all planning, management, business and engineering problems, and phenomena require statistics as an analytical tool to help in subsequent decision making. The users of these and other sciences need statistics equally. To most of the users, the applied parts and not the fine mathematics of the subject are of great importance. They have little opportunity and time in going to the sophisticated theories of statistics. At the same time, they must be in a position to interpret the technical details of the subject in order to correctly use the tool. This is important to guard against misuse of statistics. It is recognized that the users of statistics come largely from various disciplines, some having no strong mathematical background. Emphasis of such users is on the application of statistics in the real and practical problems.

My engineering background, training in planning and management, and teaching of statistics in the Asian Institute of Technology, Bangkok, have helped me in identifying the need of the subject in a wide variety of applications in the real-world perspective and the problems the students face in handling the statistical techniques. To overcome these, a lot of techniques and ideas were used and found successful in the classroom. The purpose of writing this text book is to transfer these practical aspects into a reference book from which a wider range of readers can benefit. While I started teaching statistics to the students coming from various disciplines, the students were found fearful of the subject. My constant efforts were, therefore, to make the subject interesting to them rather than a fearful one. This book will reflect to some extent these efforts.

The coverage of the book is evident from the Table of Contents. The chapter "Index Numbers" with their construction techniques, applications, and interpretations is not normally available in ordinary statistics books. Yet the planning and management students need it frequently and to them it is of a great help. This has been added for their benefit.

Bangkok   Abdul Quader Miah
April 2015

Contents

Chapter 1
Basics

Abstract Statistics deals with data. The basics of collection, summarization, presentation, description, and analysis of data are explained. Knowledge of the rules and the levels of measurement are important because each statistical technique is appropriate for data measured only at certain levels. So levels of measurement are introduced. Analysis of data that may vary will depend on the variables. So types of variables according to suitability are introduced.

Keywords Statistics · Data · Measurements · Variable

1.1 History

The subject "statistics" as we find it today has emerged from three different sciences. The first science was "Staatenkunde" which involved a collection of information on history, resources, and military expertise of nations. This science developed from the felt need of gathering information on resources and other aspects of the states. For example, Aristotle gathered and compiled information on 158 city states.

The second science was the "political arithmetic" dealing with population estimates and mortality. The present-day demography derived its roots from this science. This science developed gradually but its refined methodology has developed only recently.

The third science was "a calculus of probability" dealing with mathematical theorems and techniques for problems involving uncertainty. The theory of probability is the foundation of modern statistics.

© Springer Science+Business Media Singapore 2016   1
A.Q. Miah, *Applied Statistics for Social and Management Sciences*,
DOI 10.1007/978-981-10-0401-8_1

1.2   Statistics

Statistics may be defined in several ways. Some authors say that statistics are classified facts, especially numerical facts, about a particular class of objects. Others say that statistics is the area of science that deals with the collection of data on a relatively small number of cases so as to form conclusions about the general case.

Another definition says that statistics is the science of collecting, simplifying, and describing data as well as drawing conclusions. Wikipedia says that statistics is a mathematical science pertaining to the collection, analysis, interpretation, and presentation of data. Statistics is used by a wide variety of academic disciplines.

Statistics has two distinct branches—descriptive statistics and inferential statistics. Descriptive statistics deals with collection, summarization, presentation, and description of data and is sometimes called the primary analysis. Inferential statistics deals with further analysis of data in order to draw conclusions and is sometimes called secondary analysis.

1.3   Contents of Statistics

In the above section the definition of statistics has been outlined. In this section the contents of statistics is summarized. Statistics deals with data relating to

- collection,
- summarization,
- presentation,
- description, and
- analysis.

It is evident that any exercise in statistics concerns data. Thus data are the central requirement of any statistical work. Hence comes the question of data and their collection.

Data may be collected from secondary sources such as census reports, other documents, previous studies, etc. Data that are not readily available need to be collected from the field. There are recognized ways/techniques of data collection from the field. The important ones include

- structured/semi-structured interviews,
- standardized questionnaires,
- observation schedules,
- direct measurements, and
- experiments.

In the structured/semi-structured interview, the points of interest on which information is sought, are noted on a piece of paper. Then questions are asked to the respondents or a guided discussion is held with the respondent. The points of discussion are noted while discussing. Afterwards the information is summarized. Sometimes, group discussions are arranged and the outcomes of the discussions are recorded.

A more formalized way of collecting data from the field is the use of standardized questionnaires. A wide variety of information can be collected through this technique. Questions on selected topics are previously formulated. With the questionnaire in hand, the survey personnel go to the field and the preset questions are asked to the respondent exactly in the same manner they were set. The answers are recorded in the manner also previously prescribed in the questionnaire.

Some of the information can be obtained by simple observations, for example, housing conditions. In such cases an observation schedule may be used. The prescribed schedules (different for different purposes) are used in order to have uniformity in the recordings. This facilitates subsequent data processing and analysis.

Sometimes direct measurements of data are possible. If possible, this gives the most accurate information. Measurements of plot size, house size, road width, etc., are examples where direct measurements are possible. Direct measurements are usually done for data generated in laboratory experiments as well as in some field experiments. Devices such as scales, tapes, surveying equipments, etc., are used in measurements. Although relatively little judgment is involved in this technique, the accuracy of measurements depends on the skills and efficiency of the person recording the measurements.

Considerable data is also generated from experiments conducted by researchers in the fields, laboratories, and manufacturing processes (see Chap. 17).

1.4   Data

Based on the source, data may be classified as secondary and primary data. Secondary data are those that are obtained from the available reports, records, and documents. Primary data are those that are not readily available and as such are collected from the field or experiments. In any statistical problem when secondary data are used, care should be taken to see their relevancy and accuracy.

On the basis of the use of units of measurement or type of measurement, data may be classified as

– categorical,
– ranked, and
– metric.

Categorical data are those in which individuals are simply placed in the proper category or group, and the number in each category is counted. Each item must fit into exactly one category.

Example 1

Sex	Number
Male	706
Female	678

In the above example, male and female are categories. Only their frequencies are counted. No other measurement units are used to identify these.

Ranked data is also categorical data. But this type of data has order among the categories. In addition to categories, ranking or ordering is inherent in the data.

Example 2

People are categorized on the basis of income levels such as low, lower middle, middle, upper middle, and high income. The low income people represent the lowest category. The high income people represent the highest category. The intermediate ones follow accordingly. These are ranked data. Although low, lower middle, middle, upper middle, and high income are categories, some ranking is inherent in the categories.

Metric data are those that need certain units of measurements. These data have values that are continuous over a certain range, and are expressed with the help of standard units of measurements.

Example 3

Expenditures of AIT students on food Bahts 2021, 1850; agricultural productivity 2000 kg/rai, velocity 10.52 m/s, etc.

Often, metric data are converted to rank data. Suppose the individual incomes of the people of a city are known to us. We can categorize them in the following way (Table 1.1).

Table 1.1 Income category

Income range (US \$/month)	Category
Up to 200	Poor income
More than 200 but up to 500	Middle income
More than 500	High income

1.5 Level of Measurement

Measurement constitutes the process of assigning a value to the data and the rules defining the assignment of an appropriate value determine the level of measurement. The levels of measurement are distinguished on the basis of ordering or distance properties inherent in the measurement rules. Knowledge of the rules and the levels of measurement is important because each statistical technique is appropriate for data measured only at certain levels.

On the basis of traditional classification, four levels of measurements are identified. These are

– nominal,
– ordinal,
– interval, and
– ratio.

Nominal level is the lowest in the levels of measurements. Each value is a distinct category. The value itself serves merely as a label or name (hence, "nominal" level) for the category. No assumption regarding ordering or distances between categories is made. The real number properties (addition, subtraction, multiplication, division) are not applicable to the nominal level of measurement. Categorical data fall under this level of measurement.

Example 4
Names of cities—Bangkok, Manila, Dhaka.

Ordinal level of measurement is derived from nominal level with the addition that in the ordinal level of measurement, it is possible to rank-order all the categories according to certain criterion. Although ordering property is present in the ordinal level of measurement, the distance property is absent. Consequently, the properties of real number system cannot be used to summarize relationships of an ordinal level variable. Ranked data fall under this level of measurement.

Example 5
Education levels are measured as primary, secondary, higher secondary, and tertiary.

In this example primary, secondary, higher secondary, and tertiary are categories. But there is a meaningful ordering also. So the values would have ordering property. But we cannot say what is the distance between primary and secondary, and between secondary and higher secondary, and so forth. This means that there is no distance property.

Interval level is the third in the level of measurement. In addition to ordering, the interval level measurement has the property that the distances between the categories are defined in terms of fixed and equal units. It is important to note that in the interval level measurement we study the difference between things and not their proportionate magnitude. The interval level measurement has ordering and distance properties but the inherently determined zero point is not available.

It would be noted that in social science, a true interval level measurement is difficult to be found. If the distances between categories can be measured, a zero point can also be established. Another point is to be kept in mind. Statistics developed for one level of measurement can always be used with higher level variables, but not with variables measured at a lower level. An appropriate example in this case is the median. The median assumes an ordinal level of measurement. But it can be used with interval level or ratio level scales also. However, it cannot be used with variables measured at nominal level. Metric data fall under this level of measurement.

Example 6
Readings in a thermometer. The difference between 400 and 41 °F is the same as the difference between 80 and 81 °F. But 80 °F does not mean the double of 40 °F.

The ratio level measurement has all the properties of an interval level measurement. It has an additional property, i.e., well-defined zero point. The zero point is inherently defined by the measurement scheme. Consequently, the distance comparisons as well as ratio comparisons can be made. Any mathematical manipulation appropriate for real numbers can be applied in ratio-level measurements. Also, all statistics requiring variables measured at interval level are appropriate for use with variables at ratio level. Metric data fall under this level of measurement.

Example 7
Height of student—155, 160 cm; income Baht 4000, 20,000 per month. In all cases measurements start from zero.

In this example, the distance between 155 and 160 cm, and between Baht 4000 and Baht 20,000 is well defined. In each case zero point is specified. It can precisely be said that the income of Baht 20,000 is exactly five times the income of Baht 4000. This type of comparison cannot be made for temperature measured in Fahrenheit.

There are other typologies for measurements which we come across quite frequently. The following are important.

– Quantitative—a fixed unit of measurement is defined (interval and ratio levels).
– Qualitative—a fixed unit of measurement is not defined (nominal and ordinal levels).

1.6 Variable

The term "variable" is used to mean something that varies. But this definition is an oversimplification. The variable itself does not vary. Rather its values vary. In other words, a variable may have some values and at a certain point of time or at a certain situation it may assume a specific value. For example, "age" is a variable. It can assume any value. But the age of a particular person at a certain point of time is a fixed value. After some time this value is changed. Another example of a variable is the strength of bricks. A particular brick may have, for example, strength of 200 psi, another brick may have strength of 250 psi, and so on.

Three types of variables can be distinguished in statistical problems. These are

– numerical variable,
– categorical variable,
– rank variable.

A numerical variable is a variable whose possible values are numbers. Again, in numerical variables two types can be distinguished—a discrete variable and a continuous variable. Discrete variables can assume only integer values. There is a definite jump from one value to another. An example of a discrete variable is the number of occupants in houses. The number of occupants can be 2, 3, 5, 7, and so forth, but it cannot be a fractional number such as 3.52. The mean for several houses may however, be expressed as a fractional number. A continuous variable is one that can take any possible value within a certain range. The values are on a continuous scale of measurement. An example of a continuous variable is the speed of motor cars. The speed can be 90, 95 km/h or even 95.35 km/h.

Careful distinction should be made between a discrete variable and a continuous variable. The scale of measurement of a discrete variable is discontinuous. But the scale of measurement of a continuous variable is continuous. The scale of measurement of the number of occupants is discontinuous between 2 and 3, between 3 and 4, and so on. The scale of measurement of the speed of motor cars is continuous. There are all possible values between 95 and 96 km/h. It may be argued that when the speeds are expressed as 95.32 and 95.33 km/h., there is discontinuity between 95.32 and 95.33 because the intermediate values could be 95.321, 95.322, 95.323 … 95.329, 95.33. The point is that while recording we record usually up to two figures after the decimal point. This does not mean that the intermediate values are absent. A measured variable is discrete, because we cannot be infinitely precise in measuring the values of a variable.

A categorical variable is a variable whose values are expressed as a few categories, usually stated in words rather than in numbers. A categorical variable with only two values is called a dichotomous variable. A categorical variable with more than two values is called a polytomous variable. A categorical variable may come from two sources. A variable can be naturally categorical or a numerical variable can be converted to a categorical variable. Examples of natural categorical variables are satisfaction (with values highly satisfied, satisfied, neutral, dissatisfied, highly dissatisfied), response (yes, no, no answer). Examples of variables converted to categorical variables are exam scores (0–100) converted to grades A, B, C, D; years of education converted to primary, secondary, higher secondary, bachelor, doctoral degrees.

Rank variables are those whose values are ranks. Rank variables can also come from two sources—naturally rank variable and numerical variable converted to rank variable. Examples of natural rank variables are class ranking (first, second, third, fourth), priority (first priority, second priority, third priority). Example of a

numerical variable converted to rank variable is achievement test scores of 15
subjects reduced to ranks from 1 to 15.

1.7   Notation

Several notations are used in statistics. Some Greek letters are universally used in
statistics. These are not explained here but will be shown in places of their
occurrences later. A complete list of symbols together with their meanings is shown
in the appendix. But one notation is considered here. This is summation notation. Its
uses are more frequent and the symbol is $\sum$ (summation). As an example, if we
want sum of n observations, i.e., $x_1 + x_2 + x_3 + x_4 + \ldots + x_n$, we express it as $\sum x_i$.
This means

$$\sum x_i = x_1 + x_2 + x_3 + \cdots + x_n$$

This can be generalized. If $f(x_i)$ is a function of x, then

$$f(x_i) = f(x_1) + f(x_2) + f(x_3) + \cdots + f(x_n)$$

Suppose the set of observations for two variables x and y are given as shown
below:

X	Y
3	2
4	1
1	0
2	1
5	−3

(a)
$$\sum x_i = 3 + 4 + 1 + 2 + 5$$
$$= 15$$

(b)
$$\sum y_i = 2 + 1 + 0 + 1 - 3$$
$$= 1$$

(c)
$$\sum x_i^2 = (3)^2 + (4)^2 + (1)^2 + (2)^2 + (5)^2$$
$$= 9 + 16 + 1 + 4 + 25$$
$$= 55$$

$$\sum (y_i - 2)^2 = (2 - 2)^2 + (1 - 2)^2 + (0 - 2)^2 + (1 - 2)^2 + (-3 - 2)^2$$

(d) $\qquad = 0 + 1 + 4 + 1 + 25$

$\qquad = 31$

$$\sum (x_i y_i) = 3 * 2 + 4 * 1 + 1 * 0 + 2 * 1 + 5 * (-3)$$

(e) $\qquad = 6 + 4 + 0 + 2 - 15$

$\qquad = -3$

Problems

1.1 Name the type of variables in the following cases:
Income of people; area of plots; fish production; quantity of water; attitude toward a statement (agree or disagree); attitude toward a statement (strongly agree, agree, neutral, disagree, strongly disagree); attitude (agreed, neutral, not agreed); sex; outcome of a task (pass/fail); result of exam (pass/fail); years of schooling; proportion; ranking in participation; attitude measured on a ten-point scale.

1.2 Let $x_1 = 5$; $x_2 = 6$; $x_3 = 2$; $x_4 = -2$; $x_5 = -1$;
Find

(a) $\sum x_i$; (b) Σx_i^2; (c) $\{\sum x_i\}^2$;

(d) $\sum (x_i - 1)^2$; (e) $\left\{\sum (x_i - 2)^2\right\}^2$;

1.3 Two variables are y_1 and y_2. When can the following be true?

$$\sum y_i^2 = \left\{\sum y_i\right\}^2$$

1.4 Three variables are x_1, x_2, and x_3. Find out a simple relationship between the three variables, given

$$\sum x_i^2 = \left\{\sum x_i\right\}^2$$

If $x_1 = -10$, $x_3 = 30$; find x_2.

1.5 Five sets of observations for two variables x and y are given

Observation	x	y
1	10	15
2	12	18
3	5	7
4	20	25
5	30	40

Compute

(a) $\sum x_i y_i$; (b) $\Sigma x_i \, \Sigma y_i$ (c) $\sum x_i^2 y_i^2$ (d) $\left\{ \sum (x_i y_i)^2 \right\}$

1.6   Four sets of observations for three variables x, y, and z are shown:

Observation	x	y	z
1	5	3	0
2	6	2	5
3	4	0	8
4	10	8	8

Compute

(a) $\sum x_i y_i z_i$;
(b) $\sum x_i \sum y_i \sum z_i$
(c) $\sum (x_i - 1) \sum (y_i - 3) \sum (z_i - 2)$

1.7   Data of two variables x and y are given:

x	y
10	15
20	30
25	5
30	0

Express the following in notation form:

(a) $15 + 30 + 5 + 0$
(b) $10 * 15 + 20 * 30 + 25 * 5 + 30 * 0$
(c) $10^2 * 15^2 + 20^2 * 30^2 + 25^2 * 5^2 + 30^2 * 0^2$
(d) $(10^2 + 20^2 + 25^2 + 30^2)(15 + 30 + 5 + 0)$

Answers

1.2   (a) 10; (b) 70; (c) 100; (d) 55; (e) 2500
1.3   if, either $y_1 = 0$ or $y_2 = 0$
1.4 $x_1 x_2 + x_2 x_3 + x_3 x_1 = 0$; 15
1.5   (a) 2101; (b) 8085; (c) 1,760,381; (d) 4,414,201
1.6   (a) 700; (b) 59; (c) 286
1.7   (a) $\sum y_i$; (b) $\sum x_i y_i$; (c) $\sum x_i^2 y_i^2$; (d) $(\sum x_i^2)(\sum y_i)$

Chapter 2
Presentation of Statistical Data

Abstract Data are collected often in raw form. These are then not useable unless summarized. The techniques of presentation in tabular and graphical forms are introduced. Some illustrations provided are real-world examples. Graphical presentations cover bar chart, pie chart, histogram, frequency polygon, pareto chart, frequency curve and line diagram.

Keywords Presentation · Table presentation · Graph presentation · Types of presentation

Data are often collected in raw form. These are then not useable unless summarized. There are certain guidelines for data summarization such as
   summarization

– should be as useful as possible,
– should represent data fairly, and
– should be easy to interpret.

After collection of data (primary or secondary), it is necessary to summarize them suitably and present in such forms as can facilitate subsequent analysis and interpretation. There are two major tools/techniques for presentation of data as follows:

– Presentation in tabular form
– Presentation in graphical form.

2.1 Tabular Presentation

Data may be presented in the form of statistical tables. In one table only simple frequencies can be shown. Also, in the same table cumulative frequencies, relative frequencies, and cumulative relative frequencies can be shown. Relative frequencies and cumulative frequencies are defined as follows:

Relative frequency: It means the ratio of the frequency in the category of concern to the total frequency in the reference set.

$$\text{Relative frequency of } X_i = \frac{\text{actual frequency of } X_i}{\text{sum of all frequencies}}$$

$$= \text{proportion}$$

Relative frequency of X_i (%) = proportion * 100

Cumulative frequency of X_i = sum of all frequencies of all values up to and including X_i.

In the same table the simple frequencies combined with one or more but not all, of the cumulative frequencies, relative frequencies, and cumulative relative frequencies may be shown. Table 2.1 serves as an example of tabular presentation of simple frequency distribution. In the same table all have been shown together for illustrative purposes.

In Table 2.1 only one dimension has been shown. But intelligently more than one dimension may also be shown in the same table. This is demonstrated in Table 2.2.

There are advantages in such table presentation. One advantage is that more than one parameter (here education and occupation) can be shown in the same table. This serves as a concise presentation. Another advantage is easy comparison and interpretation. In Table 2.2 both fathers' and mothers' situations against each of the variables of the parameters can be readily compared and interpreted.

In the examples provided in Tables 2.1 and 2.2, frequencies have been shown against each categories. The categories served as groups, which were predetermined on the basis of certain criteria. More specifically, these are presentations of categorical data. But in practice it often becomes necessary to group the metric data to form some groups or categories or classes. Here the point of interest is to see how well the data can be grouped or classed. There are certain guidelines that help in grouping the data. The guidelines are

Table 2.1 Types of organizations in which AIT alumni (1960–1987) are working

Type of organization	Freq.	Cumul. freq.	Relative freq.	Cumul relative freq.
Govt. office at central level	16	16	13.01	13.01
Govt. office at regional level	5	21	4.07	17.08
Public state enterprise	12	33	9.76	26.84
Private enterprise	32	65	26.02	52.86
Educational institution	31	96	25.20	78.06
Nongovt. organization	8	104	6.50	84.56
International organization	9	113	7.31	91.87
Others	10	123	8.13	100.00
Total	123		100.00	

Table 2.2 Educational and occupational status of AIT alumni (1960–1987) parents

Status and levels	Father		Mother	
	f	%	f	%
Education				
Formal education	56	10.6	115	21.8
Primary education	105	19.9	182	34.5
Secondary education	119	22.5	112	21.2
Post secondary schooling	67	12.7	43	8.1
College education	70	13.2	38	7.2
University education	108	20.5	29	5.2
NA	3	0.6	9	1.7
Total	528	100.0	528	100.0
Occupational				
Farming	71	13.4	63	11.8
Commerce	119	22.5	97	18.4
Industry	16	3.0	9	1.7
Public service	207	39.2	45	8.5
Private service	75	14.2	36	6.8
Teaching	4	0.8	4	0.8
Multiple job	13	2.5	1	0.2
Housework	0	0	204	38.6
Other	15	2.8	9	1.7
NA	8	1.5	60	11.4
Total	528	100.0	528	100.0

Source AIT, AIT Alumni 1961–1987

- every score must fit into exactly one class,
- intervals should be nice, and
- classes should preferably be of the same width.

A frequency distribution is a more compact summary of data than the original observations. To construct a frequency distribution, we need to divide the range of the data into intervals known as classes. As already mentioned, the class intervals, whenever possible, should be of equal width, to enhance the visual information in the frequency distribution. We need to apply our judgment in selecting the number of classes in order to give a reasonable display. The number of classes used depends on the number of observations and the amount of dispersion in the data. Too few or too many classes are not very informative. It has been found that the number of classes between 5 and 20 is satisfactory in most of the cases. Also, the number of classes should increase with the number of observations. The number of classes may be chosen to be approximately equal to the square root of the number of observations. Thus, no. of classes = $\sqrt{n}$ (approx.).

It is convenient to use a single nonoverlapping type of class for all types of data (discrete or continuous). Look at the following example (data in Table 2.3).

Table 2.3 Crushing strength (psi) of bricks

215	147	296	230	215	150	171	215	211	228
155	236	267	192	204	185	126	212	198	200
213	224	192	210	231	196	198	221	210	215
257	193	208	271	244	278	213	195	224	220
170	181	226	178	173	246	181	251	287	248
218	217	250	200	210	226	284	230	200	207
210	231	158	249	258	214	250	224	228	160
184	215	137	208	185	219	215	203	204	230
249	164	214	217	233	185	222	237	224	219
165	268	221	243	227	240	233	208	225	201

No. of observations	100
Approx. no. of classes	$\sqrt{100} = 10$
The highest value	296
The smallest value	126
Range	$296 - 126 = 170$

If the lowest and the highest values in the frequency distribution are chosen to be 120 and 300, respectively, the range becomes 180 (i.e., 300 – 120). So, nine classes are chosen and consequently the class width is 180/9 = 20. The results of this classification are shown in Table 2.4.

Another tabular presentation often used is cross tabulation. Cross tabulation is a joint frequency distribution of different values of two (or more) variables. Table 2.5 is an example of cross tabulation of two variables, namely "Academic Divisions" and "Levels of Satisfaction with Selected Aspects." The figures in the table are indexes of satisfaction.

Very frequently, statistical test results are also presented in the form of tabular presentation. Table 2.6 is an example.

Table 2.4 Crushing strength (psi) of bricks (classes)

Class interval	Frequency
$120 \le x < 140$	2
$140 \le x < 160$	4
$160 \le x < 180$	7
$180 \le x < 200$	13
$200 \le x < 220$	32
$220 \le x < 240$	24
$240 \le x < 260$	11
$260 \le x < 280$	4
$280 \le x < 300$	3
Total	100

Table 2.5 AIT alumni's satisfaction with selected aspects of thesis research conducted at AIT

Academic division/period	Level of satisfaction with selected aspects		
	Flexibility in topic selection	Practical applicability	Support service
Division			
AFE	0.78	0.71	0.75
CA/CS	0.58	0.64	0.63
CRD/SE	0.57	0.52	0.50
EE	0.75	0.70	0.73
ET	0.76	0.76	0.74
GTE	0.76	0.70	0.67
HSD	0.76	0.71	0.66
IEM	0.75	0.63	0.65
SEC	0.66	0.64	0.69
WRE	0.67	0.67	0.70
Period			
Before 1965	0.58	0.73	0.55
1966–1970	0.58	0.56	0.60
1971–1975	0.70	0.63	0.65
1976–1980	0.72	0.67	0.67
1981–1985	0.74	0.69	0.72
1986–1997	0.72	0.71	0.70
Average	0.71	0.68	0.68

Source AIT, AIT Alumni 1961–1987

The main purpose of tabular presentation, in fact of all aspects of presentation, is that summary or presentation should be informative and meaningful. It should facilitate interpretation and subsequent analysis. To highlight this aspect, data in Table 2.7 are used.

During the several years from 1790 to 1984, both the urban and rural population of the United States maintained a steady increase. From the figures in column 4, it is also clear that the percentage of urban population continued to increase steadily. This indicates, together with the figures in columns 2 and 3 that the urban population grew at a faster rate compared to the rural population.

The underlying purpose in making the analysis of the data presented in Table 2.7 is to demonstrate that every table prepared should convey some interpretation message. Mere presentation of data in some tables is meaningless unless such purpose is served.

Table 2.6 Attitudes toward necessity of education for men—results of multiple regression analysis

X	B	$\alpha(\beta)$	T-value	C.l. (%)
A	3.9745	0.0963	41.270	100.00
HH	−0.0171	0.0081	− 2.104	94.00
Occupation				
Agr	−0.0404	0.0491	− 0.822	59.00
Trd	−0.1678	0.0819	− 2.046	96.00
Srv	−0.0146	0.0518	− 0.282	22.00
Age years	0.0017	0.0014	1.202	77.00
Sch years	0.0089	0.0054	1.645	90.00
Test statistics				
R^2 = 184 df: Regression = 6				
F-value = 2.48 Residual = 66				
Confidence level = 97.00 %				

Notes
X = independent variables
a = constant term
β = coefficient
$\alpha(\beta)$ = standard error of β
C.l = confidence level
HH = household member
Agr = agriculture
Trd = trade
Srv = service
Rs = respondent
sch = schooling

Table 2.7 Urban population growth in the United States

Year	Urban population (million)	Rural population (million)	Urban as a percentage of total population
1790	0.2	3.7	5
1810	0.5	6.7	7
1830	1.1	11.7	9
1850	3.5	19.6	15
1870	9.9	28.7	25
1890	22.1	40.8	35
1910	42.0	50.0	46
1930	69.0	53.8	56
1950	96.5	54.2	64
1970	149.8	53.9	73
1975	155.9	57.3	73
1980	170.5	56.1	74
1984	179.9	56.2	76

Source U.S. Bureau of the census. The census bureau as cities and other incorporated places, which have 2500 or more inhabitants define "Urban areas"

2.2 Graphical Presentation

Data presented in the form of tables give good information in concise form. Tables provide all relevant information of the data. Apart from tabular presentation, graphical presentation of data has also become quite popular. It gives visual information in addition to magnitudes. Furthermore, comparisons and changes in the data can be well visualized when presented in graphical form. A very useful part of graphical presentation is the interpretation of the graphs. In every graph we should try to interpret the data.

With the help of computer software packages such as Harvard Graphics, Lotus 123, Energraphics, etc., graphical presentation of data can be made in a variety of ways. But these may broadly be categorized into the following:

- Bar chart
- Pie chart
- Histogram
- Frequency polygon
- Pareto chart
- Frequency curve
- Line diagram.

2.2.1 Bar Charts

Bar charts are used for categorical data or metric data that are transformed into categorical data. Categories are shown on the horizontal axis. Frequency, percentage, or proportion is shown on the vertical axis. Bars are separated from each other to emphasize the distinctness of the categories. The bars must be of the same width. The length of each bar is proportional to the frequency, percentage, or proportion in the category. Levels ought to be provided on both axes.

In one figure only one variable can be depicted. This is illustrated in Figs. 2.1 and 2.2. Two or more variables can also be depicted in the same figure for ease of comparison. Figures 2.3 and 2.4 show presentation of two bars and Fig. 2.5 shows triple bar presentation.

2.2.2 Pie Charts

Like bar charts, pie charts are also used for categorical data. A circle is divided into segments, the areas of which are proportional to the values in the question. But the areas are proportional to the angles the corresponding segments make at the center

Fig. 2.1 Foreign debt/foreign reserve of some countries (2008)

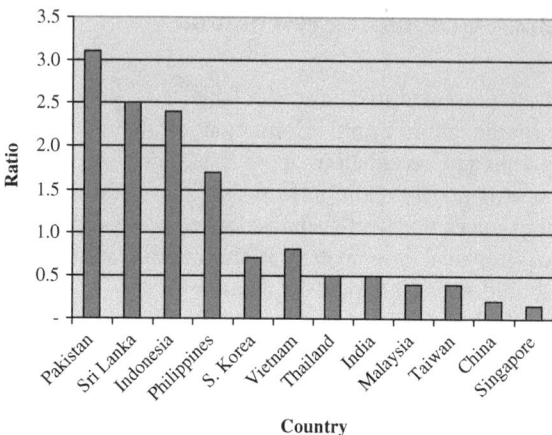

Fig. 2.2 Foreign tourist arrivals in Southeast Asia (Q1 2009)

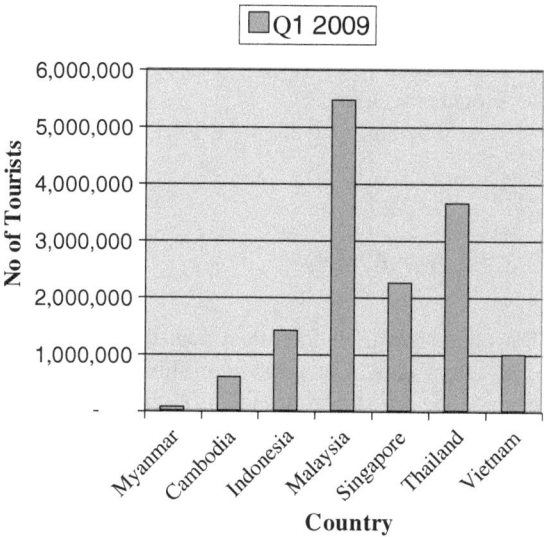

of the circle. Thus, segments of the circle are cut in such a way that their values are proportional to the angles.

In one pie chart only values of one variable can be shown. However, two or more pie charts may be constructed side by side for comparison or to study the change over time. In Fig. 2.6 Thailand population (2009) is shown. Figure 2.7 is another example of pie chart presentation (Table 2.8).

Table 2.8 Newly constructed dwellings in Bangkok and adjoining provinces

Type	1987			1991		
	Units	%	Angle	Units	%	Angle
Individual homes	34,679	65.0	234	35,604	27.6	99
Town house	16,326	30.6	110	52,116	40.4	146
Flat and condominium	1,707	3.2	12	39,861	30.9	111
Twin house	641	1.2	4	1,419	1.1	4
Total	53,353	100	360	129,000	100	360

2.2.3 Histogram

Histograms are used for metric data but converted to categories. These are somewhat similar to bar charts. However, there are some important features in histograms. The blocks in histograms are placed together one after another. These are
not separated. Classes are ordered on the horizontal axis, with scores increasing
from left to right. Areas of the blocks are proportional to the frequencies. If the class
intervals are of equal width, the heights of the blocks/rectangles are proportional to
the frequencies. If the class intervals are of unequal width, the blocks/rectangles are
drawn in such a way that the areas of the blocks/rectangles are proportional to the
frequencies. However, it is easier to interpret the histograms, if the class intervals
are of equal width.

Data of Table 2.4 are used here to construct a histogram shown in Fig. 2.8.

2.2.4 Frequency Polygon

It is also a graphical presentation of frequency distribution. It is more convenient
than the histogram. The midpoints of the upper extremes of the blocks of the
histogram are joined by straight lines. The first and the last parts of the polygon are
to be brought to the horizontal axis at a distance equal to half of the class width.

Data of Table 2.4 are used here for construction of the frequency polygon shown
in Fig. 2.9.

2.2.5 Pareto Chart

A pareto chart is a bar chart for count (discrete) data. It displays the frequency of
each count on the vertical axis and the count type on the horizontal axis. The count
types are always arranged in descending order of frequency of occurrence. The most
frequent occurring type is on the left, followed by the next–most frequently
occurring type, and so on. Bars are placed side by side with no gap between the

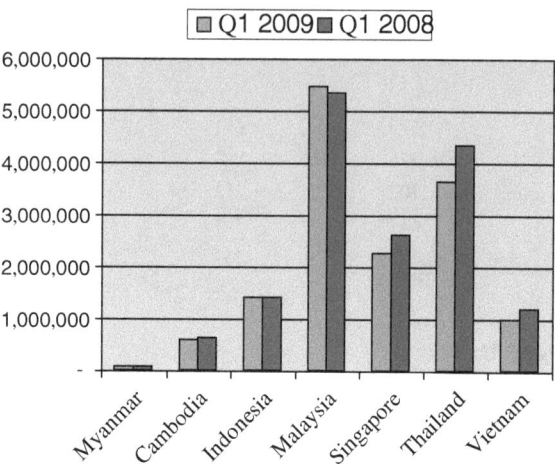

Fig. 2.3 Comparison of foreign tourist arrivals (Q1 2008 and Q1 2009). *Data Source* Pacific Asia Travel Association

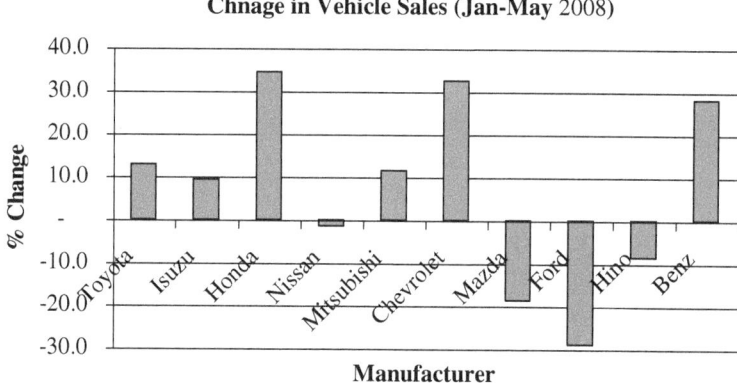

Fig. 2.4 Change in vehicles sales. *Data Source* Toyota Motors Thailand

adjacent ones. A segmented line is also drawn to depict the relative cumulative frequency distribution.

Pareto charts are useful, among other uses, in the analysis of defect data in manufacturing system, construction management, and others, and is an important part of quality improvement program since it allows the management and engineers to focus attention on the most critical defects in a production or process.

Data in Table 2.9 are used to construct the pareto chart shown in Fig. 2.10.

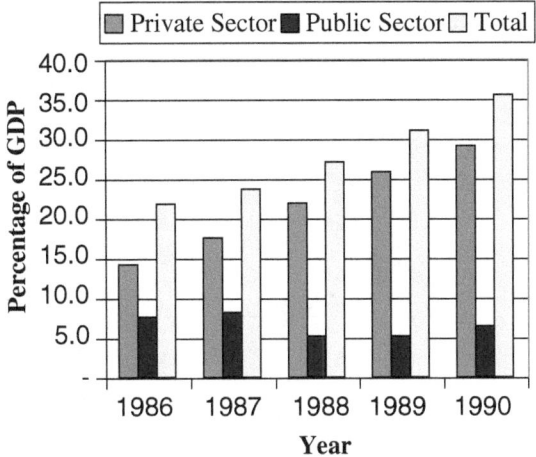

Fig. 2.5 Investment as a percentage of GDP. *Data Source* NESDB

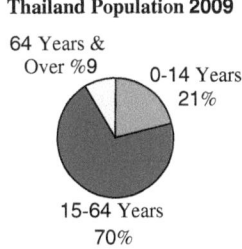

Fig. 2.6 Thailand population (2009). *Data Source* National Statistical Office

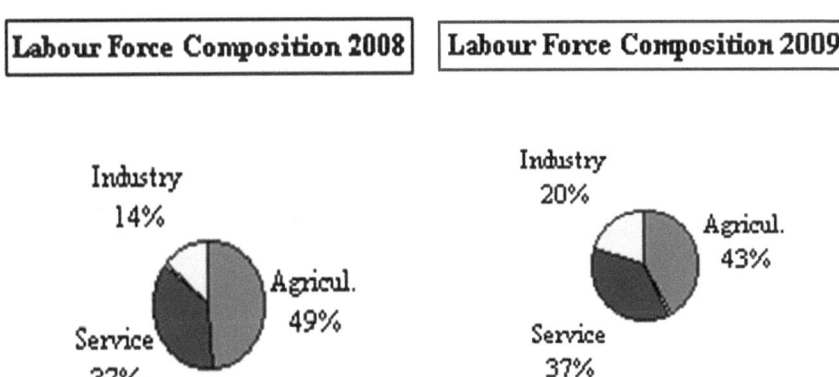

Fig. 2.7 Thai labor force by occupation. *Data Source* National Statistical Office

Fig. 2.8 Crushing strength of bricks (histogram)

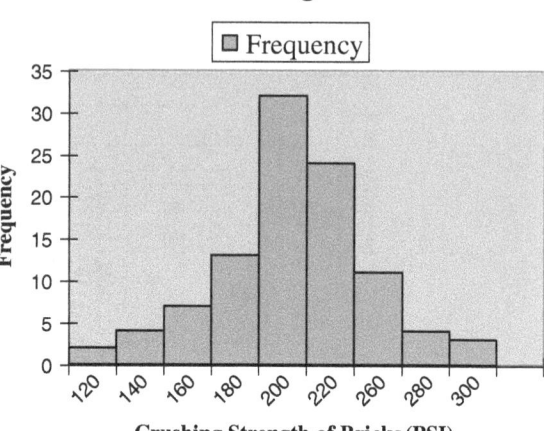

Fig. 2.9 Crushing strength of bricks (frequency polygon)

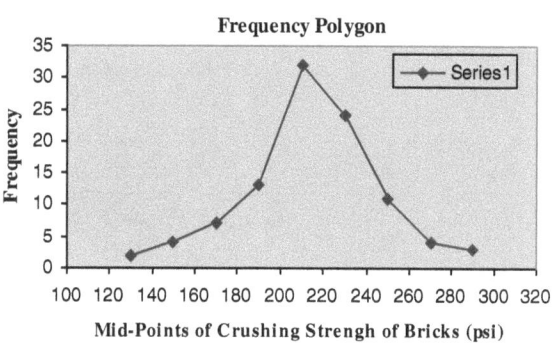

Table 2.9 Defects in building construction

Code defects	Frequency	Relative frequency (%)	R.C.F. (%)
Plaster	60	37.5	37.5
Curing	42	26.2	63.7
Flooring	12	7.5	71.2
Door	12	7.5	78.7
Distemper	10	6.3	85.0
Power line	10	6.3	91.3
Plumbing	8	5.0	96.3
Others	6	3.7	100.0
Total	160	100.0	

Fig. 2.10 Defects in building construction (pareto chart)

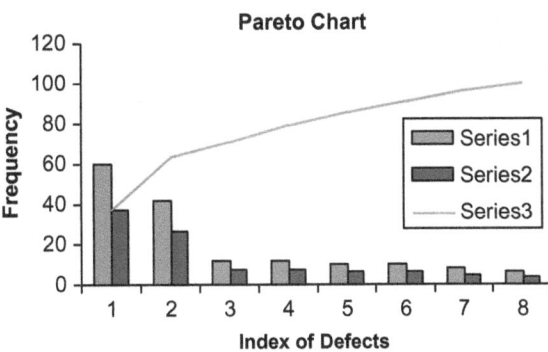

2.2.6 Line Diagram

Line diagrams are drawn by plotting the values of two continuous variables. These show trends or changes in one variable resulting from changes in the other. One important application of the line diagram is to study the changes of various economic indicators over time. Line diagrams may be presented in the form of continuous lines or segmented lines depending on the phenomenon under study. Figures 2.11, 2.12 and 2.13 will serve as examples.

2.2.7 Frequency Curve

Frequency curve is a smoothed frequency polygon. It is produced by plotting the absolute frequency of an infinitesimally small range of a continuous variable. It is a theoretical distribution.

An example of frequency polygon is given in Fig. 2.14.

Fig. 2.11 Thailand debt situation. *Data Source* Finance ministry

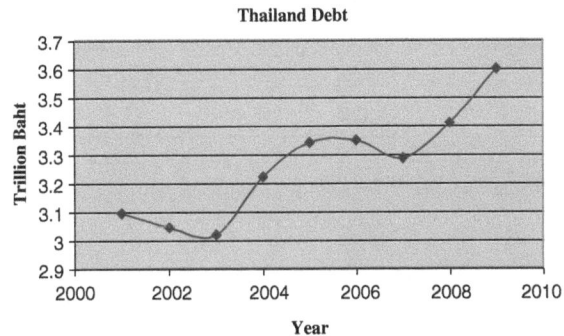

Fig. 2.12 Thailand debt as
percentage of GDP. *Data
Source* Ministry of Finance

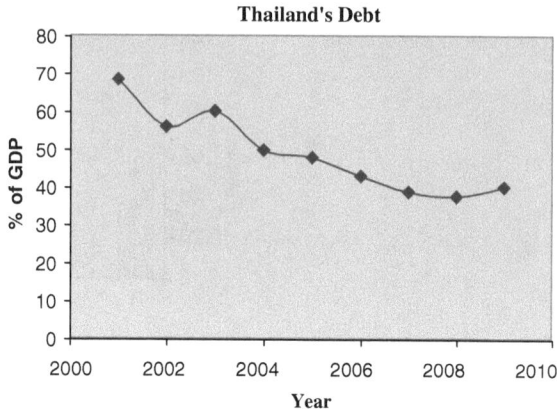

Fig. 2.13 Comparison of
economic outlook. *Data
Source* International
Monetary Fund

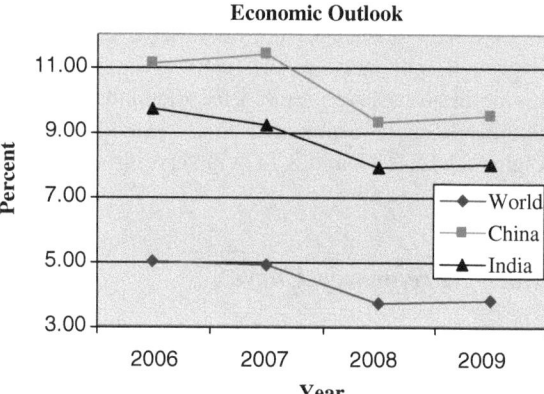

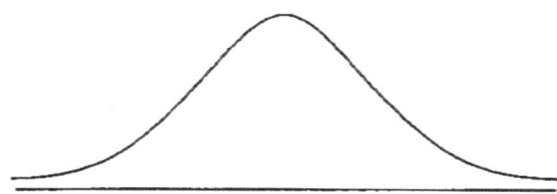

Fig. 2.14 Frequency curve

Problems

2.1 The Canadian International Development Agency (CIDA) has been providing financial support to AIT. The overall financial status as of 31 March 1991 is shown below.

Project activity	Planned expend	Actual expend	Actual/planned expend
Scholarship	4400	2270	51.6
Student research	387	108	27.9
Exploratory research	250	106	42.4
Demonstration project	250	58	23.2
Dissemination project	450	134	29.8
Seminar/workshop	400	75	18.8
Project support	150	59	39.3
Total	6287	2810	44.7

Note: The expenditure figures are in 1000 Canadian dollars

Construct bar charts for the expenditures as well as the proportions. Interpret the charts in comparing the planned and actual expenditures.

2.2 The number of enrolled students (as of May 1987) in the College of Medical Technology and Nursing, University of Tsukuba, Japan is as follows:

Level	Total no. of students	Female students
Undergraduate	7969	2194
Master degree program	1140	260
Doctoral degree program	1228	204
Laboratory school	4536	0
Medical technology	361	352
Total	15,234	3010

Construct pie charts to depict the distributions. Interpret the charts

2.3 The actual expenditures under CIDA activities in AIT during the four consecutive periods are shown:

Project component	Actual expenditures in 1000 CDN$			
	2nd quarter 1990	3rd quarter 1990	4th quarter 1990	1st quarter 1991
AGP	82	74	94	162
EPM/NRP	133	88	24	99
HSD	148	169	45	100
Project support	6	6	7	5
Total	369	337	170	366

Draw segmented line diagrams for all the four project components in one chart. Interpret the results.

2.4 A quality control manager obtained samples to check the number of defective products. The number of defective products noted was as follows:

4	7	5	8	7	8
3	11	2	7	5	9
7	16	16	12	14	5
6	12	14	11	4	9
4	10	6	13	9	6
13	12	16	15	12	17
3	5	10	20	4	19
10	8	12	9	7	12
2	12	7	3	12	11
7	7	7	6	14	8
3	15	5	4	5	10
8	6	5	6	7	9

Group the data. Construct a histogram. Draw the frequency polygon. Comment on the distribution.

2.5 Electrical power demand in Dhaka was noted in two sample occasions—one in winter and the other in summer. The recorded demands were as follows:

Time (h)	Winter demand (MW)	Summer demand (MW)
16:00	972	1141
17:00	1203	1161
18:00	1519	1147
18:30	1551	1251
19:00	1549	1334
19:30	1475	1343
20:00	1475	1318
21:00	1317	1344
22:00	1170	1281
23:00	908	1227
24:00	841	1184

(a) Draw smooth line charts to depict the trend of power demand over time.
(b) Compare the two demands.

References

Asian Institute of Technology, *AIT Alumni 1961–1987 Tracer Study*, Bangkok: AIT, 1990, p.9

International Monetary Fund

Ministry of Finance, Royal Thai Government, Thailand

NESDB, National Economic and Social Development Board, Thailand

National Statistical Office, Thailand

Pacific Asia Travel Association

Toyota Motors, Thailand

U.S. Bureau of the Census. The Census Bureau as cities and other incorporated places, which have 2500, or more inhabitants define "Urban areas"

Chapter 3
Descriptive Statistics

Abstract Descriptive statistics and inferential statistics are distinguished. Concepts of descriptive statistics are presented. The techniques of using descriptive statistics dealing with mean (arithmetic, geometric), median, and mode are explained. The technique for calculation of the "growth rates" and future projection is shown. Measures of dispersion of data are explained. Descriptive statistics entails central tendency of data.

Keywords Descriptive statistics · Inferential statistics · Central tendency · Dispersion · Growth rate

Statistical applications can be viewed as having two broad main branches. These are descriptive statistics which deals with the description of the data elements and inferential statistics which deals with the inferences. In this chapter, we shall discuss the descriptive statistics. This again will be divided into two broad areas as follows.

© Springer Science+Business Media Singapore 2016
A.Q. Miah, *Applied Statistics for Social and Management Sciences*,
DOI 10.1007/978-981-10-0401-8_3

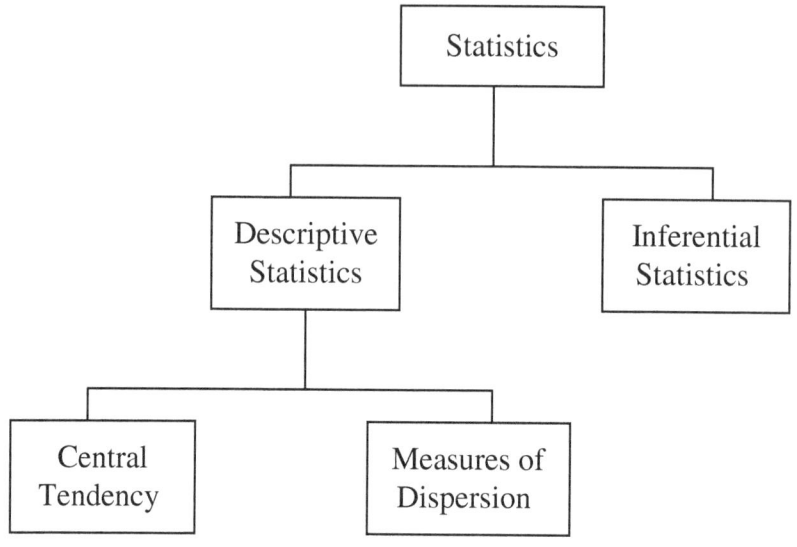

3.1 Central Tendency

Previously, we have seen how to summarize data and present them in the form of tables and graphs. Now we are in a position to analyze the data. In this attempt we should study first the measures of central tendency. There are several measures to indicate the central tendency. But the most common ones are

– mean (arithmetic; geometric)
– median and
– mode.

3.1.1 Mean

There are two types of means, namely arithmetic and geometric means. Although the term "mean" should be used for any of the two "arithmetic mean" or "geometric mean," in practice it is used only for arithmetic mean (average). Thus, whenever the term "mean" is used, it would indicate arithmetic mean (average). If it is a case of geometric mean, it would be expressed in that way.

(A) Arithmetic mean

The mean (arithmetic) of a set of scores is the sum of the scores divided by the number of scores. It is the most common measure of central location.

For assorted data, if the observations are $x_1, x_2, x_3, \ldots, x_n$, then the mean is given as

$$\bar{X} = \frac{\sum X_i}{N}$$

where
$\bar{X}$ mean,
X_i score of ith case,
N total number of cases.

Example 1
At the end of the third term final examination, the GPA of 10 randomly selected students in AIT was found to be 3.83, 3.49, 4.00, 3.50, 3.33, 3.54, 3.13, 3.88, 3.50, and 3.71. Calculate the mean GPA.
    The mean GPA is given as

$$\bar{X} = (3.83 + 3.49 + 4.00 + 3.50 + 3.33 + 3.54 + 3.13 + 3.88 + 3.50 + 3.71)/10$$
$$= 3.59$$

For weighted data, simple averages or mean do not reflect the true situation. It becomes necessary to attach different weightages to different sets of observations. The method involves weighted mean. The following example will illustrate the concept.

Example 2
One student took three tests. First test—time taken is half an hour; earned grade is 50. Second test—time taken is one hour; earned grade is 80. Third test—time taken is one hour and a half; earned grade is 70.

$$\text{Weighted mean} = \frac{\sum X_i w_i}{\sum w_i}$$

where,
X_i score of ith case
w_i weightage of ith case.

In the above example

$$X_1 = 50; \quad X_2 = 80; \quad X_3 = 70$$

$$w_1 = 1; \quad w_2 = 2; \quad w_3 = 3$$

$$\text{Weighted mean} = \frac{X_i w_i + X_2 w_2 + X_3 w_3}{w_1 + w_2 + w_3}$$
$$= \frac{50 * 1 + 80 * 2 + 70 * 3}{1 + 2 + 3}$$
$$= \frac{50 + 160 + 210}{6}$$
$$= \frac{420}{6} = 70$$

Sometimes, especially when a large number of observations are involved, each individual observation is not readily available. Instead, frequency distribution is given. In such a case the formula for calculating the mean is to be modified as follows:

$$\bar{X} = \frac{X_i f_i}{N}$$

where
X_i individual score of ith case
f_i frequency of ith case
N total number of cases
      $= \Sigma f_{i.}$

Example 3
The high-rise buildings and built spaces (average per building) of Bangkok for 4 years from 1987 to 1990 are shown in columns 2 and 3 in Table 3.1. Calculate the mean space per building over the 4-year period.

$$\bar{X} = \frac{17,242,940}{720}$$
$$= 23,949 \text{ m}^2 \text{ per building over the four - year period.}$$

Often the continuous data are transformed into groups by forming certain classes and frequencies of respective classes are provided. We need to calculate the mean of the whole dataset. In such a case, the procedure is similar to that applied to the

Table 3.1 High-rise buildings and built space in Bangkok (1987–1990)

Year	Average space (m²) x_i	No. of buildings f_i	$x_i f_i$
1987	18,868	16	301,888
1988	23,192	86	1,994,512
1989	20,346	291	5,920,686
1990	27,602	327	9,025,854
Total		720	17,242,940

previous example. The difference is only that we need to find the midpoint of the classes and assume that all the observations in the class are centered on the midpoint or are perfectly symmetrical about the midpoint. Since this assumption will not be 100 % valid in almost all practical cases, the mean calculated using this method will be approximate.

The formula for mean when frequency distribution is given for grouped data is as follows:

$$\bar{X} = \frac{\sum m_i f_i}{n}$$

where
m_i midpoint of ith class
f_i frequency of ith class
n total number of cases.

Example 4
In a sample survey of 640 households in a city the household income distribution was found as shown (first and second columns) in Table 3.2.

$$\bar{X} = \frac{13,075 * 1000}{640}$$
$$= \$20,429.69 \text{ per household.}$$

Sometimes it becomes quite inconvenient to deal with large figures as in the previous example. A short method of calculation of the mean can be applied in such situations. The technique is demonstrated as follows:

For ungrouped data

$$\bar{X} = a + \frac{\sum d_i}{n}$$

Table 3.2 Household income distribution

HH income class (000 $)	No of HHs f_i	Midpoint m_i	$m_i f_i$
$5 \leq x < 10$	100	7.5	750
$10 \leq x < 15$	110	12.5	1375
$15 \leq x < 20$	120	17.5	2100
$20 \leq x < 25$	130	22.5	2925
$25 \leq x < 30$	90	27.5	2475
$30 \leq x < 40$	60	35.0	2100
$40 \leq x < 50$	30	45.0	1350
Total	640		13,075

where

a a constant (assumed mean)
d_i deviation
      $= X_i - a$
n total no. of cases.

For grouped data

$$\bar{X} = a + \frac{\sum f_i d_i}{\sum f_i} * c$$

where,

a a constant (assumed midpoint of the middle class or assumed mean)
d_i deviation$=(m_i - a)/c$; c is class width
m_i midpoint of ith class
f_i frequency of ith class.

To illustrate the use of the formula, an example hereafter is given for grouped data.

Example 5
Scores of 50 students in a subject are transformed into several classes and the distribution is given in Table 3.3. Calculate the mean score using the short method of calculation.

a (assumed mean) = 65; c (class width) = 10

$$\bar{X} = a + \frac{\sum f_i d_i}{\sum f_i} * c$$
$$= 65 + \frac{3}{50} * 10$$
$$= 65 + 0.6$$
$$= 65.6$$

Table 3.3 Scores distribution

Class	m_i	$m_i - a$	f_i	d_i	$f_i d_i$
$30 \le x < 40$	35	−30	4	−3	−12
$40 \le x < 50$	45	−20	6	−2	−12
$50 \le x < 60$	55	−10	8	−1	−8
$60 \le x < 70$	65	0	12	0	0
$70 \le x < 80$	75	+10	9	+1	9
$80 \le x < 90$	85	+20	7	+2	14
$90 \le x < 100$	95	+30	4	+3	12
Total			50		3

(B) Geometric Mean

In certain circumstances geometric mean is preferred to arithmetic mean. It is suitable for rate of change, time series data, and ratios, and also for data in geometric progression. The geometric mean is defined as

$$\text{Geometric mean (GM)} = \sqrt[n]{(X_1 * X_2 * X_3 * \cdots X_n)}$$

The calculations can be simplified by taking the logarithm of both sides. Thus,

$$\log \text{GM} = \log\left\{ \sqrt[n]{(X_1 * X_2 * X_3 * \cdots X_n)} \right\}$$
$$= \log\{(X_1 * X_2 * X_3 * \cdots X_n)\}^{1/n}$$
$$= 1/n\{\log X_1 + \log X_2 + \log X_3 + \cdots \log X_n\}$$

Example 6
The yearly sales of a company for 5 years from 1986 to 1991 are given in Table 3.4. Calculate the mean percentage change over the 5-year period.

$$\text{GM} = \sqrt[5]{(111 * 150 * 133 * 120 * 104)}$$
$$= (2.7636336 * 10^{10})^{1/5}$$
$$= 122.55\,\%$$

Therefore, mean percentage change $= 122.55 - 100 = 22.55$ per year. The calculations can be simplified by using logarithm. Thus, taking logarithm of both sides we can get

$$\log \text{GM} = 1/5(\log 111 + \log 150 + \log 133 + \log 120 + \log 104)$$
$$= 1/5(2.0453 + 2.1761 + 2.1239 + 2.0792 + 2.0171)$$
$$= 1/5(10.4415)$$
$$= 2.0883$$
$$\text{Therefore, GM} = \text{anti - log}(2.0883)$$
$$= 122.55$$

Table 3.4 Yearly sales ($)

Year	Sales ($)	Percentage compared with a year earlier
1986	4500	–
1987	5000	111
1988	7500	150
1989	10,000	133
1990	12,000	120
1991	12,500	104

For frequency distribution the formula should be modified in the following way:

$$GM = \sqrt[n]{\left(X_1^{f1} * X_2^{f2} * X_3^{f3} * \cdots X_n^{fn}\right)}$$

$$\Rightarrow \log GM = 1/n\{f_1 * \log X_1 + f_2 * \log X_2 + f_3 * \log X_3 + \cdots f_n * \log X_n\}$$

An important application of the concept of the geometric mean is growth rate calculation. There are certain phenomena showing continuous increase/decrease and the increase/decrease is measured as a percentage or fraction of the previous year's figure. Such increase/decrease is known as growth rate. Growth rate is always measured on a per year basis.

Suppose, population of a country during the present year = P_0. Growth rate (rate of increase of population per year) = 5 % = 0.05. Therefore, at the end of year 1 population will be

$$P_1 = \text{base year population} + \text{increase in one year}$$
$$= P_0 + 0.05P_0$$
$$= P_0(1 + 0.05)$$

Population at the end of year 2 will be

$$P_2 = \text{population of year 1} + \text{increase in one year}$$
$$= P_1 + 0.05P_1$$
$$= P_0(1 + 0.05) + 0.05\,P_0(1 + 0.05)$$
$$= P_0(1 + 0.05)(1 + 0.05)$$
$$= P_0(1 + 0.05)^2$$

Population at the end of year 3 will be

$$P_3 = \text{population of year 2} + \text{increase in one year}$$
$$= P_2 + 0.05P_2$$
$$= P_0(1 + 0.05)^2 + 0.05 * P_0(1 + 0.05)^2$$
$$= P_0(1 + 0.05)^2 * (1 + 0.05)$$
$$= P_0(1 + 0.05)^3$$

If the growth rate is r, then the population at the end of n years will be

$$P_n = P_0(1 + r)^n$$

Example 7
Exports of Thailand during 1984 and 1989 were 175,237 and 515,745 million Baht.

(a) Calculate the growth rate during the period from 1984 to 1989.
(b) Assuming the same growth rate for another 5 years, estimate the export during 1994.

Solution
Here, $P_0 = 175{,}237$; $P_n = 515{,}745$; $n = 5$ years.

$$P_n = P_0(1+r)^n$$
$$\Rightarrow 515{,}745 = 175{,}237(1+r)^5$$

Therefore, $2.9431 = (1+r)^5$

Therefore, $(1+r) = (2.9431)^{1/5}$
$$= 1.2410$$

Therefore, $r = 1.2410 - 1$
$$= 0.2410$$
$$= 24.10\%$$

$$P_{1994} = P_{1989}(1+r)^5$$
$$= 515{,}745(1+0.2410)^5$$
$$= 515{,}745(1.2410)^5$$
$$= 1{,}518{,}077 \text{ million Baht}$$

Example 8
Import of Thailand in million Bahts during 3 years is shown hereafter:

Year	Import
1983	236,609
1986	241,358
1989	656,428

Calculate the average growth rate during 1983–1989.

Solution
For 1983–1986

$$241{,}358 = 236{,}609(1+r_1)^3$$
$$\Rightarrow 1.02007 = (1+r_1)^3$$
$$\Rightarrow (1+r_1) = (1.02007)^{1/3}$$
$$- 1.0066$$
$$\Rightarrow r_1 = 0.0066$$
$$= 0.66\%$$

For 1986–1989

$$656,428 = 241,358(1+r_2)^3$$
$$\Rightarrow 2.7197 = (1+r_2)^3$$
$$\Rightarrow (1+r_2) = (2.7197)^{1/3}$$
$$= 1.39586$$

For 1983–1989

$$656,428 = 236,609(1+r_3)^6$$
$$\Rightarrow 2.7743 = (1+r_3)^6$$
$$\Rightarrow (1+r_3) = (2.7743)^{1/6}$$
$$= 1.1854$$
$$\Rightarrow r_3 = 0.1854$$
$$= 18.54\%$$
$$\text{Average growth rate} = \frac{r_1 + r_2 + r_3}{3}$$
$$= \frac{0.0066 + 0.3959 + 0.1854}{3}$$
$$= 0.1960$$
$$= 19.60\%$$

3.1.2 Median

It is another measure of central tendency. When it is desirable to divide data into two groups, each group containing exactly the same number of values, the median is the appropriate measure. It is defined to be the middle value in a set of numbers arranged according to magnitude. The median has the property that half the scores are less than (or equal to) the median and half the scores are greater than (or equal to) the median. If n is odd, the middle value is one. So position of M_d is at $(n + 1)/2$. If n is even, there are two mid-values. M_d lies halfway in between these two values.

Examples 9
The following two sets of observations (ungrouped) show how to calculate the median.

(1)	(2)
415	415
480	480

(continued)

(continued)

(1)	(2)
525	525
608	608
719	719
1090	1090
2059	2059
4000	4000
6000	
No. of cases = 9	No. of cases = 8
M_d = 719	M_d = (608 + 719)/2 = 663.5

When the observations/measurements are given in classes, we need to use the following formula to calculate the median.

$$M_d = L_m + C_m * \frac{(n+1)/2 - T}{f_m}$$

where

L_m lower limit of median class
C_m class width of the median class
f_m frequency of the median class
n total number of observations
T cumulative frequency corresponding to the class preceding the median class.

Example 10
Table 3.5 shows some hypothetical data. Find the median.

Table 3.5 Hypothetical data

Data	Frequency	Cumulative frequency
$x < 10.0$	1	1
$10.0 \leq x < 10.5$	7	8
$10.5 \leq x < 11.0$	13	21
$11.0 \leq x < 11.5$	23	44
$11.5 \leq x < 12.0$	47	91
$12.0 \leq x < 12.5$	39	130
$12.5 \leq x < 13.0$	17	147
$13.0 \leq x < 13.5$	4	151
$13.5 \leq x < 14.0$	3	154
$14.0 \leq x < 14.5$	1	155
$14.5 \leq x < 15.0$	1	156
15.0 and above	1	157
Total	157	

The median class is the class in which the middle value lies. In this example, the middle observation is $(n + 1)/2 = (157 + 1)/2 = 79$th observation. This lies in the class in which the cumulative frequency is 91. The class is 11.5–12.0. Therefore,

$$M_d = 11.5 + 0.5 * \frac{79 - 44}{47}$$
$$= 11.5 + 0.37$$
$$= 11.87$$

3.1.3   Mode

The mode is defined to be that value which occurs most frequently. This is another measure of central tendency.

Example 11 (ungrouped data)

$$X : 1, 1, 1, 2, 2, 2, 2, 2, 3, 3, 3, 3, 3, 3, 4, 4, 5, 6, 8$$
$$M_o = 3; \quad \text{(unimodal)}$$
$$Y : 1, 1, 1, 2, 2, 2, 2, 2, 3, 3, 4, 4, 4, 4, 4, 5, 6, 6$$
$$M_o = 2, 4; \quad \text{(bimodal)}$$

In the first example, 3 occurs six times, the highest frequency. So the mode is 3. Since only one value has the highest frequency, the set of observations is unimodal. In the second example, both 2 and 4 occur five times—equal frequencies. So there are two modes 2 and 4. The set of observations is bimodal.

If the datasets are available in classes, we need to modify our technique of calculating the mode. In this case the following formula may be used.

$$M_o = L_m + C_m * \frac{\mathbf{a}}{\mathbf{a} + \mathbf{b}}$$

where
L_m lower limit of modal class
C_m class width of the modal class
a   absolute value of the difference in frequency between the modal class and the preceding class
b   absolute value of the difference in frequency between the modal class and the following class.

The modal class is the class with the highest frequency. The following example illustrates the use of the formula.

Table 3.6 Housing expenditure of slum dwellers

Housing expenditure class ($/month)	No of owners	No of renters
$0 \leq x < 6$	3	186
$6 \leq x < 12$	12	220
$12 \leq x < 18$	20	65
$18 \leq x < 24$	23	31
$24 \leq x < 30$	18	8
$30 \leq x < 45$	51	1
$45 \leq x < 60$	45	2
$60 \leq x < 90$	32	0
$90 \leq x$	16	0
Total	220	513

Data source: Miah (1990)

Example 12

Calculate the modes of the owner and renter categories from the data given in Table 3.6

The modal class for owners is 30–45 and that for renters is 6–12.

For owners:

$$M_o = L_m + C_m * \frac{a}{a+b}$$
$$= 30 + 15 * \frac{51 - 18}{(51 - 18) + (51 - 45)}$$
$$= 30 + 15 * \frac{33}{33 + 6}$$
$$= 30 + 12.69$$
$$= \$42.69$$

For renters:

$$M_o = L_m + C_m * \frac{a}{a+b}$$
$$= 6 + 6 * \frac{220 - 186}{(220 - 186) + (220 - 65)}$$
$$= 6 + 6 * \frac{34}{34 + 155}$$
$$= 6 + 1.08$$
$$= \$7.08$$

If there are two adjacent classes having the same highest frequency, the two classes may be combined together to form one single class and the calculation for mode can be done in the usual process.

Table 3.7 Housing expenditure of slum dwellers

Housing expenditure class ($/month)	No. of owners	No. of renters
$0 \leq x < 6$	3	186
$6 \leq x < 12$	12	220
$12 \leq x < 18$	20	65
$18 \leq x < 24$	23	31
$24 \leq x < 30$	18	8
$30 \leq x < 45$	50	1
$45 \leq x < 60$	50	2
$60 \leq x < 90$	28	0
$90 \leq x$	16	0
Total	220	513

Example 13

To illustrate the technique, we can use the previous example with slight modification of owners' frequency distribution as shown in Table 3.7.

Here, two classes 30–45 and 45–60 in case of owners having the same and the highest frequency of 50 each. So these two classes can be combined together to form one single class 30–60 with a total frequency of 50 + 50, i.e., 100 and the calculations can be carried out in the following manner:

$$M_o = L_m + C_m * \frac{a}{a+b}$$
$$= 30 + 30 * \frac{100 - 18}{(100 - 18) + (100 - 28)}$$
$$= 30 + 30 * \frac{82}{82 + 72}$$
$$= 30 + 15.97$$
$$= \$45.97$$

If there are two classes having the same and the highest frequency, but not adjacent to each other, then the dataset would be bimodal. Mode for each class can be computed independently.

3.1.4 Comparison of Mean, Median, and Mode

When the distribution is symmetrical, the mean, median, and mode coincide. It is illustrated in Fig. 3.1.

When the distribution is skewed to the right, Mean > Median > Mode. This is illustrated in Fig. 3.2.

Fig. 3.1 Symmetrical distribution

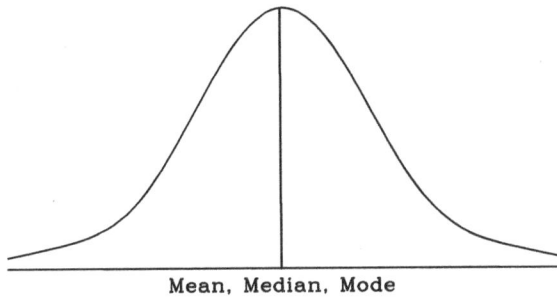

Fig. 3.2 Right-skewed distribution

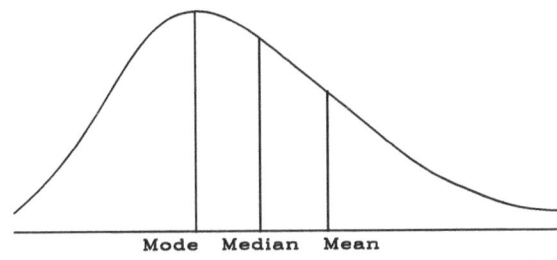

Fig. 3.3 *Left*-skewed distribution

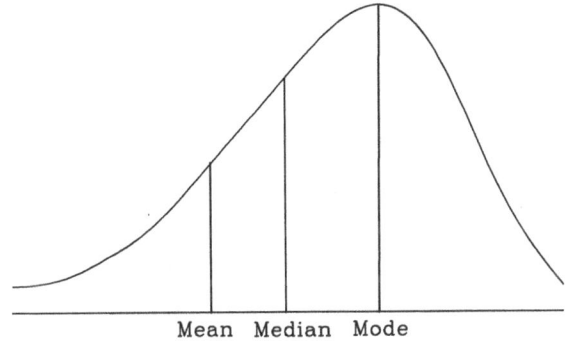

When the distribution is skewed to the left, Mean < Median < Mode. This is illustrated in Fig. 3.3.

3.2 Measures of Dispersion

In the previous chapter, we have studied measures of central tendency. Although measures of central tendency are quite useful, a measure of central tendency, by itself, is not sufficient to provide an adequate summary of the characteristics of a dataset. For example, consider the following three sets of observations in income:

$$
\begin{array}{lccccc}
x(\$) & 14,000 & 17,000 & 20,000 & 23,000 & 26,000 \\
y(\$) & 4000 & 12,000 & 20,000 & 28,000 & 36,000 \\
z(\$) & 2000 & 10,000 & 30,000 & 38,000 \\
\end{array}
$$

The mean of the income in all the three distributions is $20,000. But the distributions are not the same. Apparently, they differ. In order to study such situations, we need to study another measure called a measure of dispersion. There are several measures to study the dispersion in the datasets. Broadly these may be enumerated as follows:

- Range
- Interquartile range (IQR)
- Mean deviation
- Variance
- Standard deviation
- Coefficient of variation
- Stem-and-leaf diagram
- Other measures of dispersion.

3.2.1   Range

It is the simplest measure of dispersion. It is the absolute difference between the maximum value and the minimum value of the dataset. In the income example cited above, the range of x income set is $26,000 − 14,000 = $12,000 and range of the z income set is $38,000 − 2000 = $36,000.

3.2.2   Interquartile Range (IQR)

IQR is the difference between the third quartile and the first quartile. A general rule is necessary to set out for locating the first, second, and third quartiles in a set of data. If there are n observations arranged in ascending order, then the location of the first quartile is $(n + 1)/4$, the location of the second quartile, i.e., the median is $(n + 1)/2$ and the location of the third quartile is $3(n + 1)/4$. If $(n + 1)$ is not an integer multiple of 4, then the quartiles are to be found out by interpolation. Look at the following two examples.

Examples 1

(1)	(2)
515	450
510	449

(continued)

(continued)

(1)	(2)
$500 \Rightarrow Q_3$	445
490	440
447	438
$445 \Rightarrow Q_2$	432
441	428
438	420
$435 \Rightarrow Q_1$	
430	
420	

For set 1:

Range = 515 − 420 = 95

Location of Q_1 = $(n + 1)/4$ = $(11 + 1)/4$ = 3;

So Q_1 = third observation = 435;

Location of Q_2 = $(n + 1)/2$ = $(11 + 1)/2$ = 6;

So Q_2 = sixth observation = 445

Location of Q_3 = $3(n + 1)/4$ = $3(11 + 1)/4$ = 9;

So Q_3 = ninth observation = 500

IQR = Q_3 − Q_1 = 500 − 435 = 65

For set 2:

$$\text{Range} = 450 - 420 = 30$$

$$\text{Location of } Q_1 = (n+1)/4 = (8+1)/4 = 2.25$$

$$\text{So } Q_1 = 2\text{nd} + 0.25 \text{ from } 428 \text{ to } 432$$

$$= 428 + 0.25 * (432 - 428)$$

$$= 428 + 1.0$$

$$= 429$$

$$\text{Location of } Q_2 = (n+1)/2 = (8+1)/2 = 4.5;$$

$$\text{So } Q_2 = 4\text{th} + 0.50 \text{ from 4th to 5th}$$

$$= 438 + 0.50 * (440 - 438)$$

$$= 438 + 1$$

$$= 439$$

$$\text{Location of } Q_3 = 3(n+1)/4 = 3(8+1)/4 = 6.75;$$

$$\text{So } Q_3 = 6\text{th} + 0.75 \text{ from 6th to 7th}$$

$$= 445 + 0.75 * (449 - 445)$$

$$= 445 + 3$$

$$= 448$$

$$\text{Interquartile Range} = Q_3 - Q_1 = 448 - 429 = 19$$

For grouped data the same rule as given for the median may be used to calculate Q_1 and Q_3. The position of Q_1 is given by $(n + 1)/4$ and that for Q_3 by $3(n + 1)/4$.

Example 2

Use the dataset shown in Table 3.8 and calculate the IQR.

$$\text{Location of } Q_1 = \frac{157 + 1}{4}$$

$$= 39.5; \text{ the class is } 8.4 - 8.6$$

$$\text{So } Q_1 = 8.4 + 0.2 * \frac{39.5 - 21}{23}$$

$$= 8.4 + 0.16$$

$$= 8.56$$

$$\text{Location of } Q_3 = \frac{3(157 + 1)}{4}$$

$$= 118.5; \text{ the class is } 8.8 - 9.0$$

$$\text{So } Q_3 = 8.8 + 0.2 * \frac{118.5 - 91}{39}$$

$$= 8.8 + 0.14$$

$$= 8.94$$

$$\text{The interquartile range} = Q_3 - Q_1$$

$$= 8.94 - 8.56$$

$$= 0.38$$

Table 3.8 Hypothetical data

Data class	Frequency	Cumulative frequency
$x < 8.0$	1	1
$8.0 \leq x < 8.2$	7	8
$8.2 \leq x < 8.4$	13	21
$8.4 \leq x < 8.6$	23	44
$8.6 \leq x < 8.8$	47	91
$8.8 \leq x < 9.0$	39	130
$9.0 \leq x < 9.4$	19	149
$9.4 < x < 10.0$	5	154
10.0 and above	3	157
Total	157	

3.2.3 Mean Deviation

Mean deviation is another measure of dispersion. It is the average of the absolute deviations from some central value, usually mean. If $X_1, X_2, X_3, \ldots X_n$ are the observations, then

$$\text{Mean deviation} = \frac{\sum \{X_i - \bar{X}\}}{n}$$

Example 3

Six students have scores 50, 55, 60, 70, 75, 80. Calculate the mean deviation (Table 3.9).

$\bar{X} = 65$; Mean deviation $= 60/6 = 10$. Thus, the average absolute discrepancy of the student scores is 10.

3.2.4 Variance

Variance is the average (mean) of the squared deviations about the mean. There is an inherent problem in ordinary deviations about the mean, i.e., sum of all the deviations about the mean is zero. Variance avoids this problem. It is a good measure of the spread and is the traditional method of measuring the variability of a dataset. If $X_1, X_2, X_3, \ldots X_n$ are the observations, then the variance is given as

$$\text{Variance } S^2 = \frac{\sum \{X_i - \bar{X}\}^2}{n}$$

Table 3.9 Scores of students

Grades (X_i)	$X_i - \bar{X}$	$[X_i - \bar{X}]$
50	−15	15
55	−10	10
60	−5	5
70	+5	5
75	+10	10
80	+15	15
Total		60 ignoring sign

3.2.5 *Standard Deviation*

Variance is a measure of dispersion. But there remains a problem. The deviations are squared and hence the unit of the variable is also squared. Thus, there is an inadequacy in measuring the dispersion. In order to correct the inadequacy and to reduce the variance into the original unit of measurement, the square root of the variance is taken. This is called standard deviation.

$$\text{Standard deviation } S = \sqrt{(\text{Variance})}$$

$$= \sqrt{\left[\frac{\sum\{X_i - \bar{X}\}^2}{n}\right]}$$

$$= \sqrt{\left[\frac{\sum X_i^2}{n} - \bar{X}^2\right]}$$

For frequency distribution

$$\text{Standard deviation } S = \sqrt{\left[\frac{\sum f\{X_i - \bar{X}\}^2}{n}\right]}$$

$$= \sqrt{\left[\frac{\sum f X_i^2}{n} - \bar{X}^2\right]}$$

where X_i = ith observation or midpoint of the ith class.

Example 4

Farming costs per year of seven farmers in a Thai village are shown in Table 3.10. Calculate the mean deviation, variance, and the standard deviation.

$n = 7; \bar{X} = 9914$

Table 3.10 Farming cost

Cost (B)	$(X_i - \bar{X})$	$(X_i - \bar{X})^2$
10,000	+86	7396
12,500	+2586	6,687,396
9600	−314	98,596
10,000	+86	7396
8800	−1114	1,240,996
9500	−414	171,396
9000	−914	835,396
Total	5514 (ignoring sign)	9,048,572

$$\text{Mean deviation} = \frac{5514}{7} = 787.71 \text{ Baht}$$

$$\text{Variance } S^2 = \frac{9,048,572}{7} = 1,292,653$$

Standard deviation $S = \sqrt{(1,292,653)} = 1,136.95$ Baht

Example 5

Nonhousing expenditures of owners and renters recorded in sample surveys in several slums of a city are given in expenditure classes (Tables 3.11 and 3.12). Calculate the mean, variance, and standard deviation of owner and renter expenditures.

For owners

$$\bar{X} = \frac{19,072.5}{220}$$

$$= \$86.69$$

$$\text{Variance } S^2 = \frac{\sum f_i m_i^2}{n} - (\bar{X})^2$$

$$= \frac{2,043,393.8}{220} - (86.69)^2$$

$$= 9288.2 - 7515.2$$

$$= 1773.0$$

Table 3.11 Nonhousing expenditures of slum dwellers in a city (owners)

| Expenditure class ($/month) | Midpoint m_i | Owners | | |
		f_i	$f_i m_i$	$f_i m^2$
$0 \le x < 15$	7.5	00	0.0	0.0
$15 \le x < 30$	22.5	02	45.0	1,012.50
$30 \le x < 45$	37.5	24	900.0	33,750.00
$45 \le x < 60$	52.5	37	1942.5	101,981.30
$60 \le x < 75$	67.5	38	2565.0	173,137.50
$75 \le x < 90$	82.5	38	3135.0	258,637.50
$90 \le x < 120$	105.0	45	4725.0	496,125.0
$120 \le x < 150$	135.0	21	2835.0	382,725.0
$150 \le x < 180$	165.0	09	1485.0	245,025.0
$180 \le x < 240$	210.0	03	630.0	132,300.0
$240 \le x < 300$	270.0	03	810.0	218,700.0
Total		220	19,072.50	2,043,393.8

Table 3.12 Nonhousing expenditures of slum dwellers in a city (renters)

Expenditure class ($/month)	Midpoint m_i	Renters		
		f_i	$f_i m_i$	$f_i m_i^2$
$0 \leq x < 15$	7.5	2	15	112.5
$15 \leq x < 30$	22.5	74	1665.00	37,462.50
$30 \leq x < 45$	37.5	192	7200.00	270,000.00
$45 \leq x < 60$	52.5	125	6562.50	344,531.30
$60 \leq x < 75$	67.5	72	4860.00	328,050.00
$75 \leq x < 90$	82.5	24	1980.00	163,350.00
$90 \leq x < 120$	105	13	1365.00	143,325.00
$120 \leq x < 150$	135	04	540	72,900.00
$150 \leq x < 180$	165	04	660	108,900.00
$180 \leq x < 240$	210	02	420	88,200.00
$240 \leq x < 300$	270	01	270	72,900.00
Total		513	25,537.50	1,629,731.30

Data source: Miah (1990)

$$\text{Standard deviation} = S$$
$$= \sqrt{(1773.0)}$$
$$= \$42.1$$

For renters

$$\bar{X} = \frac{25,537.5}{513}$$
$$= \$49.78$$

$$\text{Variance } S^2 = \frac{\sum f_i m_i^2}{n} - (\bar{X})^2$$
$$= \frac{1,629,731.3}{513} - (49.78)^2$$
$$= 3176.9 - 2478.0$$
$$= 698.9$$

$$\text{Standard deviation} = S$$
$$= \sqrt{(698.8)}$$
$$= \$26.4$$

3.2.6 Coefficient of Variation

Sometimes the standard deviation which is expressed in absolute terms is inadequate and the relative measure of dispersion is preferred. Example of such a situation is comparison of variability of distributions with different variables. Coefficient of variation is the standard deviation expressed as a percentage of the arithmetic mean and is given as

$$\text{Co - efficient of variation} = \frac{100S}{\bar{X}}$$

Example 6

Use the data in the owner–renter example and show which group (owner or renter) has greater spread in the nonhousing expenditure.

$$\text{Owners co - efficient of variation} = \frac{100 * 42.1}{86.69}$$
$$= 48.56\%$$
$$\text{Renters co - efficient of variation} = \frac{100 * 26.4}{49.78}$$
$$= 53.03\%$$

Thus, the spread of nonhousing expenditures of renters is greater than that of the owners.

3.2.7 Stem-and-Leaf Diagram

The histogram is a useful graphic display. It can give the decision maker a good understanding of the data and is useful in displaying the shapes, location, and variability of the data. However, the histogram does not allow individual data points to be identified, since all the observations falling in a cell (class) are indistinguishable. The stem-and-leaf diagram is a new graphical display which is more informative than the histogram. Since stem-and-leaf display is often most useful at the initial stage of data analysis, it is one of the exploratory data analysis methods.

To construct a stem-and-leaf diagram, each number x_i is divided into two parts: a stem, consisting of one or more of the leading digits, and a leaf, consisting of the remaining digits.

Example 7

The data in crushing strength of bricks example are used to construct the stem-and-leaf diagram. The observations range from 126 to 296. So the values 12, 13, 14, ... 29 are selected as stems. The resulting diagrams are shown in Figs. 3.4 and 3.5.

Stem	Leaf	f
12	6	1
13	7	1
14	7	1
15	5, 8, 0	3
16	5, 4, 0	3
17	0, 8, 3, 1	4
18	4, 1, 5, 5, 5, 1	6
19	3, 2, 2, 6, 8, 5, 8	7
20	8, 0, 4, 3, 8, 0, 4, 0, 7, 1	10
21	5, 3, 8, 0, 7, 5, 4, 0, 7, 5, 0, 4, 9, 3, 5, 5, 2, 1, 0, 5, 9	21
22	4, 6, 2, 7, 6, 2, 1, 4, 4, 8, 4, 5, 8, 0	14
23	6, 1, 0, 1, 3, 3, 0, 7, 3	9
24	9, 9, 8, 3, 4, 6, 0, 8	8
25	7, 0, 8, 0, 1	5
26	8, 7	2
27	1, 8	2
28	4, 7	2
29	6	1
Total		100

Fig. 3.4 Stem-and-leaf diagram of crushing strength of bricks (not ordered)

Stem	Leaf	f_i	f'
12	6	1	1
13	7	1	2
14	7	1	3
15	0, 5, 8	3	6
16	0, 4, 5	3	9
17	0, 1, 3, 8	4	13
18	1, 1, 4, 5, 5, 5	6	19
19	2, 2, 3, 5, 6, 8, 8	7	26
20	0, 0, 0, 1, 3, 4, 4, 7, 8, 8	10	36
21	0, 0,0,0,1,2,3,3,4,4,5,5,5,5,5,5,7,7,8,9,9	21	57
22	0, 1, 2, 2, 4, 4, 4, 4, 5, 6, 6, 7, 8, 8	14	71
23	0, 0, 0, 1, 1, 3, 3, 6, 7	9	80
24	0, 3, 4, 6, 8, 8, 9, 9	8	88
25	0, 0, 1, 7, 8	5	93
26	7, 8	2	95
27	1, 8	2	97
28	4, 7	2	99
29	6	1	100
Total		100	

Note: f' = cumulative frequency.

Fig. 3.5 Stem-and-leaf diagram of crushing strength of bricks (ordered). *Note f'* = cumulative frequency

Sometimes there are too many stems with very few leaves in each stem. Also, there may be too few stems with too many leaves in a stem. This type of display does not provide a good impression. The problem of too many stems can be handled by choosing appropriate stems and leaves. Suppose there is an observation 375 after 296 in the example under study. In such a case it will be unwise to continue the stems 30, 31, 32, 33, 34, 35, 36, and 37. Instead of that only one stem can be shown after stem 29 as shown hereafter.

Stem	Leaf
29	6
HI	375

The other problem of having too many leaves in a stem can be handled by breaking the stem into 2–5 parts. Data in the crushing strength of bricks may be used to illustrate this technique. Let us suppose that we want to divide the stem 21 into 2 parts. The first part should contain the leaves from 0 to 4 and the second part of the stem should contain the leaves from 5 to 9. The resulting diagram would look like the following:

Stem	Leaf
21	0, 0, 0, 0, 1, 2, 3, 3, 4, 4
21	5, 5, 5, 5, 5, 5, 7, 7, 8, 9, 9

If we want to divide stem 21 into 5 parts, then the resulting diagram would look like the following:

Stem	Leaf
21	0, 0, 0, 0, 1
21	2, 3, 3
21	4, 4, 5, 5, 5, 5, 5, 5
21	7, 7
21	8, 9, 9

It would be noted that a stem-and-leaf diagram is a combination of tabular and graphical display and looks like a histogram. But in the histogram, the individual observations are not available, whereas in the stem-and-leaf diagram the individual observations are available.

Examples for usefulness of the stem-and-leaf diagram based on the above plot are highlighted hereafter.

1. Most of the observations lie between 180 and 250.
2. The central value is somewhere between 210 and 220.
3. $n = 100$; $n/2 = 50$. So median is in between observations with ranks 50 and 51. $M_d = (215 + 215)/2 = 215 \Rightarrow Q_2$.
4. Rank of 10th percentile observation = $(0.1)(100) + 0.5 = 10.5$ (midway between 10th and 11th observations); value of the 10th percentile = $(170 + 171)/2 = 170.5$
5. Rank of the first quartile = $(0.25)(100) + 0.5 = 25.5$, (approx.) (midway between 25th and 26th observations); value of the first quartile $Q_1 = (198 + 198)/2 = 198$
6. Rank of the third quartile = $(0.75)(100) + 0.5 = 75.5$, (approx.) (midway between 75th and 76th observations); value of the third quartile $Q_3 = (231 + 231)/2 = 231$.
7. IQR = $Q_3 - Q_1 = 231 - 198 = 33$.

3.2.8 Other Measures of Dispersion

– Skewness
– Kurtosis

While mean, median, and mode provide information regarding central location of data, variance, standard deviation, and others show the dispersion. But they do not provide any information on the shape of the distribution of values. Skewness and Kurtosis provide such information.

3.2.8.1 Skewness

A distribution is considered to be skewed when there are a considerably larger number of cases on one side of the distribution as compared to the other side. The skewness is defined as

$$\text{Skewness} = \frac{\sum \{(X_i - \bar{X})/s\}^3}{N}$$

If the result is positive, the distribution is skewed to the right. If the result is negative, the distribution is skewed to the left.

3.2.8.2 Kurtosis

Kurtosis shows the flatness or peakedness of the distribution. A flat distribution with short broad tails is called platykurtic. A very peaked distribution with long thin tails is called leptokurtic. The kurtosis is defined as

$$\text{Kurtosis} = \frac{\sum\{(X_i - \bar{X})/s\}^4}{N} - 3$$

A positive value indicates leptokurtosis (more peaked in the middle than the normal distribution), and a negative value indicates platykurtosis.

Problems on Central Tendency

3.1.1 Use the data shown in Table 2.4 and calculate the mean, median, and mode.
3.1.2 Use the data shown in Table 3.7 and calculate the mean and median housing expenditure of owners and renters.
3.1.3 Use the data presented in Table 3.5 and calculate the mean.
3.1.4 Scores of students in the written examination of statistics course are shown in table.

78.50	59.75	50.00	57.00	75.50
49.50	55.00	58.50	71.50	53.00
68.00	62.50	57.00	56.50	78.00
71.00	50.75	34.75	81.50	23.50
64.50	64.75	78.50	57.00	81.00
62.75	23.50	52.50	81.75	82.00
76.75	57.50	82.00	84.00	52.75
23.00	58.75	74.50	68.25	54.00
58.50	44.50	38.00	56.00	55.00
53.00	35.00	73.00	71.00	83.00

Calculate the mean, median, and mode of the scores based on individual scores and classes.

3.1.5 As per the report of the Bank of Thailand (vol. 29, No. 4, December 1989), Thailand's national assets continued to rise. Figures for 1982 and 1987 are 23,891 and 36,203 million US$, respectively.

(a) Calculate the growth rate.
(b) If the same growth rate continues, predict the country's assets during 1994.
(c) Assuming the same growth rate to continue, show when the country's assets will be double the figure of 1982.

3.1.6 United Nations reported the child population of Asia as follows:

1985	987.4 million
1990	1019.7 million

(a) Calculate the growth rate of child population between 1985 and 1990.
(b) Assume the future growth rate to be 150 % of this growth rate. Show when the
     child population of Asia will be 1159.5 million.

3.1.7 In Thailand the national government actual revenue from taxation for 4 years
are as follows:

Year	1983	1985	1987	1989
Revenue (million Baht)	129,062	144,947	185,690	302,057

(a) Calculate the average growth rate between 1983 and 1989.
(b) Estimate the revenue from taxation for 1993.

Problems on Measures on Dispersion

3.2.1 A traffic engineer measured the motor car speeds in a certain section of a
highway. The observations recorded in feet per second are as follows:

91	97	90	113	80	93	87
102	95	86	101	80	82	81
82	94	102	95	91	97	106
83	76	107	102	72	104	105
92	101	89	84	97	101	76
92	106	86	88	98	86	90
104	105	92	82	99	86	95
108	90	97	75	108	89	107
96	93	98	79	91	82	78
104	99	83	80	92	95	86

(a) Group the data in suitable classes.
(b) Calculate the range, IQR, variance, standard deviation, and coefficient of
     variation using the original observations and classes and compare the results.

3.2.2 In a final examination the scores of a sample of randomly selected students are
as follows:

94.59	93.58	90.44	54.03	97.01	74.59
92.61	90.07	90.41	92.22	83.15	63.67
74.63	80.33	84.57	90.22	92.97	81.12

(continued)

(continued)

85.86	82.93	84.20	70.25	96.32	74.87
83.12	81.36	90.09	90.11	65.68	90.21
53.95	83.88	83.08	91.15	97.88	15.86
85.58	63.37	80.84	66.44	85.46	94.15
92.61	96.62	82.42	77.02	79.90	80.33
77.26	90.02	92.20	76.55	94.60	50.06
90.11	84.12	92.00			

(a) Group the data in suitable classes and present in the frequency distribution table showing frequency, relative frequency, and cumulative relative frequency.
(b) Calculate the range and IQR.
(c) Calculate the variance, standard deviation, and coefficient of variation.

3.2.3 Students' scores in a written examination are shown in the following frequency distribution:

Score range	Frequency	Frequency cumulative
$40 \le x < 50$	3	3
$50 \le x < 60$	9	12
$60 \le x < 70$	13	25
$70 \le x < 80$	12	37
$80 \le x < 90$	5	42
$90 \le x < 100$	3	45
Total	45	

(a) Calculate the IQR.
(b) Calculate the coefficient of variation.

3.2.4 Use the data of problem 3.2.1 for the following questions:

(a) Draw a stem-and-leaf diagram.
(b) Calculate the IQR.
(c) Calculate the percentage of observations above the 75th percentile.

3.2.5 Using the data of problem 3.2.2 answer the following questions:

(a) Draw a stem-and-leaf diagram.
(b) Find the median score.

3.2.6 Use the data in problem 3.2.2 and

(a) Calculate the variance and standard deviation
(b) Calculate the coefficient of variation.

Answers

Sect. 3.1: *Central Tendency*

3.1.1 $\bar{X} = 214.6$; $M_d = 215.31$; $M_o = 214.07$
3.1.2   Owners $\bar{X} = \$43.39$; $M_d = \$40.35$
       Renters $\bar{X} = \$8.82$; $M_d = \$7.94$
3.1.3   11.88
3.1.4 $\bar{X} = 60.77$; $M_d = 58.63$; $M_o = 57.00$; 71.00
3.1.5   (a) 8.67 %; (b) \$64,791 million; (c) 1990
      (sd = 15.84)
3.1.6   (a) 0.65 %; (b) 2010
3.1.7   (a) 15.02 %; (b) 528,667 (million Baht)

   Sect. 3.2: *Measures of Dispersion*

3.2.1   (b) $R = 41$; IQR = 15; $s^2 = 91.71$; $s = 9.58$; CV = 10.36; $\bar{X} = 92.47$
3.2.2   (b) $R = 82.02$; IQR = 16.25
      (c) $s^2 = 214.99$; $s = 14.66$; CV = 17.54 %
3.2.3   (a) 18.48; CV = 18.71
3.2.4   (b) IQR = 0.38; (c) 24.84 %
3.2.5   (b) 87.65
3.2.6   (a) $s^2 = 200.76$; $s = 14.17$; (b) 17.28 %

Reference

Miah, Md.A.Q., Weber, K.E.: Potential for Slum Upgrading among Owners and Renters. AIT, Bangkok, p. 69, 72 (1990)

Chapter 4
Probability Theory

Abstract Inferential statistics involves drawing conclusion regarding population parameters based on sample data drawn from the same population. In the process probabilities are an inherent part. Probabilities along with their axioms are defined. Calculation of probabilities in different situations is different and is explained with examples.

Keywords Probability · Inferential · Axiom of probability · Examples in probability

Up to this stage we have studied the descriptive statistics—presentation of data, central tendency and dispersion. Now is the time to study how the statistical inferences are to be made. This refers to drawing of conclusions about the population, on the basis of samples drawn from the same population. Accurate conclusions can be made only if the whole population is studied. As such if some conclusions are drawn based on the sample information, there is likelihood of some degree of uncertainty. Probability deals with the nature of this uncertainty. Probability is of fundamental importance in statistical inference. Before going into the details of probability, a good idea of the following terminology would be necessary.

Population: Population or universe is the aggregate of all possible values of a variable or all possible objects whose characteristics are of interest in any particular investigation or enquiry.

Example 1

Incomes of all citizens of a country. It is a finite population. All possible outcomes in tossing a coin. It is an infinite population.

Sample: A sample is the part of a population about which information is gathered. A sample is a relatively small group chosen so as to represent the population. Data for the sample are collected. But the statistician is interested in the whole population.

Example 2

A doctor tests a new drug on 100 patients with malaria, chosen at random. Here the sample consists of 100 patients tested. The population is all malaria patients.

An outcome: An outcome is one particular result of an experiment.

An event: An event is a set of outcomes.

A sample space: A sample space (S) is the set of all possible outcomes of an experiment. The outcomes are also called elements of the sample space.

Example 3

A fair 4-sided die is tossed several times. The sides are numbered E_1, E_2, E_3 and E_4. Here tossing the die several times is an experiment. In any one toss any one of the 4 sides (E_1, E_2, E_3, E_4) may appear. Here possible outcomes are E_1, E_2, E_3 or E_4 (any one at time—not more than one at a time). Appearing E_1 or E_2 or E_3 or E_4 is an event. It is said like this "E_1 is the event that side one will come". The sample space is the aggregate of all sample points.

$$S = (E_1, E_2, E_3, E_4)$$

4.1 Probability Definition

The probability that something occurs is the proportion of times it occurs when exactly the experiment is repeated a very large (preferably infinite) number of times in independent trials. "Independent" here means that the outcome of one trial of the experiment does not affect any other outcome.

Example 4

Let, x = number of times A happened in an experiment,

n = total number of cases,

f = number of cases A did not happen.

Then $n = x + f$ and the probability of A occurring is given by

$$P(A) = \frac{x}{n} = \frac{x}{x+f}$$

4.2 Two Approaches in Calculating Probability

One is a priori probability. In this case we know all possible outcomes in a set of circumstances. So it is possible for us to evaluate the probability of one of the outcomes occurring. Here the key point is that the probability can be obtained in advance, before the event takes place.

The second approach is the empirical probability. In this case we must know how many times the event A occurred in the past, out of a known number of possible outcomes. Then it is assumed that the same proportion will continue to happen into the future.

4.3 Axioms of Probability

There are three axioms of probability as stated hereunder.

1. $P(S) = 1$
2. $0 \leq P(A) \leq 1$
3. If A and B are mutually exclusive events, then

$$P(A \text{ or } B) = P(A) + P(B)$$

The first axiom states that if an event is certain to occur, then its probability to occur is 1. If the event is certain not to occur, then its probability to occur is 0. The second axiom states that the answer to every probability problem will lie between 0 and 1, inclusive. The third axiom is self- explanatory.

If events A and B are the exhaustive events in a mutually exclusive set, then the probability of A occurring is one minus the probability of B occurring. This may be made clear by an example. During a period of 6 months (182 days) a train arrived late on 16 days. So the probability of its arriving late is 16/182. The probability of its arriving in time is $1 - 16/182 = 166/182 = 83/91$.

4.4 Probability in Mutually Exclusive Events

If two events, A and B, are possible outcomes from n occurrences, but cannot take place at one and at the same time, then the probability of A or B occurring is the sum of the probabilities that each will occur.

$$P(A \text{ or } B) = P(A) + P(B)$$

Example 5
A box contains 10 balls of which 2 are red, 3 are black and 5 are green. Therefore, probabilities of drawing specific colored ball are

$$P(\text{red}) = 2/10 = 0.2; \quad P(\text{black}) = 3/10 = 0.3; \quad P(\text{green}) \ 5/10 = 0.5.$$

So

$$P(\text{red}) + P(\text{black}) + P(\text{green}) = 0.2 + 0.3 + 0.5 = 1.0$$

In this example events are mutually exclusive.

The probability of drawing either a red or a black ball is

$$P(\text{red or black}) = P(\text{red}) + P(\text{black}) = 0.2 + 0.3 = 0.5$$

The probability of drawing either a red or a green ball is

$$P(\text{red or green}) = P(\text{red}) + P(\text{green}) = 0.2 + 0.5 = 0.7$$

The probability of drawing either a red or a black or a green ball is

$$P(\text{red or black or green}) = P(\text{red}) + P(\text{black}) + P(\text{green})$$
$$= 0.2 + 0.3 + 0.5 = 1.0$$

4.5   Probability in Independent Events

If two events, A and B, are independent of each other meaning that outcome of one does not affect the outcome of the other, then the probability that both of them will occur is the product of the probabilities that each will occur. Thus

$$P(A \text{ and } B) = P(A) * P(B)$$

Example 6

A manufacturer has two machines to produce similar parts. The probability of a defective item from either of these two machines is 1/100. The probability that an item from machine I is defective is not affected by whether or not another item produced in machine II is defective. So

$$P(A) = P(B) = 1/100.$$

The probability that both items (one from each machine) are defective is

$$P(A \text{ and } B) = P(A) * P(B) = (1/100) * (1/100) = 1/10,000$$

4.6   Probability in Dependent Events (Conditional/Unconditional Probability)

If two events, A and B, are related in such a way that the probability of B taking place depends on the probability of A having occurred, then the probability of both A and B occurring is the product of the unconditional probability of A occurring and the conditional probability of B occurring.

$$P(A \text{ and } B) = P(A) * P(B|A)$$

$P(B|A)$ means conditional probability of B, given A.

Example 7

A random sample is to be taken from a group of 100 houses on a new estate to examine the standard of workmanship. One house is selected randomly and surveyed. Then the second one is selected. This process is continued until the sample size is obtained. Let the sample size be 5.

The probability of first house to be chosen is 1/100. The probability of the second house to be chosen is 1/99, since there are only 99 un-surveyed houses left after selecting the first house. In this way, the probabilities of selecting the third, fourth and the fifth houses are 1/98, 1/97 and 1/96 respectively. Therefore, the probability of any particular five houses being selected is

$$(1/100) * (1/99) * (1/98) * (1/97) * (1/96) = 1/9,034,502,400$$

4.7   Probability in Non-mutually Exclusive Events

If two events, A and B, may occur separately or together, then the probability that A or B will occur is the sum of the probabilities of each occurring minus the probability of both A and B occurring.

$$P(A \text{ or } B) = P(A) + P(B) - P(A \text{ and } B)$$

In this case we cannot write $P(A \text{ or } B) = P(A) + P(B)$ because some elements are common both in event A and event B.

Example 8

We need to select a playing card from a well-shuffled pack. What is the probability of picking either a spade or a queen?

$$P(\text{spade or queen}) = P(\text{spade}) + P(\text{queen}) - P(\text{queen of spade})$$
$$= 13/52 + 4/52 - 1/52$$
$$= 16/52 = 4/13$$

4.8   Probability and Number of Possible Samples

We have studied the probabilities in different situations. In many cases, before the probability of an event can be evaluated, it is necessary to work out the total number of events which could result from the circumstances in question. In other words, we need to calculate the number of possible samples in different circumstances. Three types of circumstances which are likely to come across often, will be described hereafter.

4.8.1   Sampling with Replacement

Suppose, we have 4 different objects. We need to select a sample size of 3. How many samples are possible?

Of the 3, the first object can be selected in 4 different ways. Since the sampling is with replacement, the second object can also be selected in 4 different ways. Similarly, the third object can also be selected in 4 different ways. Thus, the total number of different possible samples will be

$$4 \times 4 \times 4 = 4^3 = 64$$

The probability of selecting a particular sample is 1/64, when sampling is with replacement.

In general, if there are n different objects and a sample size of r to be selected, then the total number of possible samples is n^r.

4.8.2   Sampling Without Replacement (Order Important)

In this case, the arrangement is different from that stated in Sect. 4.8.1. Here, if we select one thing, we do not put it back to the sample space, so that for the second choice we are left with less one object.

Suppose, we have 4 letters A, B, C and D. How many different arrangements (samples) can be made taken 2 at a time?

$$
\begin{array}{llll}
A: & AB & AC & AD \\
B: & BA & BC & BD \\
C: & CA & CB & CD \\
D: & DA & DB & DC
\end{array}
$$

The first letter can be selected in 4 different ways. When the first letter is selected, we are left with 3 letters. So the second letter can be selected in $4 - 1 = 3$

different ways. Therefore, the total number of different arrangements (samples) is $4 * 3 = 12$. This type of arrangement is known as permutation. Remember, here the order is important. AB is different from BA.

In general, the number of permutations (arrangements) of n items taken r at a time is given by

$$_nP_r = n(n - 1)(n - 2)\ldots(n - r + 1); \quad \text{the product is taken over } r \text{ terms}$$

$$= \frac{n!}{(n - r)!}$$

In the example illustrated above (letters A, B, C, D)

$$_nP_r = {_4}P_2 = \frac{4!}{(4 - 2)!} = \frac{4 * 3 * 2 * 1}{2 * 1} = 12$$

The probability of selecting a particular arrangement (sample) is 1/12.

4.8.3 Sampling Without Replacement (Order Irrelevant)

There are certain circumstances when order does not matter, contrary to the arrangement shown in Sect. 4.8.2. Here, we come across the concept of combination. Suppose, a quality control inspector takes a handful of components from a bin to check the number defective. He will select his sample in one fell swoop. In this case, there is no question of order of selection because all sample units are taken simultaneously. This is combination.

The number of combinations (arrangements) of n items taken r at a time is given by

$$_nC_r = \frac{n!}{(n - r)!r!}$$

In the example (letters A, B, C, D), how many arrangements or combinations (samples) can be made taken 3 at a time?

$$_nC_r = {_4}C_3 = \frac{4!}{(4 - 3)!3!} = \frac{4 * 3 * 2 * 1}{1 * 3 * 2 * 1} = 4$$

The arrangements are ABC, ABD, ACD and BCD. Here ABC is not different from ACB or BCA.

The probability of selecting a particular arrangement (sample) is 1/4.

Problems

4.1   Interpret

(a) $P(S) = 1$
(b) $0 \leq P(A) \leq 1$

4.2   You are interested in studying the productivity of land in a certain section of your countryside. There are 50 plots approximately of equal sizes. You want to take a sample of 5 plots.

(a) How many possible samples there can be?
(b) What is the probability of selecting a particular sample?

4.3   Students enrollments in 1991 in AIT are as shown in the following table.

Term	Division/School	Centers	Total
January	805	85	890
May	851	50	901
September	869	80	949
Total	2525	215	2740

Of all the students enrolled in 1991, one is selected by random process. Find the probability that he/she was enrolled in:

(a) Divisions/School in May term.
(b) Centers in January and Division/School in September term.

4.4   Suppose a list of eight real estate projects, each in a different location is presented to a board of management of the company. Each member may rank the four projects that the company may undertake. How many conceivable different rankings of the projects may be possible?

4.5   A real estate firm is to develop two sites (*A* and *B*). From location consideration, each site has equal weightage. From marketing consideration, each is equally likely to be profitable or non-profitable. Assume that the site location and marketing condition are independent. What is the probability that the site *A* will be profitable?

4.6   The Student Union of AIT wants to choose a committee consisting of three judges for the cultural show from among three men faculty and two female faculty. The women students want to know the probability that no women faculty will be chosen. What is the probability?

4.7   Ten graduates are applying for two positions—one for research supervisor and the other for research associate. Each graduate is equally qualified for the positions. You are one of the graduates applying for the positions. Suppose that the choices will be made at random.

(a) How many possible choices are there for the management to select the candidates?

(b) In how many choices which have been made in (a), you are likely to be included?

(c) In how many choices made in (a) you are likely to be chosen for a research associate?

(d) What is the probability that you will be chosen for an appointment?

(e) What is the probability that you will be chosen for a research associate?

4.8 A committee consisting of four gentlemen and three ladies is to be selected from among six gentlemen and four ladies. One gentleman out of the six and one lady out of the four are considered aggressive.

(a) How many different combinations for the committee are possible?

(b) What is the probability that both aggressive gentleman and the lady will be included in the committee? Assume random selection.

(c) What is the probability that the committee will contain no aggressive member?

4.9 In a survey of economic indicators in a certain city it was revealed that 20 % of all the working age people were engaged in service sector and 30 % were engaged in the industry sector. It was also found that 10 % of all the working age people were engaged in the formal sector (which includes both service and industry sector). If a worker is chosen at random, what is the probability that he is engaged in at least one of the service sector and industry sector?

4.10 In a countrywide survey it was found that 55 % of the farmers produce rice, 25 % of them produce jute and 15 % of them produce both the cash crops. If a farmer is chosen at random from the countryside, what is the probability that he produces at least one of the two cash crops?

4.11 Fifty per cent students get A grade in written exam, 50 % get A in assignment and 30 % get A both in written exam and assignment. A student is selected at random. What is the probability that he gets A grade at least in one of the categories?

4.12 Consider that data in Problem 4.11. If 60 % of those students who get A grade in at least one category, are chosen for a particular study, what is the probability that a randomly selected student will be chosen for the study?

Answers

4.2 (a) 2118760; (b) 1/2118760
4.3 (a) 0.31; (b) 0.0
4.4 1680
4.5 0.50
4.6 0.10
4.7 (a) 90; (b) 18; (c) 9; (d) 0.2; (e) 0.1

4.8   (a) 60; (b) 0.5; (c) 0.083
4.9   0.40
4.10   0.65
4.11   0.70
4.12   0.42

Chapter 5
Probability Distributions

Abstract A random variable is introduced. More commonly used probabilities are introduced. Discrete probability distributions introduced are binomial probability distribution, multinomial probability distribution, hypergeometric probability distribution, and Poisson probability distribution. Continuous probability distributions introduced are normal probability distribution, Student's t distribution, F distribution, and Chi-square distribution. Important features are presented in a tabular form. The technique of fitting the data of unknown distribution to known probability distribution is introduced.

Keywords Probability distribution · Important features · Data fitting to distributions

We have studied the probabilities associated with different outcomes. Now we ought to learn something of the distribution of probabilities, usually called probability distribution. Associated with probability distribution is a random variable.

A random variable is a well-defined rule for making the assignment of a numerical value to any outcome of the experiment. In other words, A variable that has numerical values and has probabilities associated with each value is called a random variable. A random variable is simple a numeric value that has an associated probability distribution.

A discrete random variable is a random variable whose values have gaps between them; not all values are possible in the range of values. A continuous random variable is a random variable whose values include all the numbers within a certain range (refer to discrete and continuous variables in Sect. 1.6).

The distribution of probabilities of a discrete random variable is called discrete probability distribution. The distribution of probabilities of a continuous random variable is called continuous probability distribution.

It is important to distinguish carefully between values of a random variable (such as 0, 1, 2, etc.) and the probabilities of these values (such as 0.065, 0.130 etc.). There is no restriction on the numeric values of a random variable may take, except that a discrete variable has gaps in its range and a continuous variable does not. But

© Springer Science+Business Media Singapore 2016
A.Q. Miah, *Applied Statistics for Social and Management Sciences*,
DOI 10.1007/978-981-10-0401-8_5

there are definite restrictions on the probabilities of a random variable. The probabilities in a probability distribution must each be nonnegative; must not exceed 1.0; and must sum, in total, to 1.0.

Basically there are two types of distributions. These are

1. Discrete Probability Distribution
   The distribution of a discrete random variable is called a discrete probability distribution. A discrete random variable is a random variable which has discrete measurement (not continuous).
2. Continuous Probability Distribution
   The distribution of a continuous random variable is called a continuous probability distribution. A continuous random variable is a random variable which has a continuous measurement.

5.1 Discrete Probability Distribution

There are a couple of discrete probability distributions. The more commonly used ones are

1. Binomial Probability Distribution
2. Multinomial Probability Distribution
3. Hypergeometric Probability Distribution
4. Poisson Probability Distribution.

5.1.1 The Binomial Distribution

A random experiment is carried out. It has two possible outcomes—"success" (yes) and "failure" (no). Outcomes are exclusive and exhaustive. P is the probability of success (yes) in a single trial. Total number of trials is n. The distribution of number of success x is called the binomial distribution and the experiment is called the binomial experiment. Its probability function is

$$P(x) = {}_nC_xP^x(1-P)^{n-x}$$

$$= \frac{n!}{(n-x)!x!}P^x(1-P)^{n-x}$$

The binomial experiment and distribution are very important in a variety of statistical inferences. This is simply because of the fact that the proportion of elements in a population possessing a certain characteristic of interest can be viewed as the probability of success in a binomial experiment.

The mean of the expected value of x (success), the binomial random variable is given by

$$\mu = E(x)$$
$$= n \cdot p$$

The variance of the binomial random variable is

$$\sigma^2 = n \cdot p \cdot q$$
$$= n \cdot p \cdot (1 - p)$$

Example 1

A student appeared in the examination of five papers. He/she believes that his/her probability of getting A in each of the papers is 0.40. The distribution of number papers getting A is binomial with $n = 5$ and $p = 0.40$.

The probabilities of getting A in different papers are as follows:

$$P(0 \text{ subject}) = P(0) = \frac{5!}{5!0!}(0.40)^0(0.60)^5$$
$$= (0.60)^5$$
$$= 0.078$$

$$P(1 \text{ subject}) = P(1) = \frac{5!}{4!1!}(0.40)^1(0.60)^4$$
$$= 5(0.40)^1(0.60)^4$$
$$= 0.259$$

$$P(2 \text{ subjects}) = P(2) = \frac{5!}{3!2!}(0.40)^2(0.60)^3$$
$$= 10(0.40)^2(0.60)^3$$
$$= 0.346$$

$$P(3 \text{ subjects}) = P(3) = \frac{5!}{2!3!}(0.40)^3(0.60)^2$$
$$= 10(0.40)^3(0.60)^2$$
$$= 0.230$$

$$P(4 \text{ subjects}) = P(4) = \frac{5!}{1!4!}(0.40)^4(0.60)^1$$
$$= 5(0.40)^4(0.60)^1$$
$$= 0.077$$

$$P(5 \text{ subjects}) = P(5) = \frac{5!}{0!5!}(0.40)^5(0.60)^0$$
$$= (0.40)^5(0.60)^0$$
$$= 0.010$$

Probability of getting A in no more than two subjects

$$= p(0) + p(1) + p(2)$$
$$= 0.078 + 0.259 + 0.346$$
$$= 0.683$$

Example 2

Suppose that the number of students favoring a change in curricula in AIT follows a binomial distribution. Previous reports show that 80 % of the students favored the change. If 200 students are selected at random, find

(a) the mean
(b) the variance
(c) the standard deviation of the students favoring the change.

Solution

In this case $p = 0.80$ and $n = 200$

$$\text{Mean} = E(x)$$
$$= n * p$$
$$= 200 * (0.80)$$
$$= 160$$
$$\text{Variance} = n * p * q$$
$$= n * p * (1-p)$$
$$= 200 * (0.80) * (1-0.80)$$
$$= 200 * (0.80) * (0.20)$$
$$= 32$$
$$\text{Standard Deviation} = \sqrt{32}$$
$$= 5.65$$

5.1.2   *Multinomial Probability Distribution*

The multinomial distribution is a generalization of the binomial distribution. In the binomial distribution, we have two possible outcomes each outcome having a certain probability. In the multinomial distribution, we consider more than two outcomes and each outcome is associated with a certain probability. Each outcome is a discrete value. Thus, the multinomial probability distribution is a discrete probability distribution.

Let,

n = number of trials
k = total number of possible outcomes
n_1 = number of occurrences of outcome 1
n_2 = number of occurrences of outcome 2
...
n_k = number of occurrences of outcome k
p_1 = probability associated with outcome 1
p_2 = probability associated with outcome 2
...
p_k = probability associated with outcome k

Then

$$n_1 + n_2 + \cdots + n_k = n$$
$$p_1 + p_2 + \cdots + p_k = 1.00$$

The probability function is given by

$$P(n_1, n_2, \ldots, n_k) = \frac{N!}{(n_1! * n_2! * \cdots n_k!)} * (p_1^{n_1} * p_2^{n_2} * \cdots p_k^{n_k})$$

Example 3

Labor force distribution in Thailand during 2009 shows the following composition:

Agriculture   42.6 %
Service   37.2 %
Industry   20.2 %
Total   100.00

If a sample of 12 is taken at random, find the probability that

1. The sample will contain 6 persons from agriculture sector, 4 persons from the service sector, and 2 persons from industry sector.
2. The sample will contain 8 persons from the agriculture sector and 4 persons from the industry sector.

Solution

Let us denote

Agriculture sector = 1
Service sector = 2
Industry sector = 3

Then

$$n_1 = 6; n_2 = 4; n_3 = 2 \text{ for question (1) above.}$$
$$n = n_1 + n_2 + n_3 = 6 + 4 + 2 = 12.$$
$$n_1 = 8; n_2 = 0; n_3 = 4 \text{ for question (2) above.}$$
$$n = n_1 + n_2 + n_3 = 8 + 0 + 4 = 12.$$
$$p_1 = 0.426; p_2 = 0.372; p_3 = 0.202.$$

Therefore, $p = p_1 + p_2 + p_3 = 0.426 + 0.372 + 0.202 = 1.00$
Now,

$$P(n_1, n_2, n_3) = \frac{N!}{(n_1! * n_2! * n_3!)} * (p_1^{n_1} * p_2^{n_2} * p_k^{n_3})$$

For Question (1):

$$p = \frac{12!}{6!4!2!} * (0.426)^6 (0.372)^4 (0.202)^2$$
$$= 0.0647$$

Question (2):

$$p = \frac{12!}{8!0!4!} * (0.426)^8 (0.372)^0 (0.202)^4$$
$$= 0.000894$$

5.1.3   Hypergeometric Distribution

The hypergeometric distribution describes the number of successes in a sequence of n draws from a finite population without replacement. It is unlike the binomial distribution which describes the number of successes for draws with replacement. The hypergeometric distribution is a discrete probability distribution.

The situation in the hypergeometric distribution may look like a binomial distribution since there are success and failure. But in binomial distribution, the probability of success or failure in each trial is the same. In hypergeometric distribution, probability of success or failure in each trial is not the same because the sampling is done without replacement. Consequent upon the effect of sampling without replacement, the size of the remaining population changes as we remove each unit in each of the trials.

If X is a random variable, its hypergeometric probability distribution is given by

$$P(X = k) = \frac{{}^{m}C_k * {}^{N-m}C_{n-k}}{{}^{N}C_n}$$

The parameters are N, m, and n.
The notations used in the above formula are as follows:

P = probability of X being equal to k,
N = population size,
m = the number of items successful in the whole population,
n = sample size,
k = the number of items successful in the sample,
${}^{m}C_k$ = combination of m things taken k at a time,
${}^{N-m}C_{n-k}$ = combination of $(N - m)$ things taken $(n - k)$ at a time,
${}^{N}C_n$ = combination of N things taken n at a time.

$$ {}^{N}C_n = \frac{N!}{n!(N - n)!} $$

$N!$ is factorial N. Its value is multiplication of all integers from N to 1. For example, $5! = 5 * 4 * 3 * 2 * 1 = 120$.
Sum of all the values of p in a particular event is equal to one.
It may be noted that in a population of size N taking a sample of size n, number of all possible samples is ${}^{N}C_n$. So the probability that a particular sample will be drawn is equal to $1/{}^{N}C_n$.
The notations in the formula can be summarized as in shown in the following table (Wikipedia website, 08 January 2010).

Situation	Drawn	Not drawn	Total
Successes	k	$m - k$	m
Failures	$n - k$	$N - (m - k) - n$	$N - m$
Total	n	$N - n$	N

The mean of the random variable of the distribution is given by

$$ \mu = \frac{nm}{N} $$

The variance of the random variable of the distribution is given by

$$ \sigma^2 = \frac{nm(N - m)(N - n)}{N^2(N - 1)} $$

Example 4

In a small farming community there are 12 farmers. Out of those, 5 have farming education. From the community a sample of 4 farmers are drawn at random. What is the probability that 2 farmers will have farming education?

In this example,

$$N = 12$$
$$m = 5$$
$$n = 4$$
$$k = 2$$

We have

$$P(X = k) = \frac{{}^{m}C_k * {}^{N-m}C_{n-k}}{{}^{N}C_n}$$

Putting these figures into the formula, we get as follows:

$$p = \frac{{}^{5}C_2 * {}^{7}C_2}{{}^{12}C_4}$$
$$= \frac{\frac{5!}{2!3!} * \frac{7!}{2!5!}}{\frac{12!}{4!8!}}$$
$$= \frac{210}{495}$$
$$= 0.424$$

For this distribution, we can also calculate the mean and the variance of the random variable as follows:

The mean is given by

$$\mu = \frac{nm}{N} = \frac{4 * 5}{12} = 1.667$$

The variance is given by

$$\sigma^2 = \frac{nm(N - m)(N - n)}{N^2(N - 1)} = \frac{4 * 5(12 - 5)(12 - 4)}{12^2(12 - 1)} = 0.707$$

5.1.4   Poisson Distribution

This distribution is named after French mathematician Simeon Denise Poisson (1971–1940). It is a discrete distribution. It expresses the probability of a number of events occurring in a fixed period of time if these events occur with a known average rate and independently of the time since the last event (Wikipedia website 07 January 2010). The distribution is suitable for analysis of time interval related data. It can also be used for number of events in other specified intervals such as distance, area, volume etc.

The probability is given by the following:

$$p(k, \lambda) = \frac{\lambda^k e^{-\lambda}}{k!},$$

where,

p = probability
k = number of occurrences
λ = expected (mean) number of occurrences that occur during the given interval (a non-negative, real and whole number, no fractional number)
e = base of natural logarithm = 2.71828.
$k!$ = factorial k (multiplication of all numbers starting from 1 until k).

The following assumptions are made in a Poisson distribution:

1. The probability that an event will occur during a time interval is same for all time intervals.
2. The number that occurs during one interval is independent of the number that occurs during another time interval.

Example 5
In a road intersection, the mean number of traffic light violation has been observed to be 12 per day. What is the probability that 8 traffic light violations will be observed in a particular day?
Here,

$$\lambda = 12$$
$$k = 8$$
$$e = 2.71828.$$

The formulae is

$$p(k, \lambda) = \frac{\lambda^k e^{-\lambda}}{k!}$$

Therefore,

$$p = \frac{12^8 * 2.718282^{-12}}{8!}$$
$$= 0.0655$$

5.1.5   *Important Features*

Some important features of the four discrete probability distributions discussed above are shown in the following table. These will act as a guide for identifying the distribution for a specific data set or an event.

Distribution	Outcomes	Sampling	Remark
Binomial	2	With replacement	
Multinomial	More than 2	Without replacement. If the population is large, both replacement and without replacement converge; Practically no change in probability	Generalization of binomial distribution
Hypergeometric	More than 2	Without replacement	
Poisson	1	NA	Data per unit interval (time, distance, area, volume)

5.2   Continuous Probability Distribution

1. Normal Probability Distribution
2. Student's t Distribution
3. F Distribution
4. Chi-Square Distribution

5.3   The Normal Distribution

The distribution of a continuous random variable is normal distribution. This is the most widely known and used of all distributions. The importance of the normal distribution lies in the fact that a vast number of phenomena have approximately normal distribution. Normal distribution has a wide application in statistics. Another importance of the normal distribution is that it has a number of mathematical properties. Some of the properties of the normal distribution are outlined in the following sections.

5.3.1   Properties/Characteristics of Normal Distribution

Among others the following properties/characteristics need to be remembered.
   The mathematical equation of the normal curve depicting the normal distribution is

$$f(x) = \frac{1}{\sigma\sqrt{(2\pi)}} e^{-(1/2)(x-\pi)2/\sigma 2} \quad \text{for } -\infty \leq x \leq \infty$$

$$f(x) = \frac{1}{\sigma\sqrt{(2\pi)}} e^{-(z2/2)} \quad \text{putting } z = (x-\mu)/\sigma$$

where,

$\pi = 3.1416$ (a constant)
$e = 2.71828$ (base of natural logarithm)
$\mu = $ mean of the normal distribution
$\sigma^2 = $ variance of the normal distribution
$z = $ standard variate

   The distribution has a bell-shaped symmetrical distribution. The y-axis (ordinate) shows the probability density function pdf. The x-axis can have two scales x and z. The x and/or z values can be shown along this axis. This is depicted in Fig. 5.1.

(i) The normal distributions have the characteristic "bell" shape and are symmetrical and are unimodal. Many distributions beside the normal distributions are unimodal and symmetric. But only a normal distribution has a particular shape given by the above mathematical formula. Only distributions that satisfy this formula, and hence have this particular shape, qualify as normal distributions. Other unimodal, symmetric distributions may be approximately normal, but they are not exactly normal unless they satisfy the mathematical formula.

(ii) The total area under the normal curve is given by

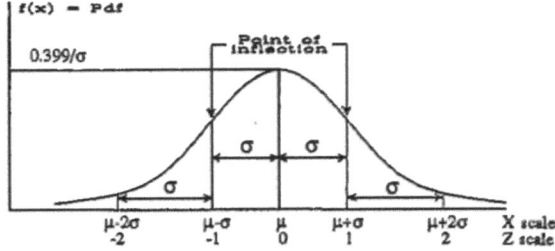

Fig. 5.1 Probability density function of normal distribution

$$\int\limits_{-\infty}^{+\infty} f(x)\mathrm{d}x = 1.00$$

Thus the total area under the normal curve is 1.00. The area represents the probability. The probability that a score is between a and b equals the area under the normal curve between a and b.

(iii)   The curve extends to infinity in both directions. The curve gets very close to the horizontal axis, but actually does not touch the axis. Most of the area under the normal curve falls between -3 and $+3$ times the standard deviation.

(iv)   For a standard normal curve, the mean is 0 and variance is 1.00.

 (v)   The distribution is unimodal; the height decreases on either side of the peak. The slope of the curve becomes steeper and steeper until a point maximum steepness is reached. Thereafter, the curve becomes less and less steep. The turning point is called the point of inflection. There are two points of inflection on either side of the peak. Each of the two points of inflection is exactly one standard deviation from the mean. Thus the distance between the two points of inflection is two standard deviations.

(vi)   All normal distributions are not identical. Some are broad with a wide range; others fall with a narrow range. But all share a valuable property: with a knowledge of the mean and the standard deviation, every characteristic can be determined. There is another more important fact than this. Measurement of the standard deviations from the mean establishes positions between and beyond which known properties of the total frequencies lie.

See the following diagram showing three normal distributions with standard deviations (spreads) of different sizes (Fig. 5.2).

5.3.2   Some Examples

(i) $P(z \leq 2.2) = ?$

z refers to the area marked shaded in the sketch.

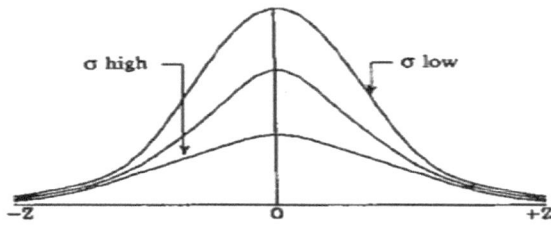

Fig. 5.2 Normal distribution with different spreads

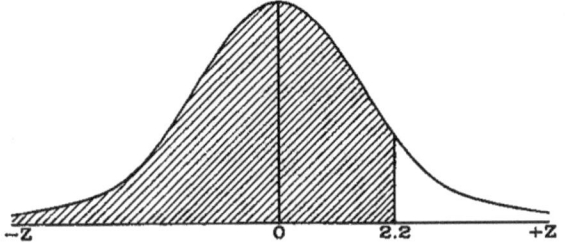

Total area under the normal curve is 1.00; area to the right of $z = 2.2$ is 0.0139 from the table.
Therefore,

$$P(z \leq 2.2) = 1 - 0.0139$$
$$= 0.9861$$
$$= 98.61\,\%$$

(ii) $P(1.38 \leq z \leq 1.42) = ?$

The probability of z lying between 1.38 and 1.42 is the area shown in the sketch.
From the table, area for $z = 1.38$ is 0.0838 and for $z = 1.42$ is 0.0778.
Therefore, the area between the two points $= 0.038 - 0.0778 = 0.006$.
Therefore, $P(1.38 \leq z \leq 1.42) = 0.006 = 0.60\ \%$.

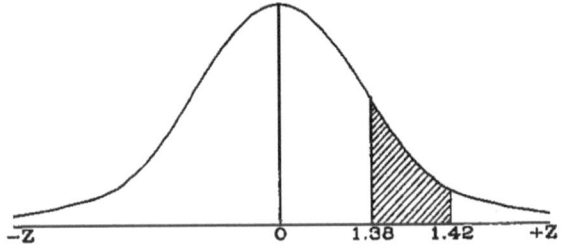

$$\mu = 15.3$$
(iii) $$\sigma\bar{x} = 0.683$$
$$P(\bar{x} \leq 14) = ?$$

In order to find out the probability of $\bar{x}$ having values less than or equal to 14, we need to convert its value to z scores as follows:

$$Z = \frac{\bar{x} - \mu}{\sigma_{\bar{x}}} = \frac{14 - 15.3}{0.683} = -1.9$$

From the table we find that the area to the left of $z = 1.9$ is 0.0287.
Therefore, $P(\bar{x} \leq 14) = P(z \leq -1.9) = 0.0287 = 2.87\,\%$.

5.4   The *t* Distribution

The "*t* distribution" is the short expression usually used for "Student's *t* Distribution". The distribution was developed by William S. Gosset. He did not use his own name, but the pen name of student and hence the name "Student's *t* Distribution".

The Z transformation statistic used in case of normal distribution is applicable when the following are true:

(i)   When the population variance σ^2 is known, irrespective of the size of sample.

(ii) The population variance σ^2 is unknown, but the sample size is large ($n > 30$). If $n > 30$, the sample estimation of the unknown population variance s_x^2 is good approximation to σ^2.

If none of these conditions is fulfilled, the Z transformation is not appropriate. But if the parent population is normal or approximately normal and the sample size is small ($n < 30$), we can apply another transformation, based on Student's *t* distribution. The transformation is

$$t = \frac{\bar{x} - \mu}{s/\sqrt{n}}$$

The *t* distribution has the following properties:

(i)   There is not just one *t* distribution, The *t* distributions are many, in fact an infinite number. Each distribution is associated with a parameter known as degree of freedom (df). In the expression

$$t = \frac{\bar{x} - \mu}{s/\sqrt{n}}$$

the degree of freedom (df) = $n - 1$.

(ii)   In appearance the *t* distribution is similar to normal distribution. It is bell shaped and symmetric about zero (mean is zero). In general, the variance is greater than 1.

(iii) It extends from minus infinity to plus infinity.

(iv)   The curve is flatter (has more spread) than the normal curve, because of larger standard deviation. However, the total area under the curve is 1.

As the sample size becomes larger, the *t* distribution approached the standard normal distribution. In fact, for $n \geq 30$, a *t* distribution is approximately standard normal. This means for $n \geq 30$, if one uses z distribution instead of t, the error will be very small.

The t curve and z curve (standard normal curve) are compared in Fig. 5.3.

Some examples will help to use the *t* distribution. The *t* table has been given in the appendix. The left most column shows the degree of freedom, the upper row

Fig. 5.3 Comparison of t and z curves

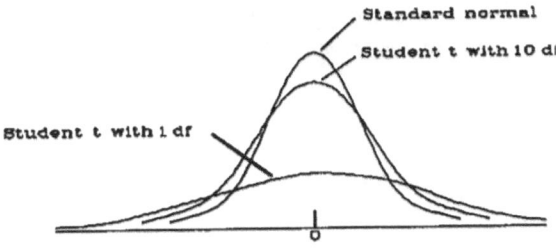

shows the alpha values. The figures inside the table are areas of the shaded part, indicating probabilities.

For 10 degrees of freedom:

$$P(t \geq 1.812) = 0.05$$

Also,

$$P(t \leq -1.812) = 0.05$$

(since the distribution is symmetrical)

For 25 degrees of freedom:

$$P(t \geq 2.060) = 0.025$$

Also,

$$P(t \leq -2.060) = 0.025$$

For 15 degrees of freedom:

$$P(-1.753 \leq t \leq 2.602) = (0.50-0.05) + (0.50-0.01)$$
$$= 0.45 + 0.49$$
$$= 0.94$$

Suppose, the students' scores are normally distributed. A sample of 14 students produced a mean of 85.78 with a standard deviation of 25.64. What proportion of the students will have scores above 95?

Here,

$$t = \frac{95.00 - 85.78}{25.64/\sqrt{14}}$$
$$= \frac{9.22}{6.829}$$
$$= 1.35$$

Therefore, $P(x \geq 95.00) = P(t \geq 1.35) = 0.10$ with 13 df.

Therefore, the proportion of students who will have scores above 95 is 0.10, i.e., 10.00 %.

5.5 The *F* Distribution

The *F* distribution was developed by a British statistician Sir Ronald A. Fisher. Suppose, there are two populations with variances of σ_1^2 and σ_2^2 and two independent random samples of sizes n_1 and n_2 are taken from the two populations producing sample variances of s_1^2 and s_2^2, respectively. Then

$$F = \frac{s_1^2/\sigma_1^2}{s_2^2/\sigma_2^2}$$

has *F* distribution with numerator degrees of freedom $n_1 - 1$ and denominator degrees of freedom $m_2 - 1$.

An *F* distribution with numerator degrees of freedom u and denominator degrees of freedom v is denoted by $F_{u,v}$.

The properties of *F* distribution are

(i)   The *F* distribution is not just one distribution. There are an infinite number of *F* distributions.

(ii) Each *F* distribution has a pair of degrees of freedom: numerator degrees of freedom $(n_1 - 1)$ and denominator degrees of freedom $(n_2 - 1)$.

(iii) An F curve starts from zero and extends to the right up to infinity, i.e., its range is 0 to infinity.

(iv) The total area under the curve is 1.

The shape of the *F* curve is shown in Fig. 5.4.

A few examples may be used to see the *F* distributions.

Example 6

Use the *F* distribution table and find out the values of

$F_{0.05,10,15}$; $F_{0.10,10,15}$; $F_{0.05,1,4}$; $P(F_{10,20} > 2.35)$; $P(F_{20,10} > 4.41)$

Fig. 5.4 The *F* distribution curve

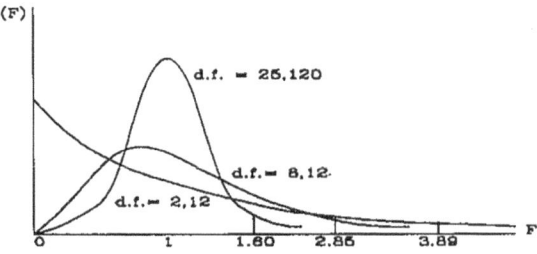

Solution

$$F_{0.05,10,15} = 7.71$$
$$F_{0.10,10,15} = 3.80$$
$$F_{0.05,1,4} = 4.24$$
$$P(F_{10,20} > 2.35) = 0.05$$
$$P(F_{20,10} > 4.41) = 0.01$$

Example 7

Two population variances are 25.32 and 20.26. Samples sizes of 21 from population1 and 17 from population2 produced variances of 10.58 and 27.60, respectively. What is the F value? What is the probability that F will be greater than this value? Here,

$$\sigma_1^2 = 25.32; \sigma_2^2 = 20.26$$
$$s_1^2 = 10.58; s_2^2 = 27.60$$
$$df_1 = 21-1; df_2 = 17-1$$
$$= 20 := 16$$
$$F = \frac{s_1^2/\sigma_1^2}{s_2^2/\sigma_2^2}$$
$$= \frac{10.58/25.32}{27.60/20.26}$$
$$= \frac{0.4178}{1.3623}$$
$$= 0.3067$$

So, $P(F_{20,16} > 0.3067) > 0.01$ from table.

The tabulated values of F distributions are upper percentage points of F. The lower percentage points of F can be calculated from the following relationship:

$$F_{1-\alpha,u,v} = \frac{1}{F_{\alpha,u,v}}$$

Thus,

$$F_{0.95,1,4} = \frac{1}{F_{0.05,1,4}} = 1/7.71 = 0.1297$$

5.6   The Chi-Square Distribution

The Chi-square distribution is another distribution useful in many statistical inferences. The Chi-square distribution is not a single distribution, but a family of distributions. For each degree of freedom, there is a member of the family.

Chi-square is a random variable and is defined to be the sum of squares of the variables $z_1, z_2, z_3, \ldots, z_d$. It is denoted by χ^2, i.e., the square of the Greek letter chi. Thus,

$$\chi^2 = z_1^2 + z_2^2 + z_3^2 + \cdots + z_d^2.$$

The Chi-square distribution is the distribution of the random variable χ^2. As an illustration, take a sample of size D from a normal distribution of z scores. Then determine

$$\chi^2 = \frac{(x_1 - \mu)^2}{\sigma^2} + \frac{(x_2 - \mu)^2}{\sigma^2} + \frac{(x_3 - \mu)^2}{\sigma^2} + \cdots + \frac{(x_d - \mu)^2}{\sigma^2}$$
$$= z_1^2 + z_2^2 + z_3^2 + \cdots + z_d^2$$

The sum we are getting as above is the first score in the Chi-square distribution for D degrees of freedom. If you repeat the process for all different samples of size D, you will get the whole Chi-square Distribution for D degrees of freedom. There is different distribution foe each different value of D. All Chi-square distributions have some common properties. These are as follows:

(i)   Every Chi-square distribution extends indefinitely to the right from zero.
(ii)   The Chi-square distribution has only one (right) tail.
(iii) The Chi-square probability density function has only one parameter, the degree of number of freedom. The number of degrees of freedom completely determines what the shape of $f(\chi^2)$ will be. When the number of degrees of freedom is small, the shape of the density function is highly skewed to the right. As the number of degrees of freedom increases, the curves become more and more bell shaped and approach the normal curve in appearance. The χ^2 curve starts from zero and extends up to infinity to the right unlike the normal curve which extends from $-\infty$ to $+\infty$.

An important application of Chi-square distribution lies in the relationship of the sample and the population variances. Let us suppose that x is a variable with normal distribution with unknown means μ and unknown variance σ^2. Let us also suppose that $x_1, x_2, x_3, \ldots, x_n$ be a random sample of size n from the same population and let the sample variance be s^2. Then

$$\frac{(n-1)s^2}{\sigma^2}$$

follows a Chi-square distribution with $n - 1$ degrees of freedom. Thus,

$$\chi^2 = \frac{(n-1)s^2}{\sigma^2}$$

Using this relationship, the confidence interval for the population variance can be constructed.

If two samples of sizes n_1 and n_2 with variances S_1^2 and S_2^2 are taken from two populations with variances σ_1^2 and σ_2^2, respectively, then

$$\chi_1^2 = \frac{(n_1 - 1)S_1^2}{\sigma_1^2}$$

$$\chi_2^2 = \frac{(n_2 - 1)S_2^2}{\sigma_2^2}$$

Dividing one by the other, we get

$$\chi_1^2/\chi_2^2 = \frac{(n_1 - 1)S_1^2}{\sigma_1^2} \Big/ \frac{(n_2 - 1)S_2^2}{\sigma_2^2}$$
$$= \frac{S_1^2/\sigma_1^2 \, \mathrm{df}(n_1 - 1)}{S_2^2/\sigma_2^2 \, \mathrm{df}(n_2 - 1)}$$

Thus, it follows that if W and Y are two Chi-square random variables,

$$F = \frac{W/u}{Y/v}$$

Follows F distribution with u df in the numerator and v df in the denominator (Fig. 5.5).

The area under the curve equals one. The area also represents the probability. A few examples will help us in the use of the Chi-square distribution.

Example 8
Find the value of $\chi^2_{0.05,4}$.

Fig. 5.5 The Chi-square distribution

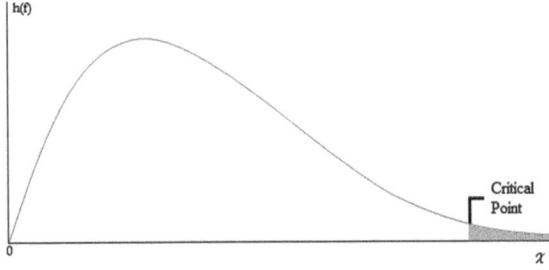

The first figure 0.05 associated with the Chi-square represents the area or the probability in the Chi-square distribution and the second figure 4 represents the degrees of freedom. From the table of Chi-square distribution, the value of the required Chi-square is 9.49. If x represents a Chi-square random variable, the present problem could be represented as

$$P(x > 9.49) = 0.05.$$

Example 9

Find the value of $P(\chi^2 > 19.02)$ for 9 degrees of freedom.

In the Chi-square distribution table, looking along the degrees of freedom, reaching at 9 and then moving to the right along the same row, we note the figure 19.02. Moving upward along the same column we find that the probability is 0.025. It may be noted that the probability (or area) denoted along the top row of the table shows the area on the right hand tail of the curve shown blank.

Example 10

A random sample of size 12 is taken from a normal population with its variance σ^2 of 5. The sample variance is S^2. Find the value of x such that

(a) $P(11S^2/5 > x) = 0.05$
(b) $P(S^2 > x) = 0.10$

Solution

(a) $P\left\{\frac{11S^2}{5} > x\right\} = 0.05$

$\Rightarrow P\left\{\frac{(12-1)S^2}{5} > x\right\} = 0.05$

Therefore, $x = \chi^2_{0.05,11}.$

$= 19.68$

(b) $P(S^2 > x) = 0.10$

$\Rightarrow P\left\{\frac{11S^2}{5} > \frac{11x}{5}\right\} = 0.10$

$\Rightarrow P\left\{\frac{(12-1)S^2}{5} > \frac{11x}{5}\right\} = 0.10$

Therefore, $\frac{11x}{5} = \chi^2_{0.10,11}$

$= 17.28$

So, $x = \frac{(17.28) * (5)}{11}$

$= 7.85$

There are three requirements for the validity of Chi-square tests. The requirements are as follows:

(i)   Random Sampling: The subjects in the sample (or samples) must be selected at random from the population(s) of interest.
(ii) Independence of Observations: Each observation must be independent of every other observations in the study.
(iii) Large Expected Frequencies: Each expected (not observed) frequency must be 5 or more in order for the Chi-square distribution to be satisfactory approximation to the normal distribution of χ^2_{obs}.

5.7 Joint Probability Distribution

Joint probability distribution is the probability distribution of two or more random variables happening together. For example, the rice production on an agricultural land depends jointly on availability of rain and use of fertilizer. Here availability of rain and use of fertilizer are two random variables.

5.7.1 Discrete Joint Probability Distribution

In case of discrete random variables, the joint probability mass function for two random variables X and Y is given by

$$P(x, y) = p(X = x, Y = y).$$

This notation indicates the joint probability of X and Y when X assumes a specific value of x and Y assumes a specific value of y. In this case, the properties of the joint probability distribution are as follows:
For all values of x and y,

$$P(x, y) \geq 0$$

$\sum_x \sum_y p(x, y) = 1$; this means summation of all probabilities in a specific case is one.

Example 11
In a study of the AIT students' performance, one researcher obtained the following distribution (number of students). Notice that there are two random variables GPA

measured by grade A, B, and C, and Gender measured by gents and ladies. The
sample size was 200.

Gender	GPA			
	A	B	C	Total
Gents	20	60	30	110
Ladies	10	20	60	90
Total	30	80	90	200

Dividing the number of each cell, we get the probability distribution as shown in
the following table:

Gender	GPA			
	A	B	C	Total
Gent	0.10	0.30	0.15	0.55
Lady	0.05	0.10	0.30	0.45
Total	0.15	0.40	0.45	1.00

If we chose a student at random, what is the probability that the student will be a
lady and her GPA will be B? The answer is 0.10. If we want to state this, we may
write as follows:

$$P(\text{lady}, B) = 0.10.$$

What is the probability that the student will have GPA of A?

$$P(A) = 0.10 + 0.05 = 0.15.$$

5.7.2   Continuous Joint Probability Distribution

In case of continuous random variables, the joint probability density function for
two random variables X and Y may be given as follows.

If the boundary of the random variable X is specified by x_1 and x_2 and that of the
random variable Y is specified by y_1 and y_2, then the joint probability density
function may be written as follows:

$$fxy(X, Y) = \text{a specific function of } X \text{ and } Y \quad \text{for} \quad \begin{array}{l} x_1 \leq X \leq x_2 \\ y_1 \leq Y \leq y_2 \end{array}$$

$$= 0, \text{otherwise}.$$

And the probability may be calculated as follows:

$$P(\text{a specific function of } x \text{ and } y) \int_{y_1}^{y_2} \int_{x_1}^{x_2} (\text{function}(x, y)dxdy$$

Example 12

Suppose bricks manufactured for construction purpose have sizes varying from 5600 cm^3 (coded as 5.6) to 7200 cm^3 (coded as 7.2). Let the size be represented by the random variable X.

Let also the weight of the bricks vary from 120 kg/m^2 (coded as 12) to 160 kg/m^2 (coded as 16). Let the random variable of weight be represented by Y.

Again let the joint probability density function of the two random variables (X and Y) be as follows:

$$fxy(X, Y) = \frac{1}{66.24}(x+y)$$

Mathematically, we can state the above problem in the following way.

$$fxy(X, Y) = \frac{1}{66.24}(x+y) \quad \text{for} \quad 5.6 \le x \le 7.2$$
$$12 \le y \le 16$$

$$= 0 \text{ otherwise.}$$

Now let us see what is the probability that a brick collected at random will have the dimension and weight as stated in the problem. The calculations can be carried as follows:

$$P\left\{\frac{1}{66.24}(x+y)\right\} = \int_{12}^{16} \int_{5.6}^{7.2} \frac{1}{66.24}(x+y)dxdy$$

$$= \frac{1}{66.24}(25.92 - 15.68 + 128 - 72),$$

by the process of double integration

$$= (1/66.24)(10.24 + 56)$$
$$= (1/66.24)(66.24)$$
$$= 1.00$$

This means that the total probability within the boundary defined by the two random variables (x and y) is always equal to one.

5.8 Data Fitting to Probability Distribution

Data obtained from natural phenomenon is analyzed using certain distributions. We have studied several distributions (discrete and continuous). Sometimes we get a set of data but we do not know which distribution can be used to analyze the data, describe the underlying characteristics and decide the action to be taken. In this situation, it is imperative to check which distribution is the best-fitted one. Certainly all data sets cannot be explained using the same or similar distributions. This exercise involves the process of fitting of data to probability distribution.

An initial examination of the data should be carried out to find out some candidate distributions. Then we can make some detailed analysis to see which distribution is the best-fitted one. In this exercise, an idea of the characteristics and properties of the candidate distributions is necessary. One way of the initial analysis is the graphical presentation of raw data sets. This may provide an idea of the distribution to be fitted. Scatter plots and histograms are useful tools in this respect.

There are many ways how to analyze the data to see the best fitting. There are a lot of software packages that can be used to find the best fit. However, one method which many statisticians suggest is to use the Chi-square Goodness-of-Fit test. How this method is used in determining the best-fit distribution for a given data set is explained with the help of an example.

Suppose we have the data shown in Table 2.3 for crushing strength of bricks in Chap. 2. Based on the type of data and having knowledge of the normal distribution, let us assume that normal distribution is a candidate distribution. Now it is our exercise to check whether the data set can be fitted well into normal distribution. We have to apply Chi-square Goodness-of-Fit test for the purpose.

First of all, we calculate the mean and standard deviation of the observations. There are 100 observations. The mean is 214.26 psi and the standard deviation is 32.18 psi. We have to group the data into a few classes with certain class width. From our knowledge of the normal distribution, we know the following:

The class width is usually one standard deviation. Starting from the mean, if we go one standard deviation toward the higher and lower sides, we find the following classes with class width of one standard deviation.

No.	Classes	Frequency (%)
1	Between −2 std. and −3 std.	2.3
2	Between −1 std. and −2 std.	13.6
3	Between mean and −1 std.	34.1
4	Between mean and +1 std.	34.1
5	Between +1 std. and +2 std.	13.6
6	Between +2 std and +3 std.	2.3
Total		100

In our example, the lowest observation is 126 and the highest observation is 296. Based on the class width principle shown in the table above, the classes and observed frequencies are shown in the table hereafter. The frequencies shown in the table below are the observed frequencies. These come from observed data.

No.	Classes	Frequency
1	Between 117.72 and 149.90	3
2	Between 149.90 and 182.08	12
3	Between 182.08 and 214.26	32
4	Between 214.26 and 246.44	38
5	Between 246.44 and 276.62	11
6	Between 276.62 and 310.80	4
Total		100

The corresponding classes in the normal distribution are shown in the following table:

To check the Goodness-of-Fit, we have to formulate the hypotheses as follows:

H_0: Data comes from normal distribution
H_a: Data does not come from normal distribution.
Let us assume the significance level = 0.05.

We now prepare the following table for the Chi-square Goodness-of-Fit test. Note that the total frequencies in both the tables above are 100 in each case. If different, these should be converted to percentages.

No.	Classes	Frequency (Oi)	Frequency (Ei)	$(Oi - Ei)^2/Ei$
1	Between 117.72 and 149.90	3	2.3	0.2130
2	Between 149.90 and 182.08	12	13.6	0.1882
3	Between 182.08 and 214.26	32	34.1	0.1293
4	Between 214.26 and 246.44	38	34.1	0.4460
5	Between 246.44 and 276.62	11	13.6	0.4971
6	Between 276.62 and 310.80	4	2.3	1.2565
Total		100	100	2.7302

Thus the calculated χ^2 = 2.7302

$$df = c-1 = 6-1 = 5$$

The critical λ^2 = 11.1 from the table for df = 5 and α = 0.05.

The calculated Chi-square value (χ^2 = 2.7302) is less than the critical Chi-square vale (χ^2 = 11.1). It falls in the acceptance region.

Decision: The null hypothesis H_0 is not rejected. This means that the data comes from normal distribution. In other words, the data set fits normal distribution.

The example shown here demonstrates how to check the fitting of data to a certain distribution using Chi-square Goodness-of-Fit test. In some cases, this procedure may be applied for two or more distributions and comparing these, the best-fitted one may be selected.

Problems

5.1 Distinguish between a variable and a random variable.

5.2 Distinguish between values and probabilities.

5.3 Assume that the number of villagers (x) opposing the construction of a dam in a certain location follows the binomial distribution. Previous reports show that 65 % of the villagers opposed the dam construction. A random sample of 100 villagers are taken.

   (a) Find the probability that 10 of those villagers will oppose the dam construction.

   (b) Find the probability that no more than 3 villagers will oppose the dam construction.

   (c) Find the probability that 3 villagers will oppose the dam construction.

5.4 In a certain manufacturing process, the number of defective items is assumed to follow the binomial distribution. Usually, 3 % of the items manufactured are found defective. A shipment is made of 150 pieces.

   (a) What is the probability that 3 pieces will be found defective ?

   (b) What is the probability that 147 pieces in the shipment will be found defective?

   (c) What is the probability that 105 of those pieces shipped will be found defective?

5.5 Find the mean, variance, and standard deviation of the number of villagers opposing the dam construction as shown in the problem 5.3

5.6 Calculate the mean, variance, and standard deviation of the number of defective items as shown in problem 5.4.

5.7 Show in the sketch $P(-0.05 \leq z \leq 1.96)$. Interpret the meaning of this.

5.8 Find the value of the following:

   (a) Find the value of $P(Z_{\alpha/2} > 1.96)$.

   (b) Find the value of $P(Z_{\alpha/2} < 1.96)$.

   (c) $P(Z_{\alpha/2} > z) = 0.025$. What is the value of z?

   (d) $P(Z_{\alpha/2} < z) = 0.025$ What is the value of z?

   (e) Find $P(0.00 < z\ 1.47)$.

   (f) Find $P(-0.44 < z\ 2.33)$.

5.9 Assume that students' scores are normally distributed with mean of 83.5 and standard deviation of 7.5.

(a) Calculate $P(80 < \text{scores} < 90)$.
(b) What is the probability of scores falling within one standard deviation from the mean?

5.10 Household income in a city is normally distributed with mean = $8000 and standard deviation = $ 1000. A household is drawn from the city at random.

(a) What is the probability that its income will fall between $7000 and $10,000?
(b) What is the minimum income needed if the household intends to be within to 5 %?

5.11 In the following problems, z is a variable that has a standard normal distribution and is a variable that is normally distributed and has a mean of 100 and standard deviation of 15.

(a) What is the probability that z is greater than 1.35?
(b) What is the probability that z is between 1.73 and 2.73?
(c) Find the interval of z scores, centered on zero, which includes 80 % of the probability of z.
(d) What is the probability that y falls between 99 and 106?
(e) Find the interval of values of y, centered on the mean of y, which includes 50 % of the probability of y.
(f) What is the probability that y is less than 115?

5.12 The mean income of a certain group of people is normally distributed with mean $2500 and standard deviation $2000. One man is taken at random.

(a) What is the probability that the income of the man will be less than $2000?
(b) What is the probability his income will be greater than $3100.
(c) What is the probability that his income will be between $2000 and $3000?
(d) What is the proportion of the people whose income will be within $2000 and $3000?

5.13 Students' scores are normally distributed with mean 68.92 and standard deviation 35.32. What should be the minimum score of a student to enable him to be within to 5 %?

5.14 In case of a normally distributed random variable,

(a) Find the proportion of scores falling within one standard deviation around the mean.
(b) Find the proportion of scores falling beyond one standard deviation above the mean.

(c) Find the proportion of scores falling below one standard deviation above the mean.

(d) What is the probability of a randomly selected score to be below 1.5 times the standard deviation lower than the mean?

5.15 Suppose that the number of hours of the RCC of AIT Functions smoothly is normally distributed.

(a) The standard deviation is 30 h. In 15 % occasion interruption takes place in less than 50 h. What is the mean time of smooth functioning?

(b) The mean time is 75 h. In 25 % occasions, the RCC runs smoothly for more than 80 h. What is the standard deviation in this case?

5.16 Find the values of the following:

(a) $t_{0.05,10}$

(b) $t_{0.01,4}$

(c) $t_{0.025,8}$

5.17 $t_{0.025,\alpha} = 1.753$. What is the value of α?

5.18 Find the values of the following:

(a) $P(t > 2.447)$ for df 6.

(b) $P(t < 2.447)$ for df 6.

(c) $P(t < -1.796)$ for df 11.

(d) $P(t > -1.796)$ for df 11.

(f) $P(-2.567 < t < 0.00)$ for df 17.

5.19 $P(t > x) = 0.05$ for 5 df What is the value of x?

5.20 $P(t < x) = 0.99$ for 10 df What is the value of x?

5.21 From a normal population, a random sample of size 15 is taken. The sample mean and standard deviation are 50.25 and 22.58, respectively. The population mean is 60.25.

(a) Find the t-value.

(b) Find $P(t > x)$, where x is the t-value calculated in (a) above.

5.22 Find the values of the following:

(a) $F_{0.05,5,15}$

(b) $F_{0.10,10,20}$

(c) $F_{0.025,8,16}$

(d) $F_{0.90,9,12}$

(e) $F_{0.95,10,15}$

5.23 Two normally distributed populations have variances of 102.59 and 95.23. Two random samples of sizes 10 from the first and 20 from the second populations, respectively, are drawn from the two populations. The sample variances are 95.26 and 90.36, respectively.

(a) Calculate the F-value.

(b) What is the probability that F-value will be greater than this value?

(c) What is the probability that the F-value will be less than the value calculated in (a) above?

5.24   Find the values of the following:

(a) $\chi^2_{0.05,20}$.

(b) $\chi^2_{0.01,15}$.

(c) $\chi^2_{0.005,25}$.

5.25   Find the values of the following:

(a) $P(\chi^2 > 15.99)$ for 10 df

(b) $P(\chi^2 > 25.00)$ for 15 df

(c) $P(\chi^2 > 25.00)$ for 10 df

(d) $P(\chi^2 > 23.50)$ for 16 df

5.26   Assume that X is a variable and has a Chi-square distribution with 5 df Find the value of x such that $P(\chi^2 > x) = 0.05$.

5.27   Assume that X is a variable and has a Chi-square distribution with 6 df Find the value of x such that $P(\chi^2 < x) = 0.90$.

5.28   The variable X has a Chi-square distribution and $P(X > 21.90) = 0.025$. What is the df?

5.29   The variable X has a Chi-square distribution and $P(X < 17.12) = 0.75$. Find the df?

5.30   From a normal population with variance 10, a sample of 18 is selected. The sample variance is S^2. Find the value of x such that

(a) $P(S^2 > x) = 0.025$.

(a) $P(S^2 < x) = 0.025$.

5.31   In the example above, ff a brick is selected at random, find the probability that the size of the brick (X) will lie between 6 and 7 and the weight (Y) will lie between 14 and 15.

5.32   Salaries of professionals in a government system vary depending on two random variables. One rv is years of education (say X) starting with 4 (Bachelor) and ending with 10 (Doctoral). Another rv is experience (say Y) varying from 0 to 20 years. With these specifications, the joint density function has been estimated to be as follows:

$$fxy(X, Y) = (1/242)(X + Y) \quad \text{for} \quad 4 \le X \le 10$$
$$0 \le Y \le 20$$

$$= 0 \text{ otherwise.}$$

If an official is chosen at random, what is the probability that his education level will be within 5 and 6 years, and experience will be within 10 and 15 years?

5.33   In a locality of rice farming, the rice yield is influenced by the seasonal rainfall and the use of fertilizer. Rainfall (say random variable X) has been recorded to vary between 10 and 30 cm. Fertilizer use (say random variable Y) has been between 2 kg/1600 and 5 kg/1600 m^2.
The joint probability function model has been established to be

$$fxy(X, Y) = \frac{1}{239}\left(\frac{x}{2} + Y^2\right)$$

Yield from a plot of land chosen at random has been noted. What is the probability that the rice yield from this plot has experienced rainfall between 15 and 20 cm and fertilizer use between 2.5 and 3 kg?

5.34   Salary of professionals in a government system is influenced by numbers of years of education from bachelor level (X), years of experience (Y) and professional training (Z). Assume the probability density function as follows:

$$fxyz(X, Y, Z) = (1/237)(X + Y + Z) \quad \text{for} \quad \begin{aligned} 4 &\leq X \leq 10 \\ 5 &\leq Y \leq 20 \\ 1 &\leq Z \leq 4 \end{aligned}$$

$$= 0 \text{ otherwise.}$$

If a professional is chosen at random from the system, find the probability that his education level will be within 5–6 years, experience within 8–10 years and professional training within 2–3 years.

Answers

5.3   (a) 2.16×10^{-30}; (b) 2.69×10^{-40}; (c) 1(approx.)
5.4   (a) 0.169; (b) 0 (approx.); (c) 0 (approx.)
5.5.   65; 22.75
5.6.   4.5; 4.37
5.8.   (a) 0.02502; (b) 0.02502; (c) 1.96; (d) −1.96; (e) 0.42922; (f) 0.64885
5.9   (a) 0.48867; (b) 0.68268
5.10   (a) 0.81859; (b) 9645
5.11   (a) 0.08849; (b) 0.03865; (c) $-1.285 < z < + 1.285$; (d) 0.18343; (e) $89.87 < y < 110.13$; (f) 0.84134
5.12   (a) 0.40129; (b) 0.38209; (c) 0.19742; (d) 19.74 %
5.13   127
5.14   (a) 68.27 %; (b) 15.87 %; (c) 34.13 %; (d) 0.06679
5.15   (a) 81.05 h; (b) 7.41 h
5.16   (a) 1.812; (b) 3.747; (c) 2.306
5.17   15
5.18   (a) 0.025; (b) 0.975; (c) 0.050; (d) 0.950; (e) 0.875; (f) 0.490
5.19   2.015
5.20   2.764

5.21   (a) −1.715; (b) 0.95 (approx.)
5.22   (a) 2.90; (b) 1.94; (c) 3.12; (d) 0.452; (e) 0.394
5.23   (a) 1.02; (b) 0.25; (c) 0.75
5.24   (a) 31.41; (b) 30.58; (c) 46.93
5.25   (a) 0.10; (b) 0.05; (c) 0.005; (d) 0.10
5.26   11.07
5.27   10.65
5.28   11
5.29   14
5.30   (a) 17.76; (b) 4.45
5.31   0.317
5.32   0.281
5.33   0.199
5.34   0.1097

Chapter 6
Statistical Inference

Abstract Parameters and statistics, estimation, estimators, and estimates are presented. Properties of estimators are explained and the central limit theorem is introduced. Point estimation and interval estimation are explained and distinguished. The techniques of calculating confidence intervals in various situations are shown.

Keywords Parameters · Estimators · Estimates · Point estimate · Interval estimate · Confidence interval

Statistical inferences may broadly be classified as (i) estimation and (ii) hypothesis testing. We shall deal with the theory of estimation first to be followed by hypothesis testing.

6.1 Parameter and Statistics

Although data are collected from samples, our main purpose is to study the population. So the problem is to estimate the population characteristics based on sample characteristics. In estimating, we shall frequently come across with such terminology as "parameter" and "statistic."

Parameter is a characteristic of population. Statistic is a characteristic of sample data.

Statistical inferences about population characteristics are called parameters, as already told. A parameter is a number that describes a population distribution. Thus, population mean and standard deviation are parameters. The population mean is a number that measures or describes the central tendency of the population distribution. The population standard deviation is number that measures or describes the variability of the population distribution. When we say statistical inference, we mean to infer the value of a population parameter such as a mean or a standard deviation.

How do we make the statistical inference? We make the inference based on a number computed from the sample data. The number is called a statistic or a sample

© Springer Science+Business Media Singapore 2016

A.Q. Miah, *Applied Statistics for Social and Management Sciences*,
DOI 10.1007/978-981-10-0401-8_6

statistic. A statistic is a number that describes a sample distribution. We must not be confused with a parameter or a statistic. A parameter is related to population and a statistic is related to a sample.

The following notations for population parameters and sample statistics should be noted.

Population	Sample
N = no. of observations	n = no. of observations
X_i = ith observation	x_i = ith observation
μ = mean	$\bar{x}$ = mean
σ^2 = variance	s^2 = variance
σ = standard deviation	s = standard deviation
Parameters = μ, σ^2, ... etc.	Statistics = $\bar{x}$. s^2. ...

6.2   Estimation

Usually in statistical studies the population parameters are unknown. Since it is almost impossible or just too much trouble because of time and expense, we need to estimate the population parameters from a sample. Here also we shall come across with the terminology "estimators" and "estimates," The random variable used to estimate the population parameter is called an "estimator." The specific value of this variable is called an "estimate" of the population parameter. The random variables $\bar{x}$ and s^2 are the estimators of the population parameters μ and σ^2, respectively. A specific value of $\bar{x}$ such as $\bar{x} = 120$ is an estimate of μ. A specific value of s^2 such as $s^2 = 237.1$ is an estimate of σ^2.

An estimate of a population parameter may be reported in two ways. If a single number is given as the estimate, it is called a point estimate. The word "point" is used to indicate that a single value is being reported as the estimate. The other way to report an estimate is to give an interval of values in which the population parameter is claimed to fall. This estimate is called an interval estimate. An example may be cited. The point for the average I.Q. of college undergraduates might be 120, implying that our best estimate of the population mean is 120. We can also say that interval estimate of the I.Q. of the undergraduates is 115–125, meaning that the population mean is expected to fall within the range of 115–125.

6.3   Properties of Estimators

When we estimate the population parameters using sample statistics, i.e., estimators, a question arises—how good are the estimators for estimating the population parameters? In other words, what are the criteria for "good" estimators? Four properties that are most relevant may be identified. These are (i) unbiasedness, (ii) efficiency, (iii) sufficiency, and (iv) consistency.

6.3.1 Unbiasedness

This property suggests that the expected value of the estimator should be very close to the population parameter being estimated. It is preferable to have the expected value of the estimator being exactly equal to the population parameter. This implies that the error term be equal to zero. An estimator is said to be unbiased if the expected value of the estimator is equal to the parameter being estimated. Thus, if we want to estimate the population mean using sample mean, then

$$E(\bar{x}) = \mu \ (\text{population mean})$$

This is the definition of the unbiased estimator. Similarly, if s^2 is an unbiased estimator of σ^2, then

$$E(s^2) = \sigma^2$$

6.3.2 Efficiency

This property suggests that an estimator should have a relatively small variance. From a population if a sample size of n is repeated, then in each case the estimator should have values close to each other. This means that if we use random sample more than once in the same population, then in each case the estimator should have values of a particular estimator close to each other. Even if we chose a random sample of size n and find a particular value of an estimator, and if another researcher uses a sample of the same size from the same population and finds a value of the estimator, then these two values should be close to each other.

The most efficient estimator among a group of unbiased estimators is the one with the smallest variance.

6.3.3 Sufficiency

This property suggests that the estimator uses all the information about the population parameter that the sample can provide. For instance, we certainly want an estimator to use all sample observations, as well as the information provided by these observations. Let us take us the case of median. It uses only the rankings of the observations and not their precise numerical values. Hence, the median is not a sufficient estimator. A primary importance of the property of sufficiency is that it is a necessary condition for efficiency.

6.3.4 Consistency

The distribution of an estimator normally changes as the sample size changes. Then it is important to see what happens when the sample size tends to be infinity ($n \rightarrow \infty$). The central limit theorem (which will be introduced hereafter) states that in the limit n approaches a very large size, the distribution of $\bar{x}$ approaches the normal distribution. In general, an estimator is said to be consistent, if it yields estimates which approach the population parameter being estimated as n becomes larger.

6.4 Central Limit Theorem

This relates to the size of the sample. When the population is not normally distributed, the sample size has an important role. When n is small, the shape of the distribution will depend mostly on the shape of the parent population. As n becomes large, one of the most important theorems in statistical inference says that the shape of the sampling distribution will become more and more like a normal distribution, no matter what the shape of the parent population is. This is called the central limit theorem which is formally stated as follows:

The distribution of means of random samples taken from a population having mean μ and finite variance σ^2 approaches the normal distribution with mean μ and variance σ^2/n as n goes to infinity.

The meaning of the theorem may be put in the following simple form.

The distribution of the sample mean, based on random samples of size n drawn from a population with mean μ and standard deviation σ, has the following characteristics:

(i) the mean $\mu\bar{x}$ is equal to the population mean μ;
(ii) the standard deviation $\sigma\bar{x}$ is exactly equal to the population standard deviation divided by the square root of the sample size, $\sigma/\sqrt{n}$;
(iii) the shape is approximately normal. The approximation of the shape to normality improves rapidly with increasingly sample size, so that for $n > 10$, the shape can be taken to be normal. Furthermore, if the population is normally distributed, then the distribution of the sample means is exactly normal, even for small sample size.

6.5   Some Examples in Estimation

We are given a population consisting of numbers 1, 2, and 3. We need to select sample size of 2 with replacement.

Q.1:   How many samples are possible?
Q.2:   List the samples.
Q.3:   Show that $E(\bar{x}) = \mu$ (mean of the sampling distribution equal to the population mean).
Q.4:   Show that $E(s^2) = \sigma^2$ (variance of the individual samples about the means of the sampling distribution equal to the population variance).
Q.5:   Show that $\sigma\bar{x}^2 = \sigma^2/n$ (variance of the means of the sampling distribution about the population mean equal to the population variance divided by the sample size).

Solutions:

Q.1:   Here $N = 3$; $r = 2$.

Therefore, no. of samples $= N^2 = 3^2 = 9$.
See the following table and calculations for answers to the rest questions:

$$\mu = (1+2+3)/3 = 2$$

Sample	$\bar{x}$	$(x_i - \bar{x})^2 = s^2$	$(\bar{x} - \mu)$	$(\bar{x} - \mu)^2$
1,1	1.0	$(1 - 1.0)^2 + (1 - 1.0)^2 = 0.0$	-1.0	1.00
1,2	1.5	$(1 - 1.5)^2 + (2 - 1.5)^2 = 0.5$	-0.5	0.25
1,3	2.0	$(1 - 2.0)^2 + (3 - 2.0)^2 = 2.0$	0.0	0.00
2,1	1.5	$(2 - 1.5)^2 + (1 - 1.5)^2 = 0.5$	-0.5	0.25
2,2	2.0	$(2 - 2.0)^2 + (2 - 2.0)^2 = 0.0$	0.0	0.00
2,3	2.5	$(2 - 2.5)^2 + (3 - 2.5)^2 = 0.5$	0.5	0.25
3,1	2.0	$(3 - 2.0)^2 + (1 - 2.0)^2 = 2.0$	0.0	0.00
3,2	2.5	$(3 - 2.5)^2 + (2 - 2.5)^2 = 0.5$	0.5	0.25
3,3	3.0	$(3 - 3.0)2 + (3 - 3.0)^2 = 0.0$	1.0	1.00
9	18.0	6.0		3.00

Q.2:   Listing of the samples is shown above.
Q.3: $E(\bar{x}) = \frac{18}{9} = 2$
$\mu = \frac{1+2+3}{3} = 2$

Therefore, $E(\bar{x}) = \mu$

Q.4:   Mean of $s^2 = E(s^2) = \frac{6}{9} = \frac{2}{3}$

Population variance

$$\sigma^2 = \frac{1}{N}(x_i - \mu)^2$$
$$= \frac{1}{3}(1-2)^2 + (2-2)^2 + (3-2)^2$$
$$= \frac{1}{3} * 2 = \frac{2}{3}$$

Therefore, $E(s^2) = \sigma^2$

Q.5: $\sigma_{\bar{x}}^2 = \frac{\sum (\bar{x} - \mu)^2}{9} = \frac{3}{9} = \frac{1}{3}$

$$\sigma^2 = \frac{(x_i - \mu)^2}{3}$$
$$= \frac{(1-2)^2 + (2-2)^2 + (3-3)^2}{3}$$
$$= \frac{2}{3}$$
$$\sigma^2/n = \frac{2}{3} * \frac{1}{2} = \frac{1}{3}$$

Therefore,
$$\sigma_{\bar{x}}^2 = \frac{\sigma^2}{n}$$
$$\sigma_{\bar{x}} = \frac{\sigma}{\sqrt{n}}$$

6.6   Point Estimation

As the terminology implies, a point estimate of a population parameter is a single numerical value corresponding to the parameter. The population parameter is not known. So we want to estimate it. We do this estimation with the help of sample data. So our interest in this section is to study the point estimates of different population parameters. The situation where often we need to estimate the population parameters are as follows:

(i) To estimate the population mean μ (single population)
     In this case the sample mean $\bar{x}$ is considered to be the point estimate of the population mean. In other words, $\mu = \bar{x}$.

(ii)   To estimate the population variance σ^2 (single population)
       Similar to the mean, the sample variance is taken to be the point estimate of
       the population variance. Symbolically, $\sigma^2 = s^2$.
(iii) To estimate the population proportion P (single population)
       The sample proportion may reasonably be taken as the point estimate of the
       population proportion. Thus, if x is the number of responses of interest in a
       sample of size n, then the sample proportion $p = x/n$ and as such the point
       estimate of the population proportion $P = x/n$.
(iv)   To estimate the difference between two population means namely, $\mu_1 = \mu_2$.
       For this purpose two independent random samples from the two populations
       are necessary. If $\bar{x}_1$ and $\bar{x}_2$ are the sample means of the two random samples
       drawn from the two populations having means μ_1 and μ_2, respectively, then
       the point estimate of the difference of the two population means is the dif-
       ference of the two sample means. Thus, $\mu_1 - \mu_2 = \bar{x}_1 - \bar{x}_2$.
(v)   To estimate the difference between two population proportions, $P_1 = P_2$

This estimation is similar to the one described in (iv) above. The difference
between the two independent random sample (drawn from the two populations)
proportions is taken to be the point estimate of the difference of the two population
proportions. Thus, $P_1 - P_2 = p_1 - p_2$.

6.7   Interval Estimation/Confidence Interval of the Mean of a Single Population

In the previous section we have seen how the point estimation of population
parameters is made. But it should be agreed that a point estimate does not provide
enough information regarding population parameter. For example, if we want to
estimate the mean income of a certain group of people, a single value of the
population parameter may not be very meaningful. We would rather be interested in
estimating the range or interval within which the population mean is expected to lie.
Thus, the interval estimate of the form $L \leq \mu \leq U$ could be more useful. Here in this
expression L and U are the two statistics showing the lower and the upper bounds of
the parameter. The two pints L and U are random variables, since they are the
functions of the same data.

The two end points L and U are set such that the probability of the population
parameter lying between the two end points is $(1 - \alpha)$. Thus,

$$P(L \leq \mu \leq U) = (1 - \alpha).$$

The interval $L \leq \mu \leq U$ is called the $100(1 - \alpha)$ % confidence interval of the
parameter μ. Here the population mean has been used as an example of any population parameter. In specific cases, it should be replaced by the parameter of
interest. The interpretation of the confidence interval is simple. It implies "we are

$100(1 - \alpha)$ % confident that the population parameter will lie between these end points L and U."

There can be several cases of sampling distribution of $\bar{x}$ and accordingly there will be equal number of ways of constructing confidence interval for interval estimation of the population mean. These are highlighted in the following illustrations.

(i) First case: Population has normal distribution; σ^2 (population variance) known.

Let us suppose that $\bar{x}$ is a random variable. Its mean μ is unknown, but the variance σ^2 is known. Le us also assume that a random sample size of n with values $x_1, x_2, x_3, \ldots, x_n$ is taken from this population. The mean of this sample is $\bar{x}$. The confidence interval of the population mean can be obtained by considering sampling distribution of $\bar{x}$. If the population is normal, the sampling distribution of $\bar{x}$ is also normal. The mean of $\bar{x}$ is μ and the variance of $\bar{x}$ is σ^2/n. The distribution of statistic

$$Z = \frac{\bar{x} - \mu}{\sigma/\sqrt{n}}$$

is a standard normal distribution.

Now,

$$P\left(-Z_{\alpha/2} \leq Z \leq Z_{\alpha/2}\right) = 1-\alpha$$

Using the substitution of Z we get,

$$P\left(-Z_{\alpha/2} \leq \frac{\bar{x}-\mu}{\sigma/\sqrt{n}} \leq Z_{\alpha/2}\right) = 1 - \alpha$$

On simplification, it gives

$$P\left(-Z_{\alpha/2} * \sigma/\sqrt{n} \leq \bar{x} - \mu \leq Z_{\alpha/2} * \sigma/\sqrt{n}\right) = 1 - \alpha$$

This can be rearranged as

$$P\left(\bar{x} - Z_{\alpha/2} * \sigma/\sqrt{n} \leq \mu \leq \bar{x} + Z_{\alpha/2} * \sigma/\sqrt{n}\right) = 1 - \alpha$$

Thus, the $100(1 - \alpha)$ confidence interval is given by

$$\bar{x} - Z_{\alpha/2} * \sigma/\sqrt{n} \leq \mu \leq \bar{x} + Z_{\alpha/2} * \sigma/\sqrt{n}$$

(ii) Second case: Population is unknown (does not have to be normal); σ^2 (population variance) known; $n \geq 30$

In this case, use of the central limit theorem is of relevance. If the sample size is large ($n \geq 30$) then the distribution of $\bar{x}$ may be assumed to be normal. Furthermore, if the population variance σ^2 is known, then the Z-transformation is also relevant and the distribution of Z statistic can be taken to be standard normal. Therefore, the confidence interval is given by

$$\bar{x} - Z_{\alpha/2} * \sigma/\sqrt{n} \leq \mu \leq \bar{x} + Z_{\alpha/2} * \sigma/\sqrt{n}$$

(iii) Third case: Population is normal; σ^2 (population variance) unknown

(iv) In the previous two cases, the population variance σ^2 was known and a confidence interval of the population mean was constructed. But if the population variance σ^2 is not known, a difficulty arises. One possibility could be to replace the population variance σ^2 by the sample variance s^2. If the sample size is large ($n \geq 30$), this could be acceptable. But if the sample size is small ($n < 30$), this would not be acceptable. However, if the population is normal, then another alternative is available. We can accomplish the task using the t distribution. The statistic

$$t = \frac{\bar{x} - \mu}{s/\sqrt{n}}$$

has standard normal distribution with $(n - 1)$ degrees of freedom. Applying the same justification as in the case of Z distribution, we can write

$$P\left(-t_{\alpha/2,n-1} \leq t \leq t_{\alpha/2,n-1}\right) = 1 - \alpha$$

Substituting t in this expression we get

$$P\left(-t_{\alpha/2,n-1} \leq \frac{\bar{x} - \mu}{s/\sqrt{n}} \leq t_{\alpha/2,n-1}\right) = 1 - \alpha$$

On simplification, it gives

$$P\left(-t_{\alpha/2,n-1} * s/\sqrt{n} \leq \bar{x} - \mu \leq t_{\alpha/2,n-1} * s/\sqrt{n}\right) = 1 - \alpha$$

This can be rearranged as

$$P\left(\bar{x} - t_{\alpha/2,n-1} * s/\sqrt{n} \leq \mu \leq \bar{x} + t_{\alpha/2,n-1} * s/\sqrt{n}\right) = 1 - \alpha$$

Thus, the $100(1 - \alpha)$ confidence interval is given by

$$\bar{x} - t_{\alpha/2,n-1} * s/\sqrt{n} \leq \mu \leq \bar{x} + t_{\alpha/2,n-1} * s/\sqrt{n}$$

6.8   Confidence Interval of the Difference of Means of Two Normal Populations

(i)   First case: Population variance σ^2 known

Let μ_1 and μ_2 be the two population means and σ_1^2 and σ_2^2 be their respective variances. If two independent random samples of sizes n_1 and n_1 are taken, whose means are $\bar{x}_1$ and $\bar{x}_2$ respectively, then

$$Z = \frac{(\bar{x}_1 - \bar{x}_2) - (\mu_1 - \mu_2)}{\sqrt{(\sigma_1^2/n_1 + \sigma_2^2/n_2)}}$$

Here the distribution of this Z statistic is standard normal, if the two populations are normal or if the sample sizes are large ($n_1 \geq 30$, $n_1 \geq 30$) in case the populations are not normal. It may be noted that the mean of $(\bar{x}_1 - \bar{x}_2)$ is $(\mu_1 - \mu_2)$ and the standard deviation of $(\bar{x}_1 - \bar{x}_2)$ is $\sqrt{(\sigma_1^2/n_1 + \sigma_2^2/n_2)}$.

Now using the logic as before, we can write

$$P\left(-Z_{\alpha/2} \leq Z \leq Z_{\alpha/2}\right) = 1 - \alpha$$

Using the transformation of Z, we get

$$P(-Z_{\alpha/2} \leq \frac{(\bar{x}_1 - \bar{x}_2) - (\mu_1 - \mu_2)}{\sqrt{(\sigma_1^2/n_1 + \sigma_2^2/n^2)}} \leq Z_{\alpha/2}) = 1 - \alpha$$

This can be simplified as

$$P\left((\bar{x}_1 - \bar{x}_2) - Z_{\alpha/2} * \sqrt{(\sigma_1^2/n_1 + \sigma_2^2/n_2)} \right.$$
$$\leq (\mu_1 - \mu_2)$$
$$\left. \leq Z_{\alpha/2} * \sqrt{(\sigma_1^2/n_1 + \sigma_2^2/n_2)} \right)$$
$$= 1 - \alpha$$

Therefore, the $100(1 - \alpha)$ confidence interval is given by

$$P(\bar{x}_1 - \bar{x}_2) - Z_{\alpha/2} * \sqrt{(\sigma_1^2/n_1 + \sigma_2^2/n_2)}$$
$$\leq (\mu_1 - \mu_2)$$
$$\leq Z_{\alpha/2} * \sqrt{(\sigma_1^2/n_1 + \sigma_2^2/n_2)}$$

(ii) Second case: Population variances unknown

In the first case of this section the two populations were normal or the sample sizes were large. Furthermore, the population variances were known. So the Z-transformation and its distribution were used to construct the confidence interval. But the difficulty arises if the population variances are not known or if the sample sizes are small. In this case, we can overcome the difficulty using the t distribution.

Here we make two assumptions. First, two populations are normal. Second, the two population variances are equal. If the two population variances are equal, then $\sigma_1^2 = \sigma_2^2 = \sigma^2$. This σ^2 is the common variance and is estimated from the sample variances as the pooled estimator (s_p^2) of the σ^2 in the following way:

$$s_p^2 = \frac{(n_1 - 1)s_1^2 + (n_2 - 1)s_2^2}{n_1 + n_2 - 2}$$

The standard deviation of the sampling distribution of $(\bar{x}_1 - \bar{x}_2)$ is $s_p \sqrt{(1/n_1 + 1/n_2)}$ and the distribution of

$$t = \frac{(\bar{x}_1 - \bar{x}_2) - (\mu_1 - \mu_2)}{s_p \sqrt{(1/n_1 + 1/n_2)}}$$

is the t distribution with $(n_1 + n_{2-2})$ degrees of freedom. Therefore, we get as before

$$P(-t_{\alpha/2, n1 + n2-2} \le t \le t_{\alpha/2, n1 + n2-2}) = 1 - \alpha$$
$$\Rightarrow P\left(-t_{\alpha/2, n1+n2-2} \le \frac{(\bar{x}_1 - \bar{x}_2) - (\mu_1 - \mu_2)}{s_p \sqrt{(1/n_1 + 1/n_2)}} \le +t_{\alpha/2, n1+n2-2}\right) = 1 - \alpha$$

This can be simplified and rearranged as

$$P(\bar{x}_1 - \bar{x}_2) - t_{\alpha/2, n1+n2-2} * s_p \sqrt{(1/n_1 + 1/n_2)}$$
$$\le (\mu_1 - \mu_2)$$
$$\le (\bar{x}_1 - \bar{x}_2) - t_{\alpha/2, n1+n2-2} * s_p \sqrt{(1/n_1 + 1/n_2)} = 1 - \alpha$$

Therefore, the $100(1 - \alpha)$ confidence interval is given by

$$(\bar{x}_1 - \bar{x}_2) - t_{\alpha/2, n1+n2-2} * s_p \sqrt{(1/n_1 + 1/n_2)}$$
$$\le (\mu_1 - \mu_2)$$
$$\le (\bar{x}_1 - \bar{x}_2) - t_{\alpha/2, n1+n2-2} * s_p \sqrt{(1/n_1 + 1/n_2)}$$

If the two population variances are not assumed equal, then the t statistic should be calculated as follows:

$$t = \frac{(\bar{x}_1 - \bar{x}_2) - (\mu_1 - \mu_2)}{\sqrt{(s_1^2/n_1 + s_2^2/n_2)}}$$

and the degrees of freedom for the t statistic is to be calculated as

$$V = \frac{(s_1^2/n_1 + s_2^2/n_2)}{\left\{ (s_1^2/n_1)^2/(n_1 + 1) + (s_2^2/n_2)^2/(n_2 + 1) \right\}} - 2$$

and subsequently the $100(1 - \alpha)$ confidence interval will be given by

$$(\bar{x}_1 - \bar{x}_2) - t_{\alpha/2,v} * s_p \sqrt{(1/n_1 + 1/n_2)}$$
$$\leq (\mu_1 - \mu_2)$$
$$\leq (\bar{x}_1 - \bar{x}_2) - t_{\alpha/2,v} * s_p \sqrt{(1/n_1 + 1/n_2)}$$

6.9 Confidence Interval of the Variance of a Normal Population

Let x be a random variable with unknown mean μ and unknown variance σ^2. Let also n be the sample size with values $x_1, x_2, x_3, \dots x_n$ giving sample variance s^2. We have seen the statistic

$$\chi^2 = \frac{(n - 1)s^2}{\sigma^2}$$

has a chi-square distribution with $(n - 1)$ degrees of freedom. The chi-square distribution is shown in Fig. 6.1.

From the figure we can write

$$P\left(\chi^2_{1-\alpha/2,n-1} \leq \chi^2 \leq \chi^2_{\alpha/2,n-1} \right) = 1 - \alpha$$

$$\Rightarrow P(\chi^2_{1-\alpha/2,n-1} \leq \frac{(n - 1)s^2}{\sigma^2} \leq \chi^2_{\alpha/2,n-1}) = 1 - \alpha$$

This by simplification and rearrangement we can write

$$P\left\{ \frac{(n - 1)s^2}{\chi^2_{\alpha/2,n-1}} \leq \sigma^2 \leq \frac{(n - 1)s^2}{\chi^2_{1-\alpha/2,n-1}} \right\} = 1 - \alpha$$

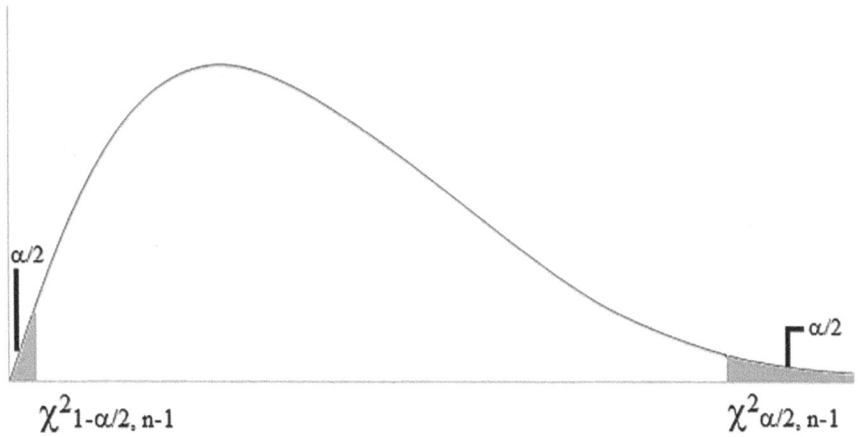

$\chi^2_{1-\alpha/2,\, n-1}$ $\chi^2_{\alpha/2,\, n-1}$

Fig. 6.1 Distribution of chi-square distribution

Therefore, the $100(1 - \alpha)$ % confidence interval of the population variance is given by

$$\frac{(n-1)s^2}{\chi^2_{\alpha/2,n-1}} \le \sigma^2 \le \frac{(n-1)s^2}{\chi^2_{1-\alpha/2,n-1}}$$

6.10   Confidence Interval of a Population Proportion

Let a sample size of n be taken from a large population in which x ($x \le n$) number of observations have the characteristic of our interest. Then the sample proportion is $p = x/n$. Here p and n are the parameters of a binomial distribution. The sampling distribution of p (sample proportion) is approximately normal with mean equaling P (population proportion) and variance P $(1 - P)/n$. This is valid if P is not very close to zero or one and if n is relatively large. Thus, the statistic

$$Z = \frac{p - P}{\sqrt{\{P(1-P)/n\}}}$$

has standard normal distribution. From the Z distribution curve, we can write

$$P\left(-Z_{\alpha/2} \le \frac{p - P}{\sqrt{\{P(1-P)/n\}}} \le Z_{\alpha/2}\right) = 1-\alpha$$
$$\Rightarrow P(-Z_{\alpha/2} * \sqrt{\{P(1-P)/n\}} \le p - P$$
$$\le +Z_{\alpha/2} * \sqrt{\{P(1-P)/n\}} = 1-\alpha$$

This can be simplified and rearranged as

$$P(p - Z_{\alpha/2} * \sqrt{\{P(1-P)/n\}} \leq P \leq p + Z_{\alpha/2} * \sqrt{\{P(1-P)/n\}}$$
$$= 1 - \alpha$$

It may be noted here that in the lower and upper limits of the above expression there is a term $P(1-P)/n$, the standard error of p (sample proportion containing an unknown parameter P). However, for the standard error, P can be estimated by p. Therefore, the $100(-\alpha)$ % confidence interval of the population proportion is given by

$$p - Z_{\alpha/2} * \sqrt{\{p(1-p)/n\}} \leq P \leq p + Z_{\alpha/2} * \sqrt{\{p(1-p)/n\}}$$

6.11   Confidence Interval of the Difference of Two Population Proportions

Let,

n_1 independent sample drawn from population 1,
x_1 no. of units in n_1 having particular characteristic of our interest,
n_2 independent sample drawn from population 2,
x_2 no. of units in n_2 having particular characteristic of our interest,
P_1 & P_2 respective populations proportions.

Now,

p_1 x_1/n_1 (sample 1 proportion),
P_2 x_2/n_2 (sample 2 proportion).

In this case too n_1, p_1 and n_2, p_2 are parameters of binomial distribution. But if the sample sizes are large and the population proportions P_1 and P_2 are not close one or zero, then the binomial distribution can be approximated by normal distribution.

Thus,

$$Z = \frac{(p_1 - p_2) - (P_1 - P_2)}{\sqrt{\{P_1(1-P_1)/n_1 + P_2(1-P_2)/n_2\}}}$$

has approximately standard normal distribution. As reasoned before, P_1 and P_2 for the standard errors can be estimated by p_1 and p_2, respectively. Thus,

$$P\left[(p_1-p_2) - Z_{\alpha/2} * \sqrt{\left\{ \frac{P_1(1-P_1)}{n_1} + \frac{P_2(1-P_2)}{n_2} \right\}} \right.$$

$$\leq (P_1-P_2)$$

$$\left. \leq (p_1-p_2) + Z_{\alpha/2} * \sqrt{\left\{ \frac{P_1(1-P_1)}{n_1} + \frac{P_2(1-P_2)}{n_2} \right\}} \right] = 1 - \alpha$$

After simplification and rearrangement, the $100(1-\alpha)\,\%$ confidence interval can be written as

$$(p_1-p_2) - Z_{\alpha/2} * \sqrt{\left\{ \frac{P_1(1-P_1)}{n_1} + \frac{P_2(1-P_2)}{n_2} \right\}}$$

$$\leq (P_1-P_2)$$

$$\leq (p_1-p_2) + Z_{\alpha/2} * \sqrt{\left\{ \frac{P_1(1-P_1)}{n_1} + \frac{P_2(1-P_2)}{n_2} \right\}}$$

6.12   Finite Population Correction Factor

In the previous sections for variance of the sampling distribution of $\bar{x}$ we have used

$$\sigma\bar{x}^2 = \frac{\sigma^2}{n}$$

If the sample size is small as compared to the population size N, then this expression holds good. But if the sample size n is not very small fraction of the population size N, then approximate expression would be

$$\sigma\bar{x}^2 = \frac{\sigma^2}{n} * \frac{N-n}{N-1}$$

$$\Rightarrow \sigma\bar{x} = \sqrt{\left\{ \frac{\sigma^2}{n} * \frac{N-n}{N-1} \right\}}$$

The term $(N-n)/(N-1)$ is called the finite population correction factor (fpc). In case of proportions also, if the sample size is not a very small fraction of the population size, the variance of p would be given by

$$\text{var}(p) = \frac{p(1-p)}{n} * \frac{N-n}{N-1}$$

$$\Rightarrow \text{Standard error}(p) = \sqrt{\left\{\frac{p(1-p)}{n} * \frac{N-n}{N-1}\right\}}$$

As N becomes larger relative to n, then n/N becomes small and so fpc approaches unity. If $n/N \leq 0.05$ or in other words, if the sample size is not more than 5 % of the population size, then the fpc may be omitted.

Using the fpc, the confidence interval for population mean would be given by

$$\bar{x} - z_{\alpha/2} * \frac{s}{\sqrt{n}} \sqrt{\left\{\frac{N-n}{N-1}\right\}} \leq \mu \leq$$

$$\bar{x} + z_{\alpha/2} * \frac{s}{\sqrt{n}} \sqrt{\left\{\frac{N-n}{N-1}\right\}}$$

Examples

Example 1
Compressive strength of concrete is normally distributed with standard deviation of 40.35 psi. A random sample of 25 specimens showed the mean strength of 3165 psi. Construct a 95 % confidence interval of the mean strength of concrete.
  Solution:

$$\sigma = 40.35 \text{ psi}; \quad \bar{x} = 3165 \text{ psi}$$
$$\alpha = 1-0.95 = 0.05; \quad \alpha/2 = 0.025$$
$$Z_{\alpha/2} = 1.96$$

The 95 % confidence interval is given by

$$\bar{x} - Z_{\alpha/2} * \sigma/\sqrt{n} \leq \mu \leq \bar{x} + Z_{\alpha/2} * \sigma/\sqrt{n}$$
$$\Rightarrow 3165 - 1.96 * \frac{40.35}{\sqrt{25}} \leq \mu \leq 3165 + 1.96 * \frac{40.35}{\sqrt{25}}$$
$$\Rightarrow 3165 - 1.96 * 8.07 \leq \mu \leq 3165 + 1.96 * 8.07$$
$$\Rightarrow 3165 - 15.82 \leq \mu \leq 3165 + 15.82$$
$$\Rightarrow 3149.18 \leq \mu \leq 3180.82$$

At 95 % confidence level, the mean strength of concrete will lie between 3149.18 and 3180.82 psi.

Example 2

From a normally distributed population a random sample of 15 is drawn. The sample mean and standard deviation are found to be 2500 and 145, respectively. Construct a 95 % confidence interval of the population mean.

Solution:

$$\bar{x} = 2500; \quad s = 145; n = 15$$
$$\alpha = 1 - 0.95 = 0.05; \alpha/2 = 0.025$$
$$Z_{\alpha/2} = 1.96$$

Here the population is normal; its variance is not given; also the sample size is small (less than 30). So we need to use the t distribution.

$$t_{\alpha/2, n-1} = t_{0.025, 14} = 2.145$$

Thus, 95 % confidence interval is given by

$$\bar{x} - t_{\alpha/2, n-1} * s/\sqrt{n} \leq \mu \leq \bar{x} + t_{\alpha/2, n-1} * s/\sqrt{n}$$
$$\Rightarrow 2500 - 2.145 * \frac{145}{\sqrt{15}} \leq \mu \leq 2500 + 2.145 * \frac{145}{\sqrt{15}}$$
$$\Rightarrow 2500 - 80.31 \leq \mu \leq 2500 + 80.31$$
$$\Rightarrow 2419.69 \leq \mu \leq 2580.31$$

The population mean will lie between 2419.69 and 2580.31.

Example 3

Incomes of two normally distributed populations are being studied. The variances are $\sigma_1^2 = \$508$ and $\sigma_2^2 = \$425$. Two randomly selected samples of sizes $n_1 = 40$ and $n_2 = 60$ produced mean $\bar{x}_1 = \$2520$ and $\bar{x}_2 = \$1950$, respectively. Construct a 95 % confidence interval of the difference of the two population incomes.

Solution:

$$\sigma_1^2 = \$508; \quad \sigma_2^2 = \$425$$
$$\bar{x}_1 = \$2520; \quad \bar{x}_2 = \$1950$$
$$n_1 = 40; \quad n_2 = 60$$
$$\alpha = 1 - 0.95$$
$$= 0.05; \quad \alpha/2 = 0.025$$
$$Z_{\alpha/2} = 1.96$$

Therefore, the $100(1 - \alpha)$ confidence interval is given by

$$(\bar{x}_1 - \bar{x}_2) - Z_{\alpha/2} * \sqrt{(\sigma_1^2/n_1 + \sigma_2^2/n_2)}$$
$$\leq (\mu_1 - \mu_2)$$
$$\leq + Z_{\alpha/2} * \sqrt{(\sigma_1^2/n_1 + \sigma_2^2/n_2)})$$
$$\Rightarrow (2520 - 1950) - 1.96 * \sqrt{\left\{ \frac{508}{40} + \frac{425}{60} \right\}}$$
$$\leq (\mu_1 - \mu_2)$$
$$\leq + (2520 - 1950) + 1.96 * \sqrt{\left\{ \frac{508}{40} + \frac{425}{60} \right\}}$$
$$570 - 1.96 * (4.45) \leq (\mu_1 - \mu_2) \leq + 570 + 1.96 * (4.45)$$
$$561.28 \leq (\mu_1 - \mu_2) \leq 578.72$$

At 95 % confidence the difference between the two population means will lie between $561.28 and $578.72.

Example 4
The diameters of pipes manufactured in two machines are assumed to be normally distributed. The engineer-in-charge of the quality control is investigating the pipe diameters. He selected two random samples of sizes $n_1 = 20$ and $n_2 = 22$ which produced means and standard deviations $\bar{x}_1 = 12.80$ mm, $s_1 = 0.55$ mm, $\bar{x}_2 = 11.50$ mm and $s_2 = 0.48$ mm, respectively. Assume that the two population variances σ_1^2 and σ_2^2 are equal.

Construct a 90 % confidence interval for the difference of mean diameters.
   Solution:

$$n_1 = 20; \qquad\qquad n_2 = 22$$
$$\bar{x}_1 = 12.80 \text{ mm}; \qquad \bar{x}_2 = 11.50 \text{ mm}$$
$$s_1 = 0.55 \text{ mm}; \qquad s_2 = 0.48 \text{ mm}$$
$$\sigma_1^2 = \sigma_2^2;$$
$$\alpha = 1 - 0.90 = 0.10; \quad \alpha/2 = 0.05$$

Here we need to use the t distribution:

$$s_p^2 = \frac{(n_1 - 1)s_1^2 + (n_2 - 1)s_2^2}{n_1 + n_2 - 2}$$
$$= \frac{(20 - 1)(0.55)^2 + (22 - 1)(0.48)^2}{20 + 22 - 2}$$
$$= \frac{5.75 + 4.84}{40}$$
$$= 0.264$$
$$s_p = 0.515$$
$$t_{\alpha/2, n1 + n2 - 2} = t_{0.05, 40} = 1.684$$

The 90 % confidence interval is given by

$$(\bar{x}_1 - \bar{x}_2) - t_{\alpha/2, n1+n2-2} * s_p \sqrt{(1/n_1 + 1/n2)}$$
$$\leq (\mu_1 - \mu_2) \leq (\bar{x}_1 - \bar{x}_2) - t_{\alpha/2, n1+n2-2} * s_p \sqrt{(1/n_1 + 1/n2)}$$
$$\Rightarrow (12.80-11.50) - 1.684 * 0.515 * \sqrt{(1/20 + 1/22)} \leq (\mu_1 - \mu_2) \leq$$
$$(12.80-11.50) + 1.684 * 0.515 * \sqrt{(1/20 + 1/22)}$$
$$\Rightarrow 1.30 - 0.268 \leq (\mu_1 - \mu_2) \leq 1.30 + 0.268$$
$$\Rightarrow 1.032 \leq (\mu_1 - \mu_2) \leq 1.568$$

The difference between the mean diameters of pipes produced from two machines will lie between 1.032 and 1.568 mm.

Example 5

The GPA of students is normally distributed. A random sample of size 10 students showed the GPA: 3.5, 3.3, 3.0, 2.8, 2.9, 3.9, 3.7, 3.8, 3.4, and 2.9.

Construct a 90 % confidence interval of σ^2.

Solution:

The sample mean is $\bar{x} = 3.32$.

The sample variance is $s^2 = 0.164$.

$$\alpha = 1-0.90 = 0.10;$$
$$\alpha/2 = 0.05$$
$$\chi^2_{\alpha/2, n-1} = \chi^2_{0.05, 9} = 16.92$$
$$\chi^2_{1-\alpha/2, n-1} = \chi^2_{0.95, 9} = 3.33$$

Therefore, the 90 % confidence interval of the population variance is given by

$$\frac{(n-1)s^2}{\chi^2_{\alpha/2, n-1}} \leq \sigma^2 \leq \frac{(n-1)s^2}{\chi^2_{1-\alpha/2, n-1}}$$
$$\Rightarrow \frac{9(0.164)}{16.92} \leq \sigma^2 \leq \frac{9(0.164)}{3.33}$$
$$\Rightarrow 0.087 \leq \sigma^2 \leq 0.443$$

Example 6

A quality control manager wants to estimate the fraction of defective items in the manufacturing process. A random sample of 6000 units contained 240 defective units. Construct a 95 % confidence interval of the fraction defective.

Solution:

$$n = 6000; \quad x = 240$$

$$p = \frac{240}{6000} = 0.04$$

$$\alpha = 1-0.95 = 0.05; \quad \alpha/2 = 0.025$$

$$Z_{\alpha/2} = 1.96$$

The confidence interval is given by

$$p - Z_{\alpha/2} * \sqrt{\{P(1-P)/n\}} \le P \le$$
$$p + Z_{\alpha/2} * \sqrt{\{P(1-P)/n\}}$$

$$\Rightarrow 0.04 - 1.96 * \sqrt{\left\{ \frac{0.04(1-0.04)}{6000} \right\}} \le P \le$$

$$0.04 + 1.96 * \sqrt{\left\{ \frac{0.04(1-0.04)}{6000} \right\}}$$

$$\Rightarrow 0.04 - 0.005 \le P \le 0.04 + 0.005$$
$$\Rightarrow 0.035 \le P \le 0.045$$

The fraction of defective items in the manufacturing process will lie between 0.035 and 0.045.

Example 7

The fraction of defective items produced by two different processes is to be estimated. Two random samples from the two processes were taken. In the process 1, 24 units were found defective in a sample of 1200. In the process 2, 15 units were found defective in a sample of 1500. Construct a 90 % confidence interval on the difference in the fraction of defective items.

$$n_1 = 1200; \quad n_1 = 1500;$$
$$x_1 = 24; \quad x_2 = 15;$$
$$p_1 = \frac{24}{100} \quad p_2 = \frac{15}{1500}$$
$$= 0.02 \quad = 0.01$$

$$\alpha = 1-0.90 = 0.10; \quad \alpha/2 = 0.05$$
$$Z_{\alpha/2} = 1.645$$

The 90 % confidence interval is given by

$$(p_1-p_2) - Z_{\alpha/2} * \sqrt{\left\{\frac{P_1(1-P_1)}{n_1} + \frac{P_2(1-P_2)}{n_2}\right\}}$$

$$\leq (P_1-P_2)$$

$$\leq (p_1-p_2) + Z_{\alpha/2} * \sqrt{\left\{\frac{P_1(1-P_1)}{n_1} + \frac{P_2(1-P_2)}{n_2}\right\}}$$

$$\Rightarrow (0.02-0.01)-1.645 * \sqrt{\left\{\frac{0.02(1-0.02)}{1200} + \frac{0.01(1-0.01)}{1500}\right\}}$$

$$\leq (P_1-P_2)$$

$$\leq (0.02-0.01) + 1.645 * \sqrt{\left\{\frac{0.02(1-0.02)}{1200} + \frac{0.01(1-0.01)}{1500}\right\}}$$

$$\Rightarrow 0.01-0.0079 \leq (P_1-P_2) \leq 0.01+0.0079$$
$$\Rightarrow 0.0021 \leq (P_1-P_2) \leq 0.0179$$

The difference in the fraction of defective items from the two processes will lie between 0.0021 and 0.0179.

Problems

6.1   Distinguish between

    (i) Parameter and statistic

    (ii) Confidence level and significance level.

6.2   In the interval estimation of the population mean, state the effect of the following:

    (a) sample size on the interval

    (b) sample standard deviation on the interval

    (c) confidence level on the interval.

6.3   Based on a simple random sampling involving a sample size of 225, the 95 % confidence interval of the mean income of the people of a city is shown to be
      $15{,}010.70 \leq \mu \leq 15{,}989.30$:

    (a) What is the sample mean?

    (b) What is the sample standard deviation?

6.4   A process produces bags of refined sugar. A random sample of 25 bags had a mean weight of 1.95 kg:

    (a) If the contents of the bags are normally distributed with std. dev. of 0.21 kg., find 95 % confidence interval of the μ.

(b) The contents of the bags are normally distributed. The population std. dev. is not known, but sample std. dev. is 0.21 kg. Find the 95 % confidence interval of μ.

(c) The distribution of the contents of bags is not known. Sample std. dev. is 0.21. Find the 95 % confidence interval of μ.

6.5 A quality control manager of a ball bearing manufacturing plant wants to estimate the mean diameter of ball bearings manufactured in the plant. He checked 61 ball bearings at random and found to have the mean diameter of 9.4 mm with a standard deviation of 3.25 mm.

Estimate the mean diameter of all ball bearings manufactured in the plant. You should be 95 % confident in your estimate.

6.6 A hospital wishes to estimate the average number of days required for treatment of patients between ages of 25 and 34. A random sample of 500 hospital patients between these ages produced a mean and standard deviation equal to 5.4 and 3.1 days, respectively.

Estimate the mean length of stay for the population of patients from which the sample was drawn.

If the total number of patients within the same age group is 5000, estimate the mean length of stay with 98 % confidence interval.

6.7 A sample of 16 persons is studied. Their scores have a mean of 32.84 and a standard of 2.08.

Construct 95 and 99 % confidence intervals for the population mean and compare the two results.

Construct a 95 % confidence interval if the sample size is 36 and the population size is 600.

6.8 Income is normally distributed in a population. The mean is Baht 15,000 and the standard deviation is Baht 2200. Total population size is 25,000.

(a) Find the range of income, centered on the mean that will include 90 % probability of income.

(b) What is the probability of income falling between Baht 11,381 and bath 18,619?

(c) How many people in the population will have income greater than Baht 18,619?

(d) How many people will have income less than Baht 11,381?

(e) How many people will have income between Baht 11,381 and Baht 18,619?

6.9 In a 95 % confidence interval estimate, the population proportion is stated to lie between 0.352 and 0.548. What was the sample size?

6.10 Construct a 90 % confidence interval for the population mean based on sample size of 36, if the sample mean is 950 and the population standard is 70. What happens when 95 % confidence interval is used?

6.11 In a bank, 25 loan applications were randomly selected for the purpose of determining the average amount required for each loan. The mean loan amount was found to be $900 with a standard deviation of $150. Estimate the 90 % confidence interval of the mean loan amount for all the applications in the bank. What is the point estimate for the mean loan amount for all the applications?

6.12   To estimate the IQ of 450 students in an institution, a sample size of 25 was chosen. The mean IQ of this sample was found to be 127 with a standard deviation of 5.4 points. Make an interval estimation of the mean IQ of the 450 students. Use 90 % confidence level. What is the point estimate in this case?

6.13   The compressive strength of concrete is normally distributed. From an experiment a civil engineer obtained a random sample of size 10 and noted the following strength measurements (in psi):
2200, 2210, 2205, 2100, 2115, 2300, 2250, 2350, 2150, 2250

(a) Construct a 90 % confidence interval for σ^2.
(b) Construct a 95 % confidence interval for σ^2.

6.14   Students scores in two universities are to be studied. The scores are normally distributed. Two independent samples produced the following results:

University	Sample size	Mean score	Std. dev.
1	41	85	8.2
2	51	76	7.5

(a) If the two population variances are 8.5 and 7.2, respectively, construct 95 % confidence interval for the difference between the two university mean scores.
(b) If the population variances are not known, construct 95 % confidence interval for the difference between the two university mean scores, assuming the two population variances equal.
(c) Repeat the problem in (b) above, assuming the two population variances unequal.

6.15   A random sample of size 41 from a normal population has a standard deviation of 0.55. Construct a 95 % confidence interval for the population variance σ^2.

6.16   From a normal population a sample of size 30 produced a mean of 950 and a standard deviation of 70. Construct a 95 % confidence interval for the population variance σ^2.

6.17   In a sample of 144 households 54 were found to own cars. Find a 95 % confidence interval of the proportion of households who have cars.

6.18   The Division of Human Settlements Development wanted to see if the work load of Statistics course is heavy. A random sample of 36 students was selected for opinion. Out of them 10 students termed the course heavy.

Make an estimation of the proportion of students who think that the course is heavy. Use 97.5 % confidence level.

6.19 To estimate the proportion of unemployed in a city, a researcher selected a random sample of 400 persons from the working class. Of them, 25 were unemployed.

Estimate the proportion of unemployed workers.

If the size of the working class is 12,000, estimate the number of persons unemployed.

6.20 A drug manufacturer wants to study the effect of a patent medicine. In a random sample of 25 he finds that 15 obtained relief.

Estimate the population proportion not obtaining relief from the medicine. If the population size is 10,000, how many people will not obtain relief from the medicine?

6.21 To estimate the proportion of passengers who had purchased tickets for more than $200 over a year's time, an airline official obtained a random sample of 80. The number of those who purchased tickets for more than $200 was 45. Make the estimation. Use 97.5 % confidence level.

6.22 A manufacturer produces spare parts in his factory. He ships the spare parts; he prepares 10,000 pieces in each shipment. He is aware of the quality control. During one shipment, he took a random sample of 500 parts for inspection. He found 10 parts to be defective. The manufacturer desires to estimate the number of defective parts with 90 % confidence level.

(a) Construct the necessary confidence interval.
(b) What is the number of defective parts in the shipment?

6.23 In a study of students' opinion on course standard of two departments, two random samples were drawn. In department A, 35 students out of a sample of 45 were found to be satisfied. In department B, 31 students out of 49 were found satisfied.

(a) Construct a 95 % confidence interval for the difference of the proportion of students satisfied with the course standard in the two departments.
(b) Construct a 99 % confidence interval for the difference of the proportion of students satisfied with the course standard in the two departments.

6.24 Population size is 5000; the sample size is 256; the sample standard deviation is 2 and the sample mean is 70. Construct a 99 % confidence interval. Use FPC. What happens if you do not use FPC?

Answers

6.3 (a) 15,500 (b) 3,744.64
6.4 (a) $1.868 \leq \mu \leq 2.032$; (b) $1.863 \leq \mu \leq 2.037$
      (c) $1.863 \leq \mu \leq 2.037$; if assumed normal
6.5 $8.58 \leq \mu \leq 10.22$ mm

6.6   Point estimate 5.4 days;
      95 % confidence estimate $5.13 \leq \mu \leq 5.67$ days;
      $5.09 \leq \mu \leq 5.71$ days
6.7 $31.73 \leq \mu \leq 33.95$; $31.31 \leq \mu \leq 34.37$; $31.85 \leq \mu \leq 33.83$
6.8   (a) $11{,}381 \leq \mu \leq 18{,}619$ Baht;
      (b) 0.90; (c) 1,250; (d) 1,250; (e) 22,500
6.9 $n = 81$
6.10 $930.81 \leq \mu \leq 969.19$; interval increased by 7.36
6.11 $848.67 \leq \mu \leq 951.33$; $900;
6.12 $125.20 \leq \mu \leq 128.80$; 127
6.13 (a) $2{,}961 \leq \sigma^2 \leq 15{,}043$; (b) $2{,}634 \leq \sigma^2 \leq 18{,}553$
6.14 (a) $7.92 \leq \mu_1 - \mu_2 \leq 10.08$; (b) $5.75 \leq \mu_1 - \mu_2 \leq 12.25$
      (c) $5.75 \leq \mu_1 - \mu_2 \leq 12.25$
6.15 $0.20 \leq \sigma? \leq 0.50$
6.16 $3108 \leq \sigma? \leq 8854$
6.17 $0.296 \leq P \leq 0.454$
6.18 $0.111 \leq P \leq 0.445$
6.19 0.0625; $0.0388 \leq P \leq 0.0862$ (95 %); 750; $466 \leq N \leq 1034$
6.20 0.40; $0.21 \leq P \leq 0.59$ (95 %); 4,000; $2{,}100 \leq N \leq 5{,}900$
6.21 $0.4383 \leq P \leq 0.6867$
6.22 (a) $0.01 \leq P \leq 0.03$; (b) $100 \leq N \leq 300$
6.23 (a) $-0.037 \leq P_1 - P_2 \leq 0.327$; (b) $-0.094 \leq P_1 - P_2 \leq 0.384$
6.24 (a) $69.69 \leq \mu \leq 70.31$; (b) $69.68 \leq \mu \leq 70.32$

Chapter 7
Hypothesis Testing

Abstract The importance of hypothesis testing lies in the fact that many types of decision-problems can be formulated as hypothesis testing problems. Simple hypothesis, composite hypothesis, null, and alternative hypotheses are introduced. One-tail and two-tail tests are explained. Errors in hypothesis testing are mentioned. A standard procedure for testing of hypotheses is set. Examples are provided for testing of various types of hypotheses. A standard procedure is set for testing of population means and proportions in many situations. The technique for drawing conclusions at the end of the test is illustrated through examples. The drawing of conclusions is the core of hypothesis testing. A highly useful and new chart called "Flow Chart for Hypothesis Testing" is provided. The technique of calculating power of hypothesis testing is illustrated with the help of examples.

Keywords Hypothesis testing · One-tail test · Two-tail test · Simple hypothesis · Composite hypothesis · Null and alternative hypothesis · Flow chart of hypothesis testing · Test procedure · Power of test

In the previous chapter, we have studied how to estimate the characteristics of population known as parameters. We dealt with the point estimation and interval estimation. Apart from this, there are many problems where we need to decide whether to accept or to reject a statement about some population parameters. The statement is called a hypothesis, and the decision-making procedure regarding the hypothesis is called hypothesis testing. Hypothesis testing is a very useful aspect of statistical inference. The importance of hypothesis testing lies in the fact that many types of the decision-problems can be formulated as hypothesis testing problems.

7.1 Introduction

There are several types of hypotheses. A thorough discussion of all those is beyond the scope of this book. Here, we shall deal with statistical hypotheses only. Statistical hypotheses concern assumed values of the population parameters. The assumptions regarding the assumed values of the parameters are referred to as

© Springer Science+Business Media Singapore 2016
A.Q. Miah, *Applied Statistics for Social and Management Sciences*,
DOI 10.1007/978-981-10-0401-8_7

statistical hypotheses. Determining the validity of an assumption of this kind is called a test of a statistical hypothesis or simply hypothesis testing.

Within the scope of statistical hypotheses, there are a few types. We can distinguish between simple hypotheses and composite hypotheses. In a simple hypothesis, only one value of the population parameter is specified. For example, the mean IQ of a group of students is 120. The exact difference between two population parameters is also a simple hypothesis. Example is $\mu_1 - \mu_2 = 0$. In a composite hypothesis, instead of specifying one value, a range of values is specified. Example of this hypothesis is $\mu \neq 120$. Another example is $\mu_1 - \mu_2 \neq 0$.

In testing hypotheses, the standard procedure is to state two conflicting hypotheses. These are mutually exclusive. These are called "null hypothesis" and "alternative hypothesis". In a null hypothesis, that value of a population parameter is specified which the researcher hopes would be rejected. The word "null" means invalid, void or amounting to nothing. In an alternative hypothesis, those values of the population parameter are specified which the researcher believes to hold true. Null and alternative hypotheses are usually denoted by H_0 and H_a, respectively. Examples are given below.

$$H_0: \mu = 120 \text{ (null hypothesis)}$$
$$H_a: \mu \neq 120 \text{ (alternative hypothesis)}$$
$$\Rightarrow$$
$$H_0: \mu_1 - \mu_2 = 120 \text{ (null hypothesis)}$$
$$H_a: \mu_1 - \mu_2 \neq 120 \text{ (alternative hypothesis)}$$

The null hypothesis and the alternative hypothesis can both be simple or composite. See the following examples.

$$H_0: \mu = 120 \text{ (simple)}$$
$$H_a: \mu = 100 \text{ (simple)}$$
$$\Rightarrow$$
$$H_a: \mu \neq 100 \text{ (composite)}$$
$$\Rightarrow$$
$$H_a: \mu < 100 \text{ (composite)}$$
$$H_0: \mu < 120 \text{ (composite)}$$
$$H_a: \mu = 100 \text{ (simple)}$$
$$\Rightarrow$$
$$H_a: \mu > 100 \text{ (composite)}$$

Here, it is useful to have an understanding of the concept of tests. A test may be one-tail or two-tail test. A statistical test in which the alternative hypothesis specifies that the population parameter lies entirely above or entirely below the value specified in the null hypothesis, is called a one-tail test. An alternative hypothesis which does not specify that the parameter lies on one particular side of the value specified in the null hypothesis is called a two-tail test. The following examples illustrate this concept.

One-Tail Test

H_0: Production of a crop = 2 tons per acre
H_a: Production of a crop > 2 tons per acre

Two-Tail Test

H_0: Production of a crop = 2 tons per acre
H_a: Production of a crop ≠ 2 tons per acre

It is important to note that hypotheses are statements regarding population parameters and not statements regarding the sample. We have defined null and alternative hypothesis. There is a question how the value in the null hypothesis is set. This may be set in any one of the three ways. First, our past experience or knowledge of the phenomenon or prior experimentation may help us in setting the value in the null hypothesis. Testing of hypothesis in this case refers to determination whether that value or situation has changed. Second, the value in the null hypothesis may also be set based on some theory or model regarding the phenomenon. Testing in this case refers to the verification of the theory or model. Third, sometimes the value in the null hypothesis is set based on external considerations, designs, specifications, etc. In this case, the objective of testing is to see whether the results conform to these considerations, designs, specifications.

When we make some statements regarding population parameters and test those, one objective is to make a decision whether the statement is true or false. The testing procedure uses the information from random samples drawn from the respective population. If the sample information is consistent with the hypothesis, then we can conclude that the hypothesis is true. On the other hand, if the sample information is inconsistent with the hypothesis, then we can conclude that the hypothesis is false.

The usual procedure in testing a hypothesis is to take a random sample from the population of our interest, compute an appropriate test statistic from the sample data, and then use this statistic to make decision whether the hypothesis is accepted or rejected. We shall detail the procedure in the subsequent sections.

Errors in Hypothesis Testing

In estimating we might have noticed that we cannot estimate the population parameters with certainty; we need to use probability distribution. In hypothesis testing also we cannot say with certainty that a particular hypothesis is to be accepted or rejected. We face the problem because of uncertainty inherent in sampling from a population. In decision to accept or to reject a hypothesis we always commit some error. This error is called error of hypothesis testing. In hypothesis testing usually our decision will focus on the null hypothesis H_0 whether to accept or to reject. Therefore, error of hypothesis testing can better be explained by use of the null hypothesis. There are chances of error in decision because the decision to accept or to reject H_0 is based on the probabilities and not on the certainty. There are two types of errors. They are called Type I error and Type II error. Based on test results we may decide to reject the null hypothesis when this hypothesis is, in fact, true. The error committed in this case is Type I error and is usually denoted by α. This is also called significance level. Again, based on the test results, we may decide to accept the null hypothesis when this hypothesis is not actually true. The error committed in this case is Type II error. This is usually denoted by β. The situation is summarized in the Table 7.1.

The probabilities of occurrences of type I and type II errors are illustrated below:

$$\alpha = P(\text{type I error}) = P(\text{reject } H_0 \,|\, H_0 \text{ is true})$$

$$\beta = P(\text{type II error}) = P(\text{not reject } H_0 \,|\, H_0 \text{ is false})$$

Associated with type I and type II errors is the power of test defined by

$$\text{Power} = 1 - \beta = P(\text{reject } H_0 \,|\, H_0 \text{ is false})$$

Thus, the power of the test is the probability that a false hypothesis is correctly rejected. The analyst or the researcher usually sets the probability of type I error, i.e., α and it is under his control. Since the significance level α is the probability of type I error when the null hypothesis is actually rejected and since the analyst/decision maker sets this value himself, he has a control over it and as such rejection of H_0 is a strong conclusion.

Unlike type I error, type II error is not a constant quantity. Type II error is a function of the true mean value of the population as well as the sample size. The probability of type II error decreases as the sample size increases. Also decreasing α would increase β and vice versa.

There is another error called type III error. This is also called type 0 error. This error occurs when a false null hypothesis is rejected but the direction is wrong. For example, suppose that the null hypothesis is H_0: $\mu_1 = \mu_2$ or $\mu_1 - \mu_2 = 0$ (two population means are equal), and the alternative hypothesis is H_a: $\mu_1 - \mu_2 > 0$ which

Table 7.1 Decision in hypothesis testing	Decision	Actual situation of H_0	
		True	False
	Reject	Type I error (α)	✓
	Not reject	✓	Type I error (β)

means $\mu_1 > \mu_2$. Based on the test result, we find the H_0 to be false and so we reject it and conclude that $\mu_1 > \mu_2$. But actually $\mu_2 > \mu_1$. This means that there is an error in decision about the direction. This error is type III error.

7.2   Test Procedure

The standard test procedure in testing hypothesis followed in this book involves identifying the critical region in the distribution concerned. This also involves identification of the acceptance and rejection regions of the null hypothesis. Then based on the sample information, a test statistic is computed. If this test statistic falls in the acceptance region, the null hypothesis is accepted (better not rejected). If this statistic falls in the rejection, the null hypothesis is rejected.

Suppose that we are interested in testing the following hypothesis:

$$H_0: \mu = \mu_0 \ (\mu_0 \text{ is a specific value of } \mu)$$
$$H_a: \mu > \mu_0$$

For identifying the acceptance and rejection regions, we look at the alternative hypothesis. In this particular example, the alternative hypothesis states that the population mean is greater than 0, the specific value set in the null hypothesis. This suggests that the critical region falls on the right side of the distribution as illustrated in Fig. 7.1.

On the basis of the set significance level α, the critical value of the test statistic (in this particular case Z_α) is identified and the acceptance and rejection regions are noted. Since the alternative hypothesis specifies/implies one side of the distribution, the α value is also set on the same side of the distribution. This is a one-tail test, since only one tail of the distribution is used to test the hypothesis.

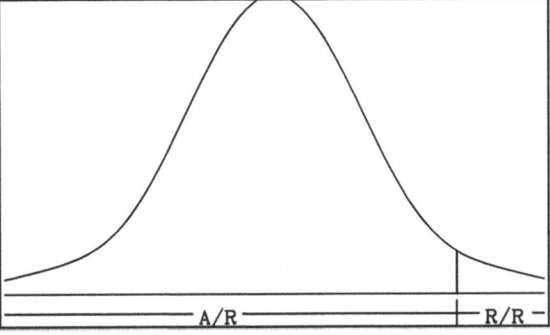

Fig. 7.1 Acceptance and rejection regions (rejection region on *right side*)

Fig. 7.2   Acceptance and
rejection regions (rejection
region on *left side*)

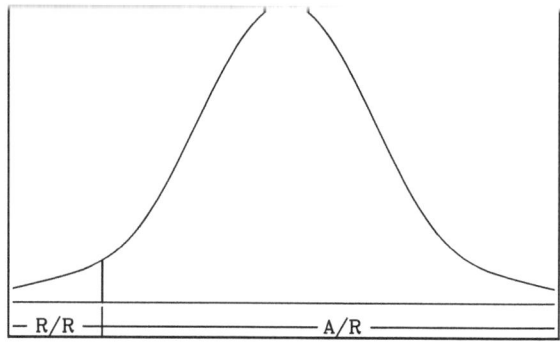

Now, if we are interested in testing the following hypothesis:

$$H_0: \mu = \mu_0$$
$$H_a: \mu < \mu_0$$

The alternative hypothesis states that the population mean is less than the specific value set in the null hypothesis. The critical region is, therefore, on the left side of the distribution as shown in Fig. 7.2.

Based on the test statistic (in this case Z_α) the rejection and acceptance regions are established. In this case also, the rejection region is on one side of the distribution and, therefore, the test is a one-tail test.

Let us now consider the third situation. Let the set of hypotheses we are testing in, be

$$H_0: \mu = \mu_0$$
$$H_a: \mu \neq \mu_0$$

In this case, the alternative hypothesis states that the population mean is not equal to μ_0. This implies that the population mean may either be less than or greater than μ_0. Here obviously, two situations are involved simultaneously. In order to incorporate this, the set α value is equally divided into two, $\alpha/2$ each. Therefore, based on these two $\alpha/2$ values, two critical regions are established. However, the acceptance region is only one as before. The regions are illustrated in Fig. 7.3.

There are two test statistics (in this particular case $-Z_{\alpha/2}$ and $Z_{\alpha/2}$). The null hypothesis would be rejected if the calculated test statistic falls in either of the two rejection regions.

The standard procedure in testing hypotheses may be summarized as follows:

(i)   Set the significance level α, if not already given.
(ii) State the appropriate hypotheses, null and alternative.
(iii) Sketch the acceptance and rejection regions.
(iv) Compute the appropriate test statistic based on sample information.

Fig. 7.3 Acceptance and rejection regions (rejection region on *both sides*)

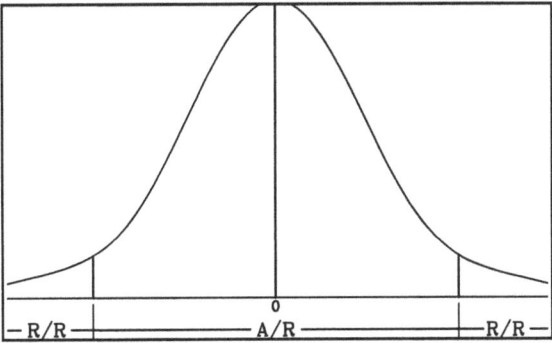

(v) Examine whether the calculated test statistic falls in the acceptance or rejection region. If it falls in the rejection region, reject the null hypothesis. If it falls in the acceptance region, do not reject the null hypothesis.

(vi) Make suitable conclusion. The nature of the conclusion may be tuned to the problem under study.

7.3 Hypothesis Testing—One Population Mean (Variance Known)

This test is valid (a) if the population is normal or (b) if the central limit theorem holds in case the population is not normal. Here, we shall study the one-tail and two-tail tests.

7.3.1 One-Tail Test

Let X be the random variable. If the population is normal or where the central limit theorem holds, the distribution is also normal. Its mean equals the population mean μ which is unknown. Its variance σ^2 is known. We are interested in testing the hypothesis

$$H_0: \mu = \mu_0 \ (\mu_0 \text{ is a specific value of } \mu)$$
$$H_a: \mu > \mu_0 \ (\text{or } \mu < \mu_0)$$

The appropriate test statistic is

$$Z_c = \frac{\bar{x} - \mu_0}{\sigma/\sqrt{n}}$$

The subscript c means the calculated value of Z to distinguish it from the table value of Z. The quantity $\sigma/\sqrt{n}$ is the standard error of $\bar{x}$. In other words, $\sigma/\sqrt{n}$ is the standard deviation of the sampling distribution of $\bar{x}$. Thus, the test statistic could also be written as

$$Z_c = \frac{\bar{x} - \mu_0}{\text{standard error of } \bar{x}}$$

If the alternative hypothesis is H_a: $\mu > \mu_0$, the rejection region is on the right side of the distribution and the null hypothesis is rejected if

$$Z_c > Z_\alpha$$

If the alternative hypothesis is H_a: $\mu < \mu_0$, the rejection region is on the left side of the distribution and the null hypothesis is rejected if

$$Z_c < -Z_\alpha$$

Example 1
The IQ of students is normally distributed. It is claimed that the mean IQ of all students in AIT is greater than 120. It is known from past record that the standard deviation of IQ of all students is 30.25. A random sample of 50 students showed a mean of 130. Test the claim at 5 % level of significance.

Solution
Here,

$$n = 50; \quad \bar{x} = 130$$
$$\sigma = 30.25; \quad \alpha = 0.05$$

The null and alternative hypotheses are formulated as follows:

$$H_0: \mu = 120$$
$$H_a: \mu > 120$$

This is a one-tail test. From Table $Z_{0.05} = 1.645$. The acceptance and rejection regions are shown in the following figure.

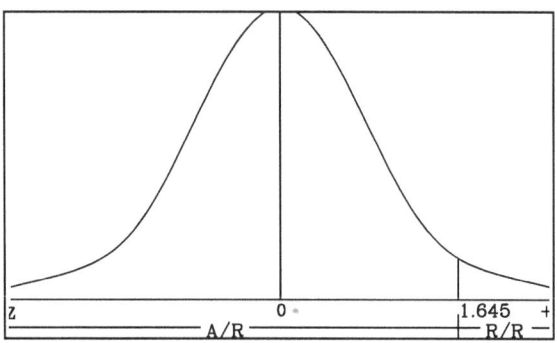

$$Z_c = \frac{\bar{x} - \mu_0}{\sigma/\sqrt{n}}$$
$$= \frac{130 - 120}{30.25\sqrt{50}}$$
$$= 2.34$$

This falls on the rejection region. The null hypothesis is, therefore, rejected at 5 % level of significance.

Conclusion: The mean IQ of all the students is greater than 120.

7.3.2 Two-Tail Test

Here, the same considerations prescribed in Sect. 7.3.1 apply except that the form of the alternative hypothesis is different. We are interested in testing the hypothesis

$$H_0: \mu = \mu_0$$
$$H_a: \mu \neq \mu_0$$

The test statistic Z_c is the same as before,

$$Z_c = \frac{\bar{x} - \mu_0}{\sigma/\sqrt{n}}$$

The set α value is divided equally into two parts, $\alpha/2$ each and the values of $-Z_{\alpha/2}$ and $+Z_{\alpha/2}$ are identified. There are two rejection regions, one located on each side of the distribution. The null hypothesis is rejected if

$$Z_c > Z_{\alpha/2} \quad \text{or} \quad \text{if } Z_c < -Z_{\alpha/2}$$

7.4 Hypothesis Testing—One Population Mean (Variance Unknown—Large Sample)

The test procedure outlined in Sect. 7.3 assumes known population variance σ^2. But if the population variance is not known, we cannot use it in calculating the test statistic. However, if the sample is large ($n \geq 30$), the sample variance s^2 can be substituted for σ^2 in the test procedure. Thus, the test statistic is

$$Z_c = \frac{\bar{x} - \mu_0}{s/\sqrt{n}}$$

The rest procedure is the same as outlined in Sect. 7.3.

Example 2

In Example 1 above, suppose that the claim is that the mean IQ of students is less than 120. Suppose also that sample mean is 115 and the sample standard deviation is 30.25 (population standard deviation not known). Test the claim.

Solution

The population variance is not known. But the sample size is large.
   Here,

$$n = 50; \quad \bar{x} = 115$$
$$\sigma = 30.25; \quad \alpha = 0.05$$

The null and alternative hypotheses are formulated as follows:

$$H_0: \mu = 120$$
$$H_a: \mu < 120$$

This is also a one-tail test. From Table $Z_{0.05} = -1.645$. The acceptance and rejection regions are shown in the following figure

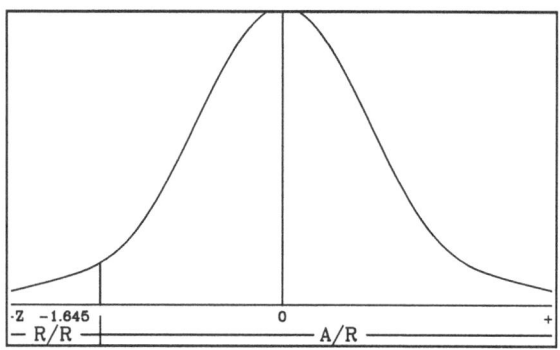

$$Z_c = \frac{\bar{x} - \mu_0}{s/\sqrt{n}}$$
$$= \frac{115 - 120}{30.25\sqrt{50}}$$
$$= -1.17$$

This falls on the acceptance region. The null hypothesis is, therefore, not rejected at 5 % level of significance.
   Conclusion: The mean IQ of all the students is not less than 120.

7.5 Hypothesis Testing—Equality of Two Population Means (Variance Known)

Both one-tail and two-tail tests may be necessary in testing hypothesis concerning equality of two population means. The test procedure is valid if both the populations are normal. It is also valid if the populations are not normal but conditions of central limit theorem apply.

Let there be two populations—population 1 and population 2. Let also X_1 and X_2 are the random variables in populations 1 and 2 respectively. Here X_1 has unknown mean μ_1 and known variance σ_1^2, and X_2 has unknown mean μ_2 and known variance σ_2^2.

Two independent samples—one from each population, are drawn. If the populations are normal (or where the conditions of central limit theorem apply) the distributions of $(\overline{x_1} - \overline{x_2})$ is normal and its standard error is

$$\sqrt{\left\{\sigma_1^2/n_1 + \sigma_2^2/n_2\right\}}$$

7.5.1 One-Tail Test

We are interested in testing the hypothesis of the form

$$
\begin{array}{ccc}
H_0\colon \mu_1 = \mu_2 & & H_0\colon \mu_1 - \mu_2 = 0 \\
& \Rightarrow & \\
H_a\colon \mu_1 > \mu_2 & & H_a\colon \mu_1 - \mu_2 > 0
\end{array}
$$

The alternative hypothesis states that the difference between the two population means is greater than zero, a specific value set in the null hypothesis. Thus, the rejection region falls on the right-hand tail of the distribution. The appropriate test statistic is

$$
\begin{aligned}
Z_c &= \frac{(\overline{x_1} - \overline{x_2}) - (\mu_1 - \mu_2)}{\sqrt{\left\{\sigma_1^2/n_1 + \sigma_2^2/n_2\right\}_1}} \\
&= \frac{(\overline{x_1} - \overline{x_2})}{\sqrt{\left\{\sigma_1^2/n_1 + \sigma_2^2/n_2\right\}_1}}
\end{aligned}
$$

since $(\mu_1 - \mu_2) = 0$ according to null hypothesis.
The null hypothesis is rejected if

$$Z_c > Z_\alpha.$$

If the alternate hypothesis is of the form

$$H_0: \mu_1 = \mu_2 \qquad H_0: \mu_1 - \mu_2 = 0$$
$$\Rightarrow$$
$$H_a: \mu_1 < \mu_2 \qquad H_a: \mu_1 - \mu_2 < 0$$

the rejection region falls on the left-hand tail of the distribution and the null hypothesis is rejected if

$$Z_c < -Z_\alpha.$$

Example 3

Two manufacturing processes are to be compared. The following information is available:

Process 1	Process 2
$\sigma_1 = 15.18$	$\sigma_2 = 18.50$
$n_1 = 40$	$n_2 = 45$
$\overline{x_1} = 150.56$	$\overline{x_2} = 141.23$

Test at 5 % level of significance whether the mean from process 1 is greater than that from process 2.

Solution

$$\alpha = 0.05$$

The null and alternative hypotheses are as follows:

$$\mu_1 - \mu_2 = 0$$
$$\mu_1 - \mu_2 > 0$$

This is a one-tail test. The rejection region falls on the right side of the distribution. From table, $Z_{0.05} = 1.645$. The acceptance and rejection regions are shown in the following figure.

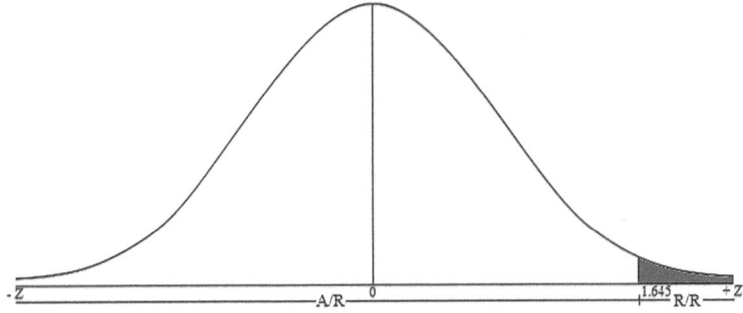

$$Z_c = \frac{(\bar{x_1} - \bar{x_2})}{\sqrt{\{\sigma_1^2/n_1 + \sigma_2^2/n_2\}}}$$
$$= \frac{150.56 - 141.23}{\sqrt{\{(15.18)^2/40 + (18.50)^2/45\}}}$$
$$= \frac{9.33}{3.66} = 2.55$$

This falls in the rejection region. The null hypothesis is, therefore, rejected at 5 % level of significance.

Conclusion: The mean from process 1 is significantly greater than that from process 2.

7.5.2   Two-Tail Test

Our interest here is to test a hypothesis of the form

$$H_0: \mu_1 = \mu_2 \qquad H_0: \mu_1 - \mu_2 = 0$$
$$\Rightarrow$$
$$H_a: \mu_1 \neq \mu_2 \qquad H_a: \mu_1 - \mu_2 \neq 0$$

The alternative hypothesis states that the difference between the two means of the two populations is not equal to zero, a specific value set in the null hypothesis. This implies that the difference could either be less than zero or greater than zero. Thus, the rejection region may fall either on the right or on the left side of the distribution. This means that there are two rejection regions. However, there is only one acceptance region. The test statistic is the same as outlined in the previous section.

$$Z_c = \frac{(\overline{x_1} - \overline{x_2})}{\sqrt{\{\sigma_1^2/n_1 + \sigma_2^2/n_2\}}}$$

The null hypothesis is rejected if

$$Z_c < -Z_{\alpha/2} \quad \text{or,} \quad \text{if } Z_c > Z_{\alpha/2}$$

Example 4

Incomes of normally distributed populations are being studied. The variances are $\sigma_1^2 = \$508$ and $\sigma_2^2 = \$425$. Two randomly selected samples $n_1 = 40$ and $n_2 = 60$ produced means $\overline{x}_1 = \$2520$ and $\overline{x}_2 = \$2350$, respectively.

Test at 90 % confidence level whether the two population means are equal.

Solution

$$\sigma_1^2 = \$508; \qquad \sigma_2^2 = \$425$$
$$\overline{x_1} = \$2520; \qquad \overline{x_2} = \$2350$$
$$n_1 = 40; \qquad n_2 = 60$$
$$\alpha = 1 - 0.90 = 0.10$$

The null and alternative hypotheses are formulated as follows:

$$H_0: \mu_1 - \mu_2 = 0$$
$$H_a: \mu_1 - \mu_2 \neq 0$$

This is a two-tail test. $Z_{\alpha/2} = Z_{0.05} = 1.645$ (from table). The acceptance and rejection regions are shown in the following figure.

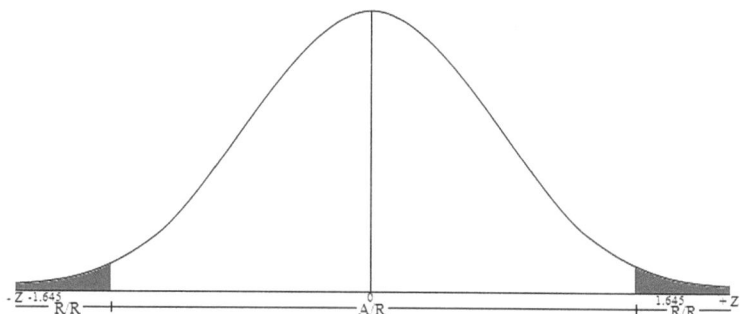

$$Z_c = \frac{(\bar{x}_1 - \bar{x}_2)}{\sqrt{\{\sigma_1^2/n_1 + \sigma_2^2/n_2\}}}$$
$$= \frac{2520 - 2350}{\sqrt{\{508/40 + 425/60\}}}$$
$$= \frac{170}{4.45} = 38.20$$

This falls in the rejection region. The null hypothesis is, therefore, rejected at 90 % confidence level.

Conclusion: The two population incomes are not equal.

7.6 Hypothesis Testing—One Population Mean (Variance Unknown)

The test procedure developed so far used Z statistic. This applies if the population is normal and if the population variance is known. Also, this applies when the sample size is large ($n \geq 30$), in case the population is nonnormal. Use of Z statistic in case of nonnormal population but with large sample size implies that the underlying population is assumed to be approximately normal. In many practical cases, this approximation is satisfactory.

The difficulty arises if the sample size is small and if the population variance is not known. In such cases, we need to use the t distribution and not the Z distribution.

7.6.1 One-Tail Test

Let X be a random variable whose mean μ is unknown and variance σ^2 is also unknown. The population is assumed to be normal. Then the sampling distribution, i.e., the distribution of $\bar{x}$ is also normal and its standard error is $s/\sqrt{n}$. We are now interested in testing the hypothesis of the form

$$H_0: \mu_1 = \mu_2$$
$$H_a: \mu_1 > \mu_2$$

The appropriate test statistic is

$$t_c = \frac{\bar{x} - \mu_0}{s/\sqrt{n}}$$

where s is sample standard deviation. The degrees of freedom is $n - 1$. As indicated by the alternative hypothesis, the rejection region falls on the right-hand tail of the distribution. The null hypothesis can be rejected if

$$t_c > t_{\alpha,n-1}$$

If we are interested in testing hypothesis of the form

$$H_0: \mu_1 = \mu_2$$
$$H_a: \mu_1 < \mu_2$$

the rejection region falls on the left-hand tail of the distribution. In this case, the null hypothesis can be rejected if

$$t_c < -t_{\alpha,n-1}$$

Example 5
In the IQ example suppose that the mean IQ of student population is assumed to be less than 120. A random sample of size 25 shows the mean of 115 and standard deviation of 20.15. Test the assumption at 97.5 % confidence level.

Solution
The population is normal. The population variance is not known. The sample size is less than 30 (small sample). So we need to use t distribution.

$$n = 25; \qquad \bar{x} = 115$$
$$s = 20.15; \quad \alpha = 1 - 0.975 = 0.025$$

The null and alternative hypotheses are formulated as shown to suit the problem.

$$H_0: \mu_1 = 120$$
$$H_a: \mu_1 < 120$$

This is a one-tail test. The t value, i.e., $t_{0.025,24} = -2.064$. The acceptance and rejection regions are shown in the figure.

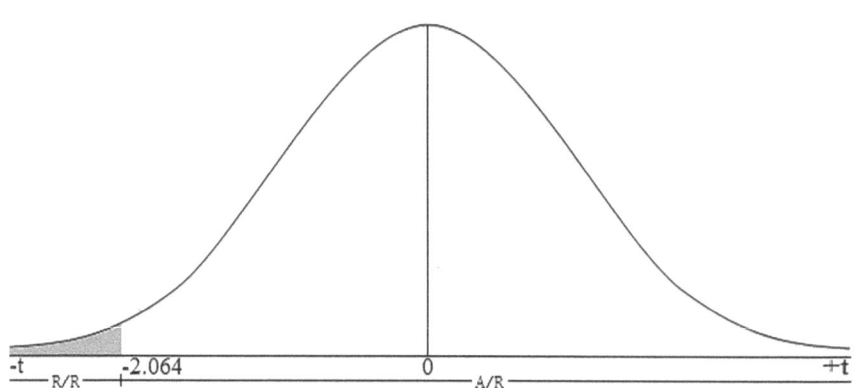

$$t_c = \frac{\bar{x} - \mu_0}{s/\sqrt{n}}$$
$$= \frac{115 - 120}{\frac{20.15}{\sqrt{25}}}$$
$$= \frac{-5}{4.02}$$
$$= -1.24$$

This falls in the acceptance region. The null hypothesis is not rejected at 97.5 % confidence level.

Conclusion: The mean IQ of the student population is not statistically less than 120.

7.6.2   Two-Tail Test

In this case we are interested in testing the hypothesis of the form

$$H_0: \mu_1 = \mu_0$$
$$H_a: \mu_1 \neq \mu_0.$$

The alternative hypothesis specifies that the value of μ may be either on the left or on the right-hand side of the distribution. The test statistic remains the same as

$$t_c = \frac{\bar{x} - \mu_0}{s/\sqrt{n}}$$

The null hypothesis is rejected if

$$t_c < -t_{\alpha/2,n-1} \quad \text{or} \quad t_c > t_{\alpha/2,n-1}$$

Example 6

In the IQ example, suppose that the mean IQ of the student population is assumed to be different from 120 (same as saying not equal to 120). A random sample of size 25 shows the mean of 115 with a standard deviation of 20.15. Test the assumption at 98 % confidence level.

Solution

The population is normal. The population variance is unknown. The sample size is small ($n < 30$). So, we should use t distribution.

$$n = 25; \quad \bar{x} = 115$$
$$s = 20.15; \quad \alpha = 1 - 0.98 = 0.02$$

The null and alternative hypotheses are formulated as shown to suit the problem.

$$H_0: \mu = 120$$
$$H_a: \mu \neq 120$$

This is a two-tail test. In this case $t_{\alpha/2,n-1} = t_{0.01,24} = 2.492$. The acceptance and rejection regions are shown.

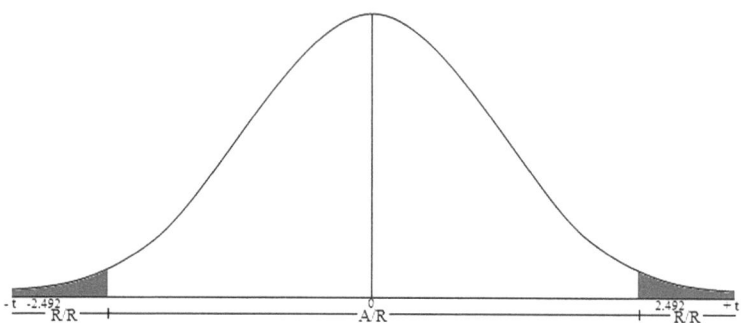

$$\begin{aligned}
t_c &= \frac{\bar{x} - \mu_0}{s/\sqrt{n}} \\
&= \frac{115 - 120}{20.15/\sqrt{25}} \\
&= \frac{-5}{4.02} \\
&= -1.24
\end{aligned}$$

This falls in the acceptance region. The null hypothesis is therefore, not rejected at 95 % confidence level.

Conclusion: The mean IQ of the student population is not different from 120.

7.7   Hypothesis Testing—Equality of Two Population Means (Variance Unknown—Small Sample)

Let μ_1 and μ_2 be two unknown means and σ_1^2 and σ_2^2 be the unknown variances of two normal populations 1 and 2, respectively. There are two situations which require different treatments. In one situation σ_1^2 and σ_2^2 are assumed to be equal. In another situation σ_1^2 and σ_2^2 are assumed not to be equal. Here we need to use the t distribution.

7.7.1 Situation 1: $\sigma_1^2 = \sigma_2^2 = \sigma^2$

Let two random independent samples with variances of s_1^2 and s_2^2 be drawn from the two populations. We need to compute one common or pooled estimated variance s_p^2 for computing the test statistic. The common or pooled estimated variance is given by

$$s_p^2 = \frac{(n_1 - 1)s_1^2 + (n_2 - 1)s_2^2}{n_1 + n_2 - 2}$$

The degrees of freedom is $n_1 + n_2 - 2$. If we are interested in testing the hypothesis of the form

$$H_0: \mu_1 = \mu_2 \qquad H_0: \mu_1 - \mu_2 = 0$$
$$\Rightarrow$$
$$H_a: \mu_1 < \mu_2 \qquad H_a: \mu_1 - \mu_2 < 0$$

implying one tail test, the appropriate test statistic is

$$t_c = \frac{(\bar{x_1} - \bar{x_2}) - (\mu_1 - \mu_2)}{s_p \sqrt{(1/n_1 + 1/n_2)}}$$
$$= \frac{(\bar{x_1} - \bar{x_2})}{s_p \sqrt{(1/n_1 + 1/n_2)}}$$

since $\mu_1 - \mu_2 = 0$ according to null hypothesis.

Here $S_p \sqrt{\{1/n_1 + 1/n_2\}}$ is the standard error of the distribution of the statistic $(\bar{x_1} - \bar{x_2})$.

If the alternative hypothesis is $H_a: \mu_1 - \mu_2 > 0$, the rejection region falls on the right-hand tail of the distribution. The null hypothesis is rejected if

$$t_c > t_{\alpha, n_1 + n_2 - 2}$$

with $(n_1 + n_2 - 2)$ degrees of freedom.

If the alternative hypothesis is $H_a: \mu_1 - \mu_2 < 0$, the rejection region falls on the left-hand tail of the distribution. The null hypothesis is rejected if

$$t_c < -t_{\alpha, n_1 + n_2 - 2}$$

with $(n_1 + n_2 - 2)$ degrees of freedom.

If the alternative hypothesis is $H_a: \mu_1 - \mu_2 \neq 0$, there are two rejection regions and one each of them falls on right-and left-hand tails of the distribution. The null hypothesis is rejected if

$$t_c < - t_{\alpha/2, n_1 + n_2 - 2}$$

or, if

$$t_c > - t_{\alpha/2, n_1 + n_2 - 2}$$

with $(n_1 + n_2 - 2)$ degrees of freedom.

Example 7

Let us consider the income example. The population variances are not known. Two randomly selected samples of sizes $n_1 = 20$ and $n_2 = 22$ produced means $\overline{x_1} = \$2520$ and $\overline{x}_2 = \$2450$ with standard deviations $s_1 = \$504$ and $s_2 = \$415$, respectively. Test at 5 % significance level whether the two population means are equal. Assume that the two population variances are equal.

Solution

$$
\begin{array}{ll}
n_1 = 20; & n_2 = 22 \\
\overline{x_1} = \$2520; & \overline{x_2} = \$2450 \\
s_1 = \$504; & s_2 = \$415 \\
\sigma_1^2 = \sigma_2^2 = s^2 & \\
\alpha = 0.05 &
\end{array}
$$

Sample sizes are small. So we need to use t distribution. The appropriate hypotheses are as follows:

$$H_0: \mu_1 - \mu_2 = 0$$
$$H_a: \mu_1 - \mu_2 \neq 0$$

This is a two-tail test. Now $\alpha/2 = 0.025$. The degrees of freedom d. f. $= n_1 + n_2 - 2 = 20 + 22 - 2 = 40$. From table $t_{0.025,40} = 2.2021$.
The acceptance and rejection regions are shown in the figure

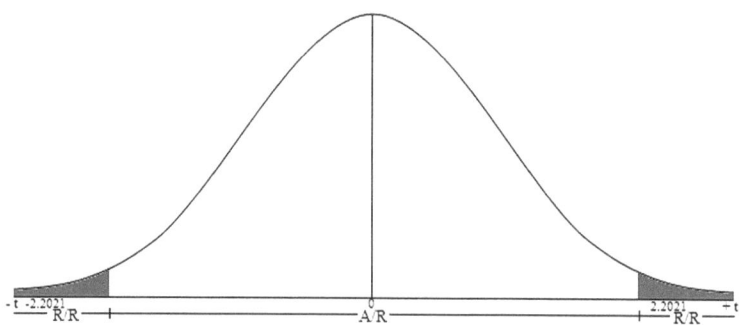

The common or pooled estimated variance is given by

$$
\begin{aligned}
s_p^2 &= \frac{(n_1 - 1)s_1^2 + (n_2 - 1)s_2^2}{n_1 + n_2 - 2} \\
&= \frac{(20 - 1)(504)^2 + (22 - 1)(415)^2}{20 + 22 - 2} \\
&= \frac{8{,}443{,}029}{40} \\
&= 211{,}075.73 \\
s_p &= 459.43 \\
t_c &= \frac{(\bar{x_1} - \bar{x_2})}{s_p \sqrt{(1/n_1 + 1/n_2)}} \\
&= \frac{2520 - 2450}{459.43 \sqrt{(1/20 + 1/22)}} \\
&= \frac{70}{141.94} \\
&= 0.49
\end{aligned}
$$

This falls in the acceptance region. The null hypothesis is, thus, not rejected at 5 % significance level.

Conclusion: Statistically, the two population mean incomes are not different.

7.7.2 Situation 2: $\sigma_1^2 \neq \sigma_2^2$

There are some situations where it is not reasonable to assume the equality of the two population variances. In such a case, there is no exact t distribution to test H_0: $\mu_1 = \mu_2$. However, the approximate statistic is

$$
t_c = \frac{(\bar{x_1} - \bar{x_2})}{s_p \sqrt{(1/n_1 + 1/n_2)}}
$$

and its degrees of freedom (v) is given by

$$
v = \frac{\{s_1^2/n_1 + s_2^2/n_2\}^2}{\frac{\left(s_1^2/n_1\right)^2}{n_1 + 1} + \frac{\left(s_2^2/n_2\right)^2}{n_2 + 1}} - 2
$$

If the alternative hypothesis is H_a: $\mu_1 - \mu_2 > 0$, the null hypothesis H_0: $\mu_1 - \mu_2 = 0$, is rejected if

$$t_c > t_{\alpha,v}$$

If the alternative hypothesis is H_a: $\mu_1 - \mu_2 < 0$, the null hypothesis is rejected if

$$t_c < -t_{\alpha,v}$$

If the alternative hypothesis is H_a: $\mu_1 - \mu_2 \neq 0$, the null hypothesis is rejected if

$$t_c < -t_{\alpha/2,v} \quad \text{or,} \quad \text{if } t_c > t_{\alpha/2,v}$$

Example 8

Solve problem 7 assuming the two population variances not equal.

Solution

$$
\begin{aligned}
&n_1 = 20; \qquad n_2 = 22 \\
&\overline{x_1} = \$2520; \quad \overline{x_2} = \$2450 \\
&s_1 = \$504; \qquad s_2 = \$415 \\
&\sigma_1^2 \neq \sigma_2^2
\end{aligned}
$$

The null and alternative hypotheses are formulated as follows:

$$
\begin{aligned}
&H_0: \mu_1 - \mu_2 = 0 \\
&H_a: \mu_1 - \mu_2 \neq 0
\end{aligned}
$$

Now, $\alpha = 0.05$; $\alpha/2 = 0.025$. The degrees of freedom v is given by

$$v = \frac{\left\{ s_1^2/n_1 + s_2^2/n_2 \right\}^2}{\frac{\left(s_1^2/n_1 \right)^2}{n_1 + 1} + \frac{\left(s_2^2/n_2 \right)^2}{n_2 + 1}} - 2$$

$$v = \frac{\left\{ (504)^2/20 + (415)^2/22 \right\}^2}{\frac{\left\{ (504)^2/20 \right\}^2}{20 + 1} + \frac{\left\{ (415)^2/22 \right\}^2}{22 + 1}} - 2$$

$$= \frac{4.21 * 10^8}{1.03 * 10^7} - 2$$

$$= 40.87 - 2$$

$$= 38.87$$

Therefore, $t_{0.025,38.87} = 2.024$. The acceptance and rejection regions are shown in the figure.

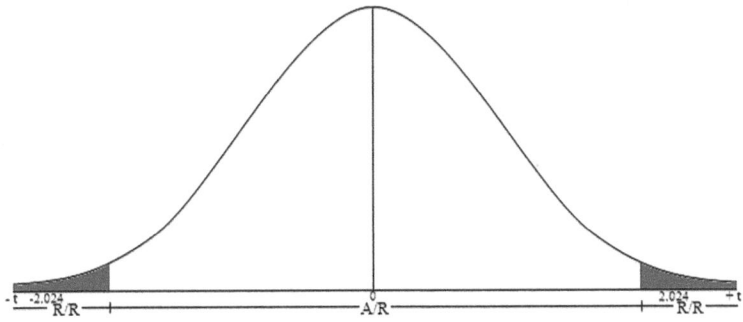

$$t_c = \frac{(\overline{x_1} - \overline{x_2})}{s_p\sqrt{(1/n_1 + 1/n_2)}}$$

$$= \frac{2520 - 2450}{\sqrt{\left\{(504)^2/20 + (415)^2/22\right\}}}$$

$$= \frac{70}{143.28}$$

$$= 0.489$$

This falls in the acceptance region. Therefore, the null hypothesis is not rejected at 5 % significance level.

Conclusion: Statistically, the two population mean incomes are the same.

7.8 Testing of Hypothesis—Population Proportion

In many planning, management and engineering decisions we are concerned with a variable which follows binomial distribution. Examples of such cases are the proportion of people below the poverty line, fraction of defective items in a manufacturing process, proportion of concrete blocks that failed in a particular test.

If the proportion is not very close to one or zero and if the sample size is large, then the binomial distribution can be approximated by normal distribution. Therefore, in testing hypothesis on proportions, test procedure based on normal distribution is developed hereafter.

7.8.1   One Population Proportion

Let X be a random variable and x be the number of observations having the characteristics of our interest in a random sample of size n. Then, the sample proportion is

$$p = \frac{x}{n}$$

If we are interested in testing a hypothesis of the form

$$H_0: P = P_0$$
$$H_a: P > P_0$$

where, P_0 is a specific value of P,
    then the appropriate test statistic is given by

$$Z_c = \frac{p - P}{\sqrt{\left\{\frac{P(1-P)}{n}\right\}}}$$

and the null hypothesis is rejected if

$$Z_c > Z_\alpha$$

If we are interested in testing the hypothesis of the form

$$H_0: P = P_0$$
$$H_a: P < P_0$$

the null hypothesis is rejected if

$$Z_c < -Z_\alpha$$

But if we are interested in testing the hypothesis of the form

$$H_0: P = P_0$$
$$H_a: P \neq P_0$$

the null hypothesis is rejected if

$$Z_c < -Z_{\alpha/2} \quad \text{or,} \quad \text{if } Z_c > Z_{\alpha/2}$$

Example 9

There has been a considerable debate on environmental problems arising out of automobiles in cities. It is claimed that more than 50 % of the people are concerned about the problem. In a random sample of 400 people, 220 people showed concern about the problem. At 99 % confidence level, test the claim that

(a) more than 50 % people are concerned about the problem.
(b) the proportion of people expressing concern about the problem is not 50 %.

Solution

$$n = 400; \quad x = 220$$
$$p = \frac{220}{400} = 0.55$$
$$a = 1 - 0.99 = 0.01$$

(a) The null and alternative hypotheses are formulated as shown.

$$H_0: P = 0.50$$
$$H_a: P > 0.50$$

This is a one-tail test. Therefore, $Z_{0.01} = 2.33$. The acceptance and rejection regions are as follows:

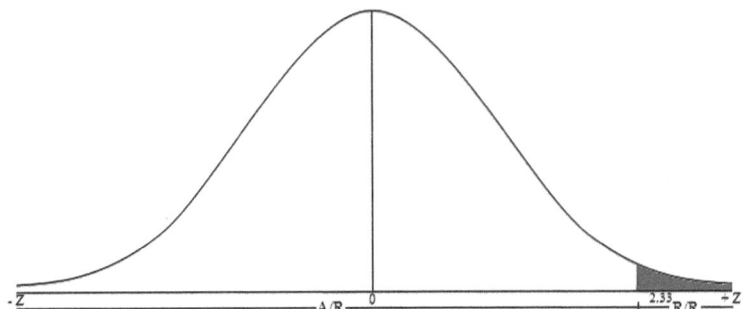

$$Z_c = \frac{p - P_0}{\sqrt{\frac{P(1-P)}{n}}}$$

$$Z_c = \frac{0.55 - 0.50}{\sqrt{\frac{0.55(1-0.50)}{400}}}$$

$$= \frac{0.05}{0.0249}$$

$$= 2.01$$

This falls in the acceptance region. Therefore, the null hypothesis is not rejected at 99 % confidence level.

Conclusion: The proportion of people expressing concern about the environmental problem is not more than 50 %.

(b) The null and alternative hypotheses are as follows:

$$H_0 : P = 0.50$$
$$H_a : P \neq 0.50$$

This is a two-tail test. Therefore, $Z_{\alpha/2} = Z_{0.005}$. From table, $Z_{0.005} = 2.575$. The acceptance and rejection regions are as follows:

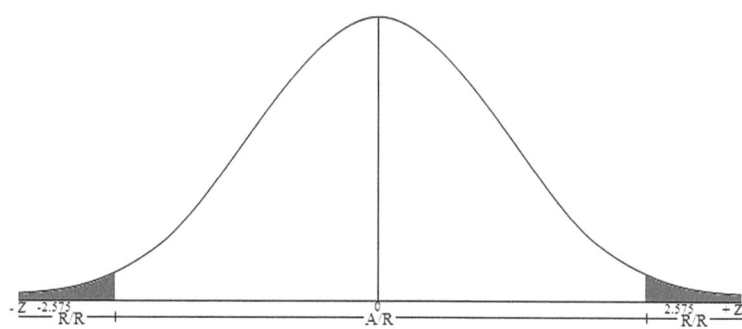

$$Z_c = \frac{p - P_0}{\sqrt{\frac{P(1-P)}{n}}}$$

$$Z_c = \frac{0.55 - 0.50}{\sqrt{\frac{0.55(1-0.50)}{400}}}$$

$$= \frac{0.05}{0.0249}$$

$$= 2.01$$

This falls in the acceptance region. The null hypothesis is, therefore, not rejected at 0.01 level of significance.

Conclusion: Statistically, the proportion of people who are concerned about the environmental problem in cities is 50 %.

7.8.2   Equality of Two Population Proportions

Let X_1 and X_2 be two random variables in population 1 and 2, respectively. Let also n_1 and n_2 be the random and independent sample sizes drawn from population 1 and 2, respectively. If x_1 and x_2 are the number of observations having the characteristics of our interest in sample 1 and 2, respectively, then

$$p_1 = \frac{x_1}{n_1}; \quad p_2 = \frac{x_2}{n_2}$$

the sample proportions. We are interested in testing whether the respective population proportions P_1 and P_2 are equal, i.e., if $P_1 = P_2$.

Thus, the null hypothesis is

$$H_0: P_1 = P_2.$$

Our concern here is the sampling distribution of $p_1 - p_2$. The distribution has the variance

$$\sigma^2_{p_1-p_2} = \sigma^2_{p_1} + \sigma^2_{p_2}$$

Therefore,

$$\sigma_{p_1-p_2} = \sqrt{\left\{\sigma^2_{p_1} + \sigma^2_{p_2}\right\}}$$

Furthermore,

$$\sigma^2_{p_1} = \frac{P_1(1 - P_1)}{n_1} \quad \text{and}$$

$$\sigma^2_{p_2} = \frac{P_2(1 - P_2)}{n_2}$$

Therefore,

$$\sigma_{p_1-p_2} = \sqrt{\left\{\frac{P_1(1-P_1)}{n_1} + \frac{P_2(1-P_2)}{n_2}\right\}}$$
$$= \sqrt{\{P(1-P)(1/n_1 + 1/n_2)\}}$$

since $P_1 = P_2$ according to null hypothesis (and taking $P_1 = P_2 = P$).
  This P in the above expression is the pooled proportion and is estimated by

$$P = \frac{p_1 n_1 + p_2 n_2}{n_1 + n_2}$$

The standard error of the distribution of $p_1 - p_2$ is thus,

$$\sqrt{\{P(1-P)(1/n_1 + 1/n_2)\}}$$

where,

$$P = \frac{p_1 n_1 + p_2 n_2}{n_1 + n_2}$$

If we are interested in testing hypothesis of the form

$$H_0: P_1 = P_2 \qquad H_0: P_1 - P_2 = 0$$
$$\Rightarrow$$
$$H_a: P_1 > P_2 \qquad H_a: P_1 - P_2 > 0$$

the appropriate test statistic is

$$Z_c = \frac{(p_1 - p_2) - (P_1 - P_2)}{\text{standard error of } p_1 - p_2}$$
$$= \frac{p_1 - p_2}{\sqrt{\{P(1-P)(1/n_1 + 1/n_2)\}}}$$

since $P_1 = P_2$ according to null hypothesis.
  The null hypothesis is rejected if

$$Z_c > Z_\sigma.$$

If we are interested in testing hypothesis of the form

$$H_0: P_1 = P_2 \qquad H_0: P_1 - P_2 = 0$$
$$\Rightarrow$$
$$H_a: P_1 < P_2 \qquad H_a: P_1 - P_2 < 0$$

the test statistic is the same and the null hypothesis is rejected if

$$Z_c < -Z_\sigma.$$

Furthermore, if we want to test the hypothesis of the form

$$H_0: P_1 = P_2 \qquad\qquad H_0: P_1 - P_2 = 0$$
$$\Rightarrow$$
$$H_a: P_1 \neq P_2 \qquad\qquad H_a: P_1 - P_2 \neq 0$$

the same statistic is used. The null hypothesis is rejected if

$$Z_c < -Z_{\alpha/2} \quad \text{or,} \quad \text{if } Z_c > Z_{\alpha/2}$$

Example 10

A quality control manager wants to compare the fraction of defective items in two manufacturing processes. In two independent random samples drawn from the two processes, he obtained the following information:

Process 1	Process 2
$n_1 = 6000$	$n_2 = 5000$
$x_1 = 240$	$x_2 = 150$

Test at 5 % significance level whether the two processes produce the same proportion of defective items.

Solution

$$p_1 = \frac{240}{6000} = 0.04; \quad p_2 = \frac{150}{5000} = 0.03$$
$$\alpha = 0.05$$

The following are the null and the alternative hypotheses:

$$H_0: P_1 - P_2 = 0$$
$$H_a: P_1 - P_2 \neq 0$$

This is obviously a two-tail test. So, $\alpha/2 = 0.025$. $Z_{0.025} = 1.96$ (from table). The following figure shows the acceptance and rejection regions.

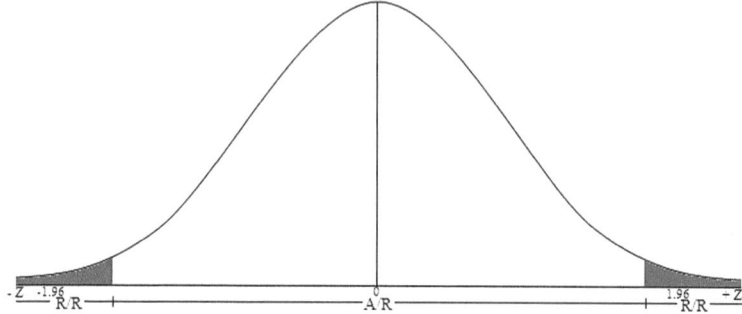

The pooled estimated P is given by

$$P = \frac{p_1 n_1 + p_2 n_2}{n_1 + n_2}$$
$$= \frac{0.04(6000) + 0.03(5000)}{6000 + 5000}$$
$$= \frac{240 + 150}{11{,}000}$$
$$= \frac{390}{11{,}000}$$
$$= 0.0355$$

Therefore, the test statistic Z_c is given by

$$Z_c = \frac{p_1 - p_2}{\sqrt{\{P(1-P)(1/n_1 + 1/n_2)}}$$
$$= \frac{0.04 - 0.03}{\sqrt{\{0.0355(1 - 0.0355)(1/6000 + 1/5000)\}}}$$
$$= \frac{0.01}{0.00354}$$
$$= 2.82$$

This calculated test statistic falls in the rejection region. The null hypothesis is, therefore, rejected at 5 % significance level.

Conclusion: The two processes do not produce the same proportion of defective items.

Flow Chart for Hypothesis Testing

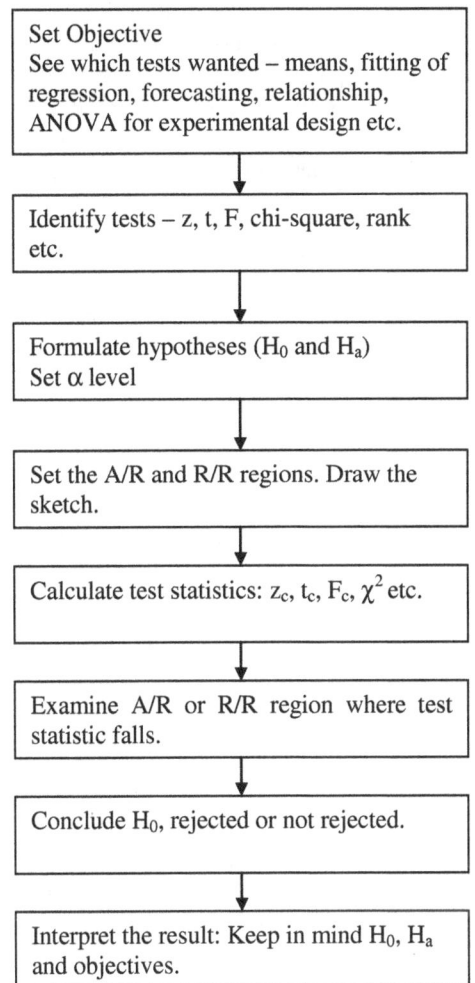

7.9   Power of Hypothesis Testing

We know

$$\alpha = P(\text{type I error}) = P(\text{reject } H_0 | H_0 \text{ is true})$$
$$\beta = P(\text{type II error}) = P(\text{not reject } H_0 | H_0 \text{ is false})$$

and the power of test defined by

$$\text{Power} = 1 - \beta = P(\text{reject } H_0 | H_0 \text{ is false})$$

Thus, it is seen that in order to calculate the power we need to have the value of β. It is not a constant quantity. It depends on α, population true mean and n. Power increases as α increases, power increases as σ (population standard deviation) increases and power increases as n increases.

For hypothesis dealing with one population mean, the hypotheses can be of the form

$$H_0: \mu = \mu_0 \text{ (a specific value)}$$
$$H_a: \mu > \mu_0$$
$$H_a: \mu < \mu_0$$
$$H_a: \mu \neq \mu_0$$

Notice that all the three forms of the alternate hypotheses are open ended. These do not refer to any specific value. In such cases, β cannot be computed, and hence power cannot be computed. To compute β and consequently power, we need to know a specific value of the population true mean in the alternate hypothesis such as $H_a: \mu = 150$.

For hypothesis dealing with two population means (comparing two population means), the hypotheses can be of the form

$$H_0: \mu_1 - \mu_2 = 0 \text{ (no difference between two population means)}$$
$$H_a: \mu_1 - \mu_2 > 0$$
$$H_a: \mu_1 - \mu_2 < 0$$
$$H_a: \mu_1 - \mu_2 \neq 0$$

Here also all the three forms of the alternate hypotheses are open ended. So we cannot compute the power of test. We need a specific value in the alternate hypothesis. It is important to note that we can put a specific value in the alternate hypothesis and compute the power of the test. But difference between two population means having a specific value has no practical use. It is only theoretical. So here we shall not work for calculation of power of test. Similar is the reasoning for comparing two population proportions.

Example 1

The following data are obtained from a study:

$$n = 50; \quad \bar{x} = 125; \quad s = 30.25; \quad \alpha = 0.05$$

Hypothesis is as follows:

$$H_0: \mu = 120$$
$$H_a: \mu > 120$$

(a) Test the hypothesis.
The A/R and R/R are shown. For $\alpha = 0.05$, $z = 1.645$.

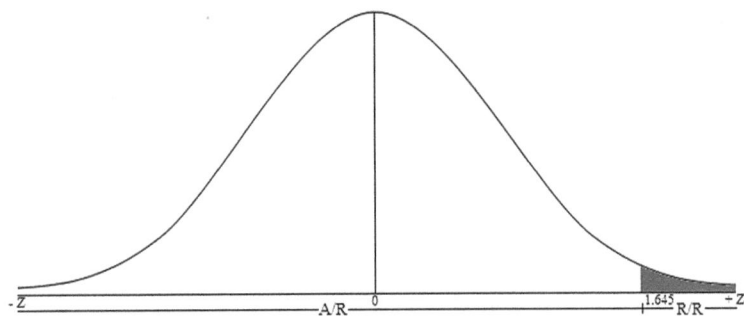

$$Z_c = \frac{\bar{x} - \mu_0}{s/\sqrt{n}} = \frac{125 - 120}{30.25/\sqrt{50}} = 1.1688$$

The test statistic Z_c falls on the acceptance region. So the H_0 is not rejected. (Note: since $n > 30$, we are using normal distribution according to central limit theorem. Population std. dev σ is not known; we are using s for σ)

(b) Calculate the type II error β
Calculate $\bar{x}$ at $z = 1.645$

$$1.645 = \frac{\bar{x} - 120}{30.25/\sqrt{50}} = \frac{\bar{x} - 120}{4.278}$$

Solving we get $\bar{x} = 127.037$
The alternate hypothesis says $\mu > 120$ which indicates many values (any value greater than 120). For calculating β, we need a specific value. Let us suppose $\mu_a = 130$.

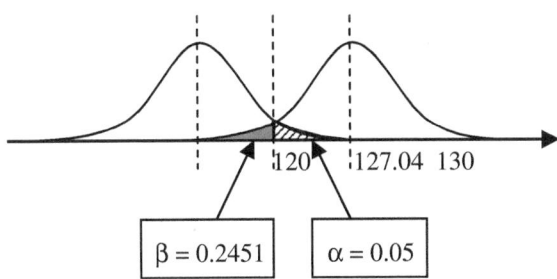

For $\mu_a = 130$, calculate z value.

$$z = \frac{127.04 - 130}{30.25/\sqrt{50}} = \frac{-2.96}{4.278} = -0.6919$$

From the normal distribution table, area of the left tail from this z value = 0.2451. Therefore, β = 0.2451.

(c) Calculate the power of the test.

$$\text{Power} = 1 - \beta = 1 - 0.2451 = 0.7549 \rightarrow 75.49\,\%$$

(d) If the power is to be 90 %, calculate the sample size.
Power = 90 % $\rightarrow$ 0.90. Therefore, $1 - \beta$ = 0.90. So β = 0.10. For β = 0.10, corresponding z score is 1.28.
For distribution with μ_0 = 120

$$z = \frac{\bar{x} - \mu_0}{30.25/\sqrt{n}} = \frac{\bar{x} - 120}{30.25/\sqrt{n}} \quad \text{or,} \quad 1.645 = \frac{\bar{x} - 120}{30.25/\sqrt{n}}$$

$$\Rightarrow$$

$$\bar{x} = 120 + 49.76/\sqrt{n}$$

For distribution with μ_0 = 130

$$z = \frac{\bar{x} - \mu_0}{30.25/\sqrt{n}} = \frac{\bar{x} - 130}{30.25/\sqrt{n}} \quad \text{or,} \quad -1.28 = \frac{\bar{x} - 130}{30.25/\sqrt{n}}$$

$$\Rightarrow$$

$$\bar{x} = 130 + 38.72/\sqrt{n}$$

This $\bar{x}$ is common to both the distribution. Therefore,

$$120 + 49.76/\sqrt{n} = 130 - 38.72/\sqrt{n}$$

On solution this gives, n = 78.29 $\approx$ 79
Using a short formula:
$n = \frac{(z_\alpha + z_\beta)^2 * \sigma^2}{(\mu_a - \mu_0)}$; σ is not known. So use s for σ.
Therefore, $n = \frac{(1.645 + 1.28)^2 * 30.25^2}{(130 - 120)^2} = 78.29 \approx 79$

$$H_0: \mu = 120$$
$$H_a: \mu > 120$$

Example 2
The following data are obtained from a study:

$$n = 50; \quad \bar{x} = 125; \quad s = 30; \quad \alpha = 0.05$$

Hypothesis is as follows:

$$H_0: \mu = 120$$
$$H_a: \mu \neq 120$$

(a) Test the hypothesis.
     The A/R and R/R are shown. For $\alpha = 0.05$, $\alpha/2 = 0.025$, $z_{\alpha/2} = 1.96$.

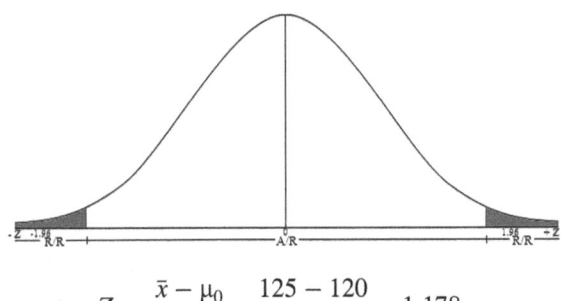

$$Z_c = \frac{\bar{x} - \mu_0}{s/\sqrt{n}} = \frac{125 - 120}{30/\sqrt{50}} = 1.178$$

The test statistic Z_c falls on the acceptance region. So the H_0 is not rejected.

(b) Calculate the Type II error β
     Let us assume $\mu_a = 130$, a specific value in the alternate hypothesis.
     Type II error is β. This is equal to the area of the left tail of the distribution
     with $\mu_a = 130$.
     Calculate $\bar{x}$ at $\alpha = 0.025$.

$$1.96 = \frac{\bar{x} - 120}{30/\sqrt{50}} = \frac{\bar{x} - 120}{4.243}$$

or, $\bar{x} - 120 = 8.3156$, or $\bar{x} = 128.32$

$$z = \frac{\bar{x} - \mu_a}{s/\sqrt{n}} = \frac{128.32 - 130}{30/\sqrt{50}} = -0.396$$

In the normal curve, the left tail area for this z score $= 0.3446$. Therefore,
$\beta = 0.3446$.
The summary is shown in the following figure.

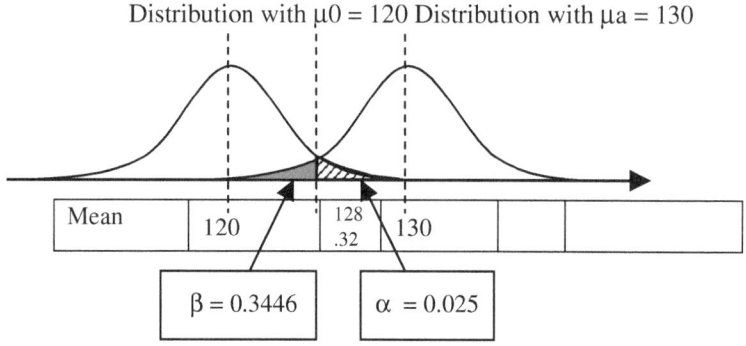

Distribution with µ0 = 120 Distribution with µa = 130

(c) Calculate the power of the test.

$$\text{Power} = 1 - \beta = 1 - 0.3446 = 0.6554 \rightarrow 65.54\,\%$$

(d) If the power is to be 85 %, calculate the sample size.
Power = 85 % → 0.85. Therefore, $1 - \beta = 0.85$. So $\beta = 0.15$. For $\beta = 0.15$,
corresponding z score is 1.035.
For distribution with $\mu_0 = 120$

$$z = \frac{\bar{x} - \mu_0}{30/\sqrt{n}} = \frac{\bar{x} - 120}{30/\sqrt{n}} \quad \text{or,} \quad 1.96 = \frac{\bar{x} - 120}{30/\sqrt{n}}$$

$$\Rightarrow$$

$$\bar{x} = 120 + 58.8/\sqrt{n}$$

For distribution with $\mu_0 = 130$

$$z = \frac{\bar{x} - \mu_0}{30/\sqrt{n}} = \frac{\bar{x} - 130}{30/\sqrt{n}} \quad \text{or,} \quad -1.035 = \frac{\bar{x} - 130}{30/\sqrt{n}}$$

$$\Rightarrow$$

$$\bar{x} = 130 - 31.05/\sqrt{n}$$

This $\bar{x}$ is common to both the distribution. Therefore,

$$120 + 58.8/\sqrt{n} = 130 - 31.05/\sqrt{n}$$

On solution this gives, n = 80.73 ≈ 81
Using the short formula
$n = \frac{(z_\alpha + z_\beta)^2 * \sigma^2}{(\mu_a - \mu_0)}$; σ is not known. So use s for σ.
Therefore, $n = \frac{(1.96 + 1.035)^2 * 30^2}{(130 - 120)^2} = 80.73 \approx 81$

Example 3

The following figures are available from a study:

$$n = 400, x = 220; \alpha = 0.01; z = 2.33$$
$$P = 220/400 = 0.55$$

$$H_0: P = 0.50$$
$$H_a: P > 0.50$$

Now,

$$z_c = \frac{p - P_0}{\sqrt{P(1-P)/n}} = \frac{0.55 - 0.50}{\sqrt{(0.55(1-0.55)/400}} = 2.01$$

This falls in the acceptance region. So the H_o is not rejected.

(b) Calculate β
   To find p at $z = 2.33$

$$z = \frac{p - P_0}{\sqrt{P(1-P)/n}} = \frac{p - 0.50}{\sqrt{(0.55(1-0.55)/400}} = \frac{(p - 0.50)}{0.02449}$$

$$2.33 = \frac{(p - 0.50)}{0.02449} \quad \text{or,} \quad p - 0.50 = 0.0570; \quad \text{so } p = 0.5570$$

The corresponding z score is given by

$$z = \frac{0.5572 - 0.60}{\sqrt{0.5572(1 - 0.5572)/400}} = \frac{-0.0428}{0.0249} = -1.72;$$

Therefore, left tail area = 0.04272.
Therefore, β = 0.04272
Therefore, Power = 1 − β = 1 − 0.04272 = 0.9572 → 95.72 %
The summary is shown below.

Distribution with μ0: P0 = 0.50 Distribution with μa: Pa = 0.60

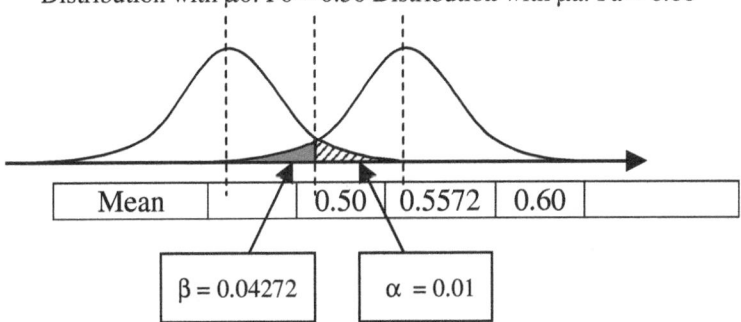

(c) If power is to be 90 %, n = ?

Power = 0.90; therefore, $1 - \beta = 0.90$. So $\beta = 0.10$
z_β (from table) = 1.28
z_α = 2.33
H_0: $\mu_0 = 0.50$
Let us assume H_a: $\mu_a = 0.60$ for finding β.

Here two distributions are involved as follows;

Distribution of H_0: with $\mu_0 = 0.50$ and
Distribution of H_a: with $\mu_a = 0.60$ as shown hereafter.

Distribution with µ0: P0 = 0.50 Distribution with µa: Pa = 0.60

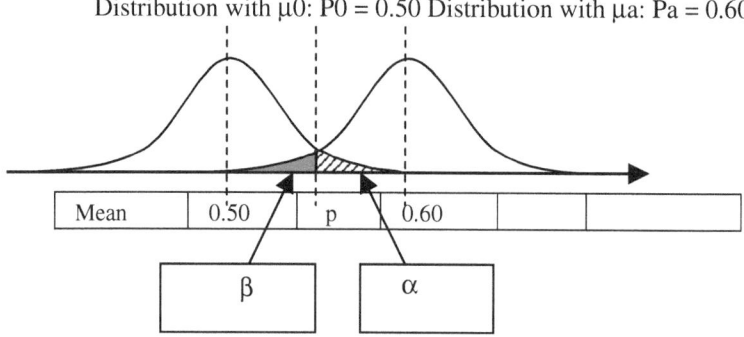

Let the p = proportion common to both the distributions. From both the distributions, find expressions for p.
From distribution of H_0: with $\mu_0 = 0.50$

$$z_\alpha = \frac{p - p_0}{\sqrt{p_0(1 - p_0)/n}} \quad \text{or,} \quad p - p_0 = z_\alpha \sqrt{p_0(1 - p_0)/n}$$

Or,

$$p = p_0 + z_\alpha \sqrt{p_0(1 - p_0)/n}$$

From distribution of H_a: with $\mu_a = 0.60$

$$z_\beta = \frac{p - p_1}{\sqrt{p_1(1 - p_1)/n}} \quad \text{or,} \quad p - p_1 = z_\beta \sqrt{p_1(1 - p_1)/n}$$

Or,

$$p = p_1 + z_\beta \sqrt{p_1(1 - p_1)/n}$$

Therefore,

$$p_0 + z_\alpha \sqrt{p_0(1 - p_0)/n} = p_1 + z_\beta \sqrt{p_1(1 - p_1)/n}$$

Putting the values we get,

$$0.50 + 2.33\sqrt{0.50(1 - 0.50)/n} = 0.60 - 1.28\sqrt{0.60(1 - 0.60)/n}$$

Or,

$$0.50 + 2.33\sqrt{0.25/n} = 0.60 - 1.28\sqrt{0.24/n}$$

Or,

$$0.50 + 1.165/\sqrt{n} = 0.60 - 0.627/\sqrt{n}$$

Or,

$$1.165/\sqrt{n} + 0.627/\sqrt{n} = 0.60 - 0.50 = 0.10$$
$$1.792/\sqrt{n} = 0.10$$
$$n = 321.13 \approx 322$$

Problems

7.1 Results of course evaluation carried out by the Students Union of AIT in respect of the course HS71: Statistics for May 1992 Term for two items of "Course and Teaching" is summarized below:

Items	Total no. of responses	No. of positive responses
Grading system clear	37	35
Appropriate text used	36	31

Are the proportion of positive responses against both the items the same? Use 5% level of significance?

7.2 It is claimed that the starting salaries (per annum) of AIT graduates from Engineering Divisions are higher than those of graduates from other Divisions. Two random samples produced the following results:

Graduates from engineering divisions	Graduates from other divisions
$n = 17$	$n = 15$
$\bar{x} = \$10{,}250$	$\bar{x} = \$9670$
$s = \$500$	$s = \$400$

(a) Using 95 % confidence level, test the claim.
(b) What is the probability that the starting salaries of the two groups are equal?

7.3   It is reported that the mean income of a population is 15,000 Baht with a std. dev of 5000 Baht. In a sample survey of 200 units the mean income was found to be 17,500. Use 2.5 % level of significance for the following questions.

(a) Test the hypothesis that the population mean income is greater than 15,000 Baht.
(b) Test if the mean population income is less than 15,000 Baht when the sample mean is 12,500 Baht.
(c) The sample mean is 14,500 Baht. Test if the population mean income is less than 15,000 Baht.
(d) The sample mean is 14,000 Baht. Test if the population mean income is equal to 15,000 Baht.

7.4   Ten years ago a survey showed that 65 of 110 employees in a factory were satisfied with their jobs. A recent questionnaire targeting the same population found 40 satisfied among a sample of 76. At the 0.10 level of significance, can we conclude that job satisfaction among the workers is declining?

7.5   It is claimed that the mean IQ of all students in a certain university is 130. It is known that the student population is normally distributed with a standard deviation of 5.4. A random sample of 25 students produced a mean IQ of 134.12 and a standard of 5.5.
Test the claim. Use 95 % confidence level.

7.6   It is claimed that the mean IQ of all students in a certain university is 130. It is known that the student population is normally distributed with a standard deviation of 5.4. A random sample of 25 students produced a mean IQ of 128.5 and a standard of 5.5.
Test the claim. Use 95 % confidence level.

7.7   The starting salaries of college graduates from university A are assumed to be the same as the starting salaries of the graduates from university B.
A random sample of 100 from university A graduates produced a mean salary of $10,250 and a standard deviation of $200. For graduates of university B a random sample of 60 yields a mean of $10,150 and a standard deviation of $180.
Test the hypothesis using 95 % confidence interval.

7.8   The starting salaries of college graduates working in city A are assumed to be the same as the starting salaries of the similar graduates working in the city B.
A random sample of 11 college graduates from city A produced a mean salary of $10,250 and a standard deviation of $200. a random sample of nine graduates from city B yields a mean of $10,150 and a standard deviation of $180.

Test the hypothesis. Use 95 % confidence interval.

7.9   Use the data given in Problem 7.8. But do not assume that the two population variances are equal. Test the claim.

7.10 In a city the mayor claimed that only 20 % of the residents were concerned with anti-noise and anti-pollution law. A group of citizens challenged this claim. They conducted a survey. In a sample survey of 81 residents 33 were found to be concerned with the law.

Test the claim. Use 99 % confidence level.

7.11 In a region, improvement in poverty was being studied. A random sample of 100 households showed that 13 households were below a defined poverty line. Does this indicate an improvement over a previously established figure of 15 %? Use 95 % confidence level to test this hypothesis?

7.12 A survey of 29 households in Bangkok, selected at random, revealed that television is watched 27 h per week, on average. The sample standard deviation was 4 h.

Do the sample data indicate that the number of hours Bangkok families watch television is greater than the national average of 25 h per week? Use 99 % confidence level.

7.13 Two types of new cars are tested for gas mileage. One group consisting of 36 cars, had an average mileage of 24 miles per gallon with a standard deviation of 1.5 miles, the corresponding figures for the other group, consisting of 72 cars, were 22.5 and 2.0, respectively.

Is there any difference between the two types of cars with respect to gas mileage, at 0.01 level of significance?

7.14 In Bangkok traffic engineers were concerned with the increased number of motor cycle accidents. They passed a law requiring all motor cyclists to wear protective helmets. Traffic police reported that before this law, motor cyclists were fatally injured in 20 % of all motor cycle accidents reported. But after the law was passed, there were only 6 fatal injuries in 120 accidents reported. Do the data indicate that the law has been helpful in reducing fatal injuries? Use 1 % level of significance.

7.15 Distinguish between Rejection region and Acceptance Region.

7.16 The manufacturer of a chemical manufacturing plant assumes that the daily average production of the chemical is 880 tons. A sample survey of 50 days observations showed a mean of 871 tons with a standard deviation of 21 tons. Test the manufacturer's assumption. Use 95 % confidence level.

7.17 In comparing the mean weight of two comparable groups of people, the following sample data were obtained:

	Group I	Group II
Sample size	40	40
Sample mean	60.50 kg	63.25 kg
Sample variance	4.25 kg^2	5.20 kg^2

Do the data provide sufficient evidence to indicate that group II have greater weight than group I? Use 5 % significance level.

7.18 A manufacturer claimed that at least 95 % of the equipment which he supplied to a factory conformed to the specification. An examination of sample of 700 pieces of the equipment revealed that 53 were faulty.

Test his claim at a significance level of 0.05.

7.19 The records of a hospital show that 52 men in a sample of 1,050 patients had heart disease; and 28 women in a sample of 1,100 had the same disease.

Do the data present sufficient evidence to indicate that there is a higher rate of heart disease among men as compared to the women?

7.20 The mean salary data of electricians and carpenters in three different work sites are given:

	Electrician	Carpenter
(a) Work Site 1		
Sample size	100	81
Mean	$1700	$1550
Standard deviation	$200	$100
(b) Work Site 2		
Sample size	25	20
Mean	$1,700	$1,650
Standard deviation	$200	$100
(assume equality of population variance)		
(c) Work Site 3		
Sample size	25	20
Mean	$1700	$1650
Standard deviation	$200	$100
(assume no equality of population variance)		

Using 0.05 level of significance, test in each situation whether the salaries of electricians and carpenters are equal.

7.21 In a random sample survey in AIT it was found that 175 married students had mean GPA of 3.34 with a standard deviation of 0.45 while 125 unmarried students had a mean GPA of 3.31 with a standard deviation of 0.31.

(a) At 0.10 level of significance, is there a difference between the GPA of married and unmarried students?

(b) At 0.05 level of significance, is the GPA of married students greater than that of unmarried students?

7.22 An agency working on environmental problems claims that not more than 50 % of the people are aware of the environmental pollution. A random sample of 255 people showed 140 to be aware of the pollution.

Carry out a suitable test to refute the claim. Use 95 % confidence level.

7.23   It is claimed that Engineering students do better in Statistics course as compared to others. Two random samples showed the following results:

Engineering students	Others
Mean score = 21.49	Mean score = 19.06
Standard deviation = 2.44	Standard deviation = 2.52
Sample size = 7	Sample size = 15

Using 2 % level of significance, test the claim.

7.24   The mean IQ of all students in a certain university is claimed to be 140. The student population is normally distributed with a standard deviation of 5.4. A random sample of 36 students produced a mean IQ of 137 and a standard of 6.5.

(a)   Test the claim.
(b)   Find the type II error.
(c)   Find the power of test.
(d)   If the power is 80 %, find the required sample size.
Use 95 % confidence level.

7.25   The claim of mean production of a manufacturing plant is 750 tons/day. A sample random survey of 50 days produced a mean of 745 tons/day with a standard deviation of 25 tons.

(a)   Test the claim. Use 0.05 % of significance;
(b)   Find the type II error;
(c)   Find the power of the test;
(d)   If the 80 % power is considered acceptable, what sample size will be ok?

7.26   An auto manufacturing plant wants to produce parts such that at least 75 % of the products meet the premium quality standards. A sample survey of 125 parts showed 95 parts that met the premium quality standard.

(a)   At 0.05 % of significance level, test the manufacturer's expectation;
(b)   Find type II error;
(c)   Find the power of the test;
(d)   If the power of the test is to be 80 %, find the required sample size.

Answers

7.1   Same
7.2   (a) Do not reject the claim; (b) 0.001
7.3   (a) Do not reject the hypothesis
       (b) Yes, less than Baht 15,000

(c) Not less than Baht 15,000

(d) Not equal to Baht 15,000

7.4   No, we cannot conclude

7.5   Not equal to 130

7.6   Yes; 130

7.7   Not same

7.8   Same

7.9   Same

7.10 More than 20 %

7.11 No.

7.12 Yes, greater than 25

7.13 Difference

7.14 Yes

7.16 Reject the claim

7.17 Yes

7.18 Reject the claim

7.19 Yes

7.20 (a) Not equal; (b) Equal; (c) Equal

7.21 (a) No difference; (b) not greater

7.22 Not refuted

7.23 Support the claim

7.24 (a) Reject the claim; (b) 0.0014; (c) 99.85 %; (d) 15 if $\mu_a = 135$

7.25 (a) Accept H_0; (b) 0.00108; (c) 99.89 %; (d) 81 if $\mu_a = 750$

7.26 (a) Accept H_0; (b) 0.1515; (c) 84.85 %; (d) 103 if $p = 0.85$

Chapter 8
The Chi-Square Test

Abstract The chi-square test is suitable for test of hypothesis dealing with categorical data. Two types of tests are provided—goodness-of-fit test and test of independence. Distributions of frequencies of different classes within two categorical variables are used to test the good fit or bad fit according to goodness-of-fit test. In the test of independence, two distributions in two categorical variables are tested whether they are related or not. Each is explained with the help of examples.

Keywords Chi-square test · Goodness-of-fit test · Test of independence

We have learnt how to estimate population parameters and test hypotheses, both in case of means and proportions. Inference about population means is limited to metric data. Inferences about population proportions can be made with categorical data. But the categorical data were restricted to two categories. Here, we will learn how to test hypothesis related to categorical data having two or more categories.

Test of hypotheses using chi-square distribution is very simple. It is easy to understand and calculate. As such it is very popular in testing hypothesis. The chi-square test makes very few assumptions about the underlying population. For this reason, it is sometimes called a nonparametric test. We shall deal with the nonparametric tests in the next chapter. Another feature may be noted in the techniques developed in the previous chapter to test hypothesis. There the hypothesis involved the parameters of the populations. In this chapter, we shall cover tests of hypotheses which do not involve population parameters.

There are a few situations in which we can use chi-square test. But we shall cover only two types of tests, namely (a) goodness-of-fit test and (b) test of independence.

8.1 Goodness-of-Fit Test

In goodness-of-fit test, each test compares a set of observed frequencies to a set of expected frequencies (calculated on the basis of null hypothesis). The test statistic is small when the two sets of frequencies are similar (good fit) and large when two sets

© Springer Science+Business Media Singapore 2016

A.Q. Miah, *Applied Statistics for Social and Management Sciences*,
DOI 10.1007/978-981-10-0401-8_8

of frequencies are quite different (bad fit). If the fit is good, the null hypothesis is supported (strictly: H_0 cannot be rejected); if the fit is bad, the alternative hypothesis is supported.

Goodness of fit may be looked into in a different way. A goodness-of-fit test is a test of one categorical variable measured in one population. The null hypothesis specifies the probability (proportion) of each possible value of the categorical variable. The proportions of each value of the categorical variable are then observed in a sample. The goodness-of-fit test determines whether the observed frequencies are close to the specified probabilities (good fit, H_0 cannot be rejected) or observed proportions are different from the specified probabilities (bad fit, H_0 is rejected). The decision between a good fit and a bad fit is made on the basis of a test statistic chi-square, which is computed from the observed and expected frequencies. The test statistic is compared to a critical value from the chi-square distribution with $df = c - 1$, where c is the number of categories of the categorical variable.

$$\text{Test statistic chi-square} = \sum \frac{(O_i - E_i)^2}{E_i}$$

where,
i no. of class
O_i observed frequencies in class i
E_i expected frequencies in class i
c number of categories
df $c - 1$

It will be noted that the goodness-of-fit test focuses on comparing two distributions: one is the observed distribution and the other the expected distribution.

Example 1
The percentage distribution of household income in a country in 1985 was as shown in Table 8.1.

A random sample of 300 families in 1990 showed the distribution shown in Table 8.2.

Table 8.1 Percent distribution of household income in 1985

Income class ($)	Percent
<1000	14
$1000 \le x < 2000$	15
$2000 \le x < 3000$	16
$3000 \le x < 4000$	18
$4000 \le x < 5000$	14
$5000 \le x < 7500$	13
7500 and above	10
Total	100

Table 8.2 Distribution of household income in 1990

Income class ($)	No of families
<1000	24
$1000 \le x < 2000$	36
$2000 \le x < 3000$	31
$3000 \le x < 4000$	33
$4000 \le x < 5000$	36
$5000 \le x < 7500$	75
7500 and above	65
Total	300

Question: Has the household income distribution changed significantly during the time? Use 5 % level of significance.

The hypotheses are formulated as follows:

H_0: Distribution of families in 1990 is similar to that of 1985
H_a: Distribution of families in 1990 is different from that of 1985

Since the figures of 1985 are given in percentages, they need to be converted to absolute frequencies. Table 8.3 shows the conversion and the subsequent calculations are done as follows:

$$df = c - 1 = 7 - 1 = 6$$

From the chi-square distribution table, we can see that at $\alpha = 0.05$ with $df = 6$, the critical value of chi-square is 12.59. The acceptance and rejection regions are shown in the figure.

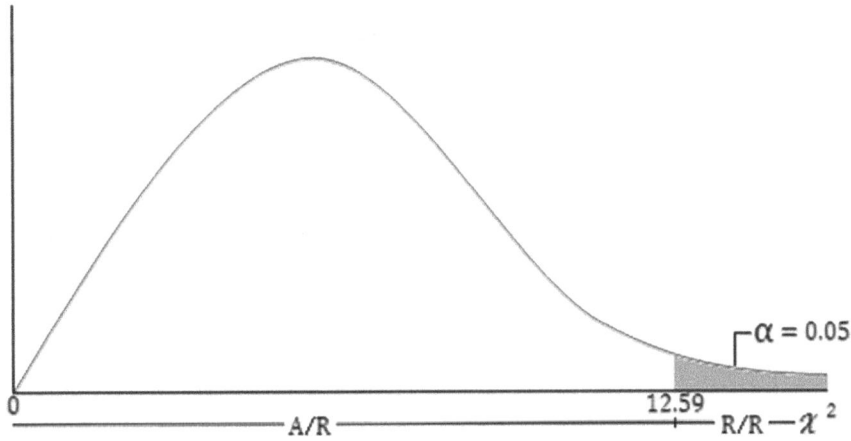

The calculated value of chi-square 98.62 falls in the rejection region. The null hypothesis is, therefore, rejected at 5 % significance level.

Table 8.3 Calculations for
the test statistic

Income class ($)	1990 O_i	1985 E_i	$O_i - E_i$	$(O_i - E_i)^2/E_i$
<1000	24	42	−18	7.71
$1000 \leq x < 2000$	36	45	−9	1.80
$2000 \leq x < 3000$	31	48	−17	6.02
$3000 \leq x < 4000$	33	54	−21	8.17
$4000 \leq x < 5000$	36	42	−6	0.86
$5000 \leq x < 7500$	75	39	+36	33.23
7500 and above	65	30	+35	40.83
Total	300	300	$\chi^2 = 98.62$	

Decision: There is a significant difference between the two distributions. This means that the income distribution has changed significantly during the time.

Another test called "test of homogeneity" is associated with goodness-of-fit test. In fact, test of homogeneity does not involve a new technique. It implies to test whether the population is homogeneous, i.e., whether all the categories have similar distribution. This means that in calculating the expected frequencies, all the categories are assumed to have equal frequencies.

8.2   Test of Independence

The second type of problem in which chi-square test is of use is testing of the difference between the observed frequencies of several classifications of two variables. The elements of a sample from the population are classified according to two different criteria. Each criterion is a variable of our interest and each class is a qualitative value. Our interest is to see whether the methods of classification, i.e., the variables are statistically independent. In this case, either classificatory or ordinal scale data may be used.

The joint frequencies of two related or independent variables may be presented in an $r \times c$ table, where r is the number of rows and c is the number of columns. This type of table is often called a contingency table and as such the test is called "Contingency Table Test". An $r \times c$ table is equivalent to saying that the first variable has r levels and the second variable has c levels.

Let O_{ij} be the observed frequency for level i of the first variable and level j of the second variable. If n is large, then the test statistic is given by

$$\text{Test statistic chi-square} = \sum\sum \frac{(O_{ij} - E_{ji})^2}{E_{ji}}$$

$$= \chi^2_{(r-1)(c-1)}$$

with $(r - 1)(c - 1)$ degrees of freedom.

The hypotheses are of the form

H_0: The two variables are independent (no relationship)
H_a: The two variables are not independent (related)

The two variables are independent is equivalent to saying that there is no relationship between the two variables. Similarly, the two variables are not independent is equivalent to saying that they are related. For this matter, the test is sometimes called a test of relationship. It may be noted here that the test does not say what sort of relationship there is or is not. It simply says that there is a relationship or there is no relationship between the two variables. More explanation, based on the joint frequencies, will have to be provided regarding the type of relationship.

The calculations for chi-square test in this case are similar to those applied to goodness-of-fit test. The expected frequencies (based on laws of probability) may be calculated as follows:

$$\text{Expected frequency} = \frac{\text{Column total} * \text{Row total}}{\text{Grand total}}$$

The number of degrees of freedom for a general contingency table with r rows and c columns is given by

$$df = (r - 1) * (c - 1)\chi^2$$

Thus, the rejection region is

$$\chi^2_{calculated} > \chi^2_{table}$$

The procedure for the test may be understood with the help of an example.

Example 2
A random sample of 1432 adults of a country showed the following distribution of educational levels by sex (Table 8.4). Is there any relationship between educational levels and sex? Use $\alpha = 0.05$

Calculations for expected frequencies are shown in the following table (Table 8.5). Note that the expected frequencies can be in fractions, but the observed frequencies cannot be in fractions.

Table 8.4 Educational levels by sex

Educational level	Sex		Total
	Male	Female	
No education	63	75	138
Primary	152	162	314
Secondary	370	350	720
Higher secondary	160	100	260
Total	745	687	1432

Table 8.5 Calculations for expected frequencies

Educational level	Expected frequencies of		Total
	Male	Female	
No education	745 * 138/1432 = 71.8	138 – 71.8 = 66.2	138
Primary	745 * 314/1432 = 163.4	314 – 163.4 = 150.6	314
Secondary	745 * 720/1432 = 374.6	720 – 374.6 = 345.4	720
Higher secondary	745 * 260/1432 = 135.2	260 – 135.2 = 124.8	260
Total	745	687	1432

The hypotheses are formulated as follows:

H_0: Educational level is independent of sex
H_a: Educational level is associated with sex

To test at $\alpha = 0.05$.

$$df = (r - 1) \times (c - 1)$$
$$\doteqdot (4 - 1) \times (2 - 1)$$
$$= 3 * 1$$
$$= 3$$

The acceptance and rejection regions are shown in the figure that follows.

From the table of chi-square distribution, the critical value of chi-square at $\alpha = 0.05$ with $df = 3$, is found to be 7.81. Observed and Expected frequencies are shown in Table 8.6.

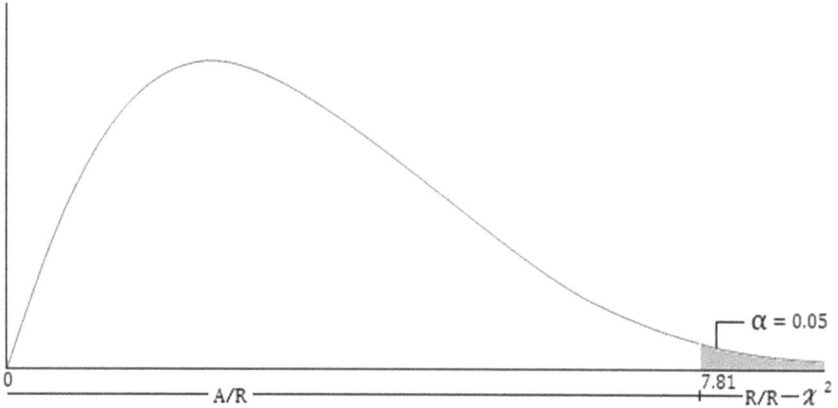

Calculations for the test statistic (chi-square) are shown in Table 8.7.

The calculated chi-square (13.51) is greater than the critical value (7.81). So the calculated chi-square falls in the rejection region.

Table 8.6 Observed and expected frequencies

Educational level	Male		Female		Row total
	O_i	E_i	O_i	E_i	
No education	63	71.8	75	66.2	138
Primary	152	163.4	162	150.6	314
Secondary	370	374.6	350	345.4	720
Higher secondary	160	135.2	100	124.8	260
Column total	745	745.0	687	687.0	1432

Table 8.7 Calculations for test statistic

Observed frequency O_i	Expected frequency E_i	$O_i - E_i$	$(O_i - E_i)^2/E_i$
63	71.8	−8.8	1.08
152	163.4	−11.4	0.80
370	374.6	−4.6	0.06
160	135.2	24.8	4.55
75	66.2	8.8	1.17
162	150.6	11.4	0.86
350	345.4	4.6	0.06
100	124.8	−24.8	4.93
Total			13.51

Calculated $\chi^2 = 13.51$

Decision: The null hypothesis is rejected. This means that there is an association between educational level and sex.

In the above example chi-square test has been applied to see if there is an association between two variables. We might have noted that this test did not give us the strength of association. In order to measure the degree of association we need to calculate the contingency co-efficient c. Fundamentally, contingency coefficient c is the same as correlation coefficient, but it may be applied to classificatory or ordinal scale data in addition to interval scale measurements. The contingency coefficient c is calculated as follows:

$$c = \sqrt{\left\{\frac{\chi^2}{N + \chi^2}\right\}};$$

where, N = grand total of observations.

In the above example,

$$c = \sqrt{\left\{\frac{13.51}{1432 + 13.51}\right\}}$$
$$= 0.097$$

Problems

8.1 A computer scientist wanted to generate random numbers. He developed a computer program to generate the random numbers. In generating 1000 random digits, he observed that the integers between zero and nine occurred according to the following frequencies:

Integers										Total
0	1	2	3	4	5	6	7	8	9	
94	93	112	101	104	96	100	99	107	94	1000
Frequency of occurrence										

The discrete uniform distribution suggests that each integer between zero and nine should occur exactly 100 times each.

Do the data indicate that the computer is generating random numbers according to the uniform discrete distribution?

8.2 The income and expenditure distributions of slum dwellers in Dhaka are shown:

Income and Expenditure of Slum Dwellers

Income/Expenditure range (Tk/month)	Frequencies			
	Income		Expenditure	
	Tenants	Owners	Tenants	Owners
$0 \le x < 500$	4	1	5	1
$500 \le x < 1000$	33	12	31	12
$1000 \le x < 1500$	25	19	28	21
$1500 \le x < 2000$	14	15	12	13
$2000 \le x$	10	39	10	39
Total	86	86	86	86

The sample survey was carried out during 1983–84. One Tk = US$0.33

(a) Are the distributions of tenants and owners across all income groups similar?

(b) Are the distributions of tenants and owners across all expenditure groups similar?

(c) Are the distributions of owners across income and expenditure groups similar?

8.3 A genetic theory predicts that the offsprings in a breeding experiment in the types a, b, c, and d will be in the proportions of 8/16, 4/16, 3/16, and 1/16, respectively.

In an experiment, 214 offsprings are observed. The frequencies of the categories a, b, c, and d were found to be 131, 52, 18, and 13, respectively.

(a) Do the data support the genetic theory?

(b) Are the distributions of the offsprings across the four categories equal/same?

8.4 The management of a company has to choose among three working schedules. The management wishes to know whether the preference for working schedules is independent of job categories. The opinions of 500 employees are shown hereafter:

Job categories	Working schedule		
	A	B	C
Salaried workers	160	140	40
Hourly workers	40	60	60

Do the necessary job.

8.5 Site condition and customer dealing in an enterprise are the subject matter of an investigator. A sample of 441 sites was investigated with the results shown in the following table. Is there evidence that site condition and customer dealing are independent?

Customer dealing	Site condition		
	Not good	Good	Very good
Unimpressive	24	52	58
Neutral	15	73	86
Impressive	17	80	36

8.6 An agricultural extension researcher wants to study the farm size and productivity relationship in certain countryside. One of the objectives was to study the distributions of farms across sizes. In a sample survey, he generated the data shown in the above table.

Show by a suitable test that the farm sizes are not evenly distributed across the sizes.

Distribution of farmers across farm sizes

Farm size (acres)	No of farms
$0 \leq x < 1.0$	66
$1.0 \leq x < 2.0$	71
$2.0 \leq x < 3.5$	48
$3.5 \leq x < 5.0$	33
$5.0 \leq x < 7.5$	15
$7.5 \leq x < 14.5$	7

8.7   A study is conducted to determine if there is a relationship between political awareness and income level. A random sample of 140 persons is surveyed. Political awareness is measured on a four-point scale from 1 (negative) to 4 (positive). Income is defined as low, medium, and high. The joint frequency distribution is shown below:

Socio-economic class	Awareness scale				
	1	2	3	4	Total
Low	17	28	6	2	53
Medium	15	14	11	7	47
High	8	12	17	3	40
Total	40	54	34	12	140

Conduct a suitable test to see if there is a relationship between the political awareness and income level. Use 0.01 level of significance. Comment on the strength of relationship.

8.8   A researcher wanted to study if there has been a change in the distribution of college, school graduates, and those who are not graduates. Previously there were 22 % college graduates, 48 % high school graduates and 30 % not graduates at all. At present a study of 1000 sample men show that there are 248 college graduates, 522 high school graduates and 230 have not finished high school. Test it at 0.01 level of significance.

8.9   A random sample of 1000 students shows that 325 are in level 1, 360 in level 2 and 315 are in level 3 according to a certain standard score-scale. Is the distribution same across all the three levels? Use $\alpha = 0.05$.

8.10   Is there a relationship between sex and smoking? A random sample of 100 persons showed the following results:

Smoking	Male	Female
Smoker	30	10
Non-smoker	20	40
Total	50	50

Test the hypothesis at 0.005 level of significance.

8.11   In a random sample survey, the age groups of drivers and the number of accidents they made are reported as follows:

No of accidents	Age of drivers		
	18–25	26–40	Over 40
0	75	115	110
1	50	65	35
2	25	20	5

Test if there is a relationship between number of accidents and the age of drivers. Use 0.01 as the level of significance.

8.12   The owner of a large company claimed that 1/10 of the personnel earned income in group1, 3/10 earned income in group2, 3/10 earned income in group3 and 3/10 earned income in group4. A random sample showed the following:

Group	Income range ($)	Frequency
1	$x < 20{,}000$	19
2	$20{,}000 \leq x < 25{,}000$	56
3	$25{,}000 \leq x < 30{,}000$	51
4	$30{,}000 \leq x$	40

Using 5 % level of significance, test the claim.

Answers

8.1   Yes (95 %)
8.2   (a)   Not similar (95 %);
       (b)   Not similar (95 %);
       (c)   Yes, Similar (95 %)

8.3   (a)   Do not support (95 %);
       (b)   Not equal (95 %)

8.4   Not independent
8.5   Not independent
8.6   Yes, there is a relationship; $c = 0.40$
8.7   Distribution same
8.8   Yes
8.9   Yes, there is a relationship
8.10   There is a relationship
8.11   Claim is not rejected ($\alpha = 0.05$)

Chapter 9
Nonparametric Test

Abstract Nonparametric tests are suitable for categorical and rank data and data having no assumption of normal distribution. Sign test, rank test (Wilcoxon rank-sum test), and Spearman rank correlation test are introduced. The technique for use of the tests is explained with the help of examples.

Keywords Nonparametric test · Rank test · Wilcoxon rank-sum test · Sign test · Spearman rank correlation test

A nonparametric test is one in which (a) it is not necessary to assume that the population is normally distributed. Even no other strong assumption regarding the population distribution is required in nonparametric test; (b) categorical or ranked data are used. In nonparametric test the computations are simple. We may notice that both the mean and standard deviation are meaningless in categorical and ranked data. In this chapter we shall deal with three tests namely, (a) sign test, (b) rank test and (c) Spearman rank correlation test.

9.1 The Sign Test

The sign test is the simplest test in the family of nonparametric tests. It is used to test a hypothesis about the median of a continuous distribution. The characteristics of a median in a continuous distribution tells us that the median is a value of the random variable such that the probability of an observed value to be less than or equal to the median is 0.50. Also, the probability of an observed value of the random variable to be greater than or equal to the median is 0.50. The procedure for the sign test is described below.

The hypotheses that may be tested are of the form

$$H_0 : M_d = M_0$$
$$H_a : M_d > M_0; M_d < M_0; M_d \neq M_0,$$

© Springer Science+Business Media Singapore 2016
A.Q. Miah, *Applied Statistics for Social and Management Sciences*,
DOI 10.1007/978-981-10-0401-8_9

where M_0 is a specific value of M_d set in the null the hypothesis and M_d is the population median.

Let a random sample of size n with observations $x_1, x_2, x_3, \ldots, x_n$ be drawn from the population of our interest. Then the differences $(x_i - M_0)$, $i = 1, 2, 3, \ldots, n$ are calculated. Let $R+$ be the number of positive differences and $R-$ be the number of negative differences. If the null hypothesis H_0: $M_d = M_0$ is true, any difference $x_i - M_0$ is equally likely to be positive or negative, i.e., its probability to be positive or negative is 0.50. The test statistic r is chosen such that r is the minimum of the two $R+$ and $R-$.

R has a binomial distribution with parameters n and 0.50. We may remember that if n is sufficiently large ($n > 20$ in this case), the binomial distribution can be approximated by the normal distribution. Some authors suggest that this approximation is valid if n is at least 10. Most of the planning researches involve a sample size of more than 20. Therefore, the test procedure developed here regarding the sign test uses the normal distribution criterion. It involves the technique of proportion. The sample proportion is given by

$$p = \frac{R}{n}$$

It may be noted here that R is the minimum of $R+$ and $R-$ and it is possible that in one or more of the difference(s) zero will be encountered. This is a case of a tie. When a tie occurs, that should be set aside and the sign test be applied to the remaining data. The n value will thus be reduced by the number of tied observations. Thus, the effective $n = n$—number of tied observations.

The test statistic is given by

$$Z_c = \frac{p - P}{\sqrt{\frac{P(1-P)}{n}}}$$

$$= \frac{p - 0.50}{\sqrt{\frac{0.50(1-0.50)}{n}}}$$

$$= \frac{p - 0.50}{\sqrt{\frac{0.50 * 0.50)}{n}}}$$

$$= \frac{p - 0.50}{\frac{0.50}{\sqrt{n}}} = \frac{p - 0.50}{0.50/\sqrt{n}}$$

If the alternative hypothesis is H_a: $M_d > M_0$, the null hypothesis is rejected if

$$[Z_c] > Z_\alpha.$$

If the alternative hypothesis is H_a: $M_d < M_0$, the null hypothesis will be rejected if

$$[Z_c] < -Z_\alpha.$$

If the alternative hypothesis is $H_a: M_d \neq M_0$, the null hypothesis will be rejected if

$$[Z_c] > Z_{\alpha/2} \quad \text{or,} \quad \text{if} [Z_c] < -Z_{\alpha/2}.$$

Example 1

A random sample of size 22 households drawn from a population had the following incomes (US $):

$$2100, 2550, 2200, 2590, 2250, 2680, 2300, 2685, 2325, 2715, 2775,$$

$$2380, 2795, 2425, 2480, 2815, 2865, 2500, 2900, 3000, 2952, 2950.$$

Test the hypothesis that the median income of the population is US $2500. Use $\alpha = 0.05$.

Solution

The hypotheses are formulated as follows:

$$H_0: M_d = 2500$$
$$H_a: M_d \neq 2500$$

This is a two-tail test. We have $\alpha = 0.05$; so, $\alpha/2 = 0.025$ and $Z_{\alpha/2} = 1.96$. The acceptance and rejection regions are shown as follows:

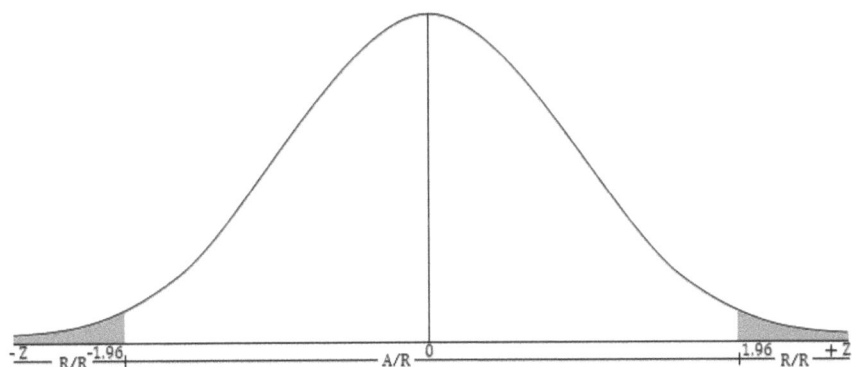

The calculations for the sign test are shown in Table 9.1.

There is one tie. Therefore, the effective $n = 22 - 1 = 21$. $R+ = 13$ and $R- = 8$. So R (min of $R+$ and $R-$) = 8. Sample proportion $p = 8/21 = 0.381$. The test statistic calculations are shown after Table 9.1.

Table 9.1 Income data for sign test

Observation	Income (x_i \$)	$x_i - 2500$	Sign
1	2100	−400	−
2	2550	+50	+
3	2200	−300	−
4	2590	+90	+
5	2250	−250	−
6	2680	+180	+
7	2300	−200	−
8	2685	+185	+
9	2325	−175	−
10	2715	+215	+
11	2775	+275	+
12	2380	−120	−
13	2795	+295	+
14	2425	−75	−
15	2480	−20	−
16	2815	+315	+
17	2865	+365	+
18	2500	0	0
19	2900	+400	+
20	3000	+500	+
21	2952	+452	+
22	2950	+450	+

$$
\begin{aligned}
Z_c &= \frac{p - 0.50}{0.50/\sqrt{n}} \\
&= \frac{0.381 - 0.50}{0.50/\sqrt{21}} \\
&= \frac{-0.119}{0.50/\sqrt{21}} \\
&= -1.09 \\
[Z_c] &= 1.09
\end{aligned}
$$

This falls in the acceptance region. Therefore, the null hypothesis is not rejected at 5 % level of significance.

Conclusion: The population median is US \$2500.

The sign test can also be applied to paired samples, if the samples are drawn from two continuous populations. The computations and test procedure are slightly different in this case.

Let (x_{1i}, x_{2i}), $i = 1, 2, 3, \ldots, n$ be the paired observations drawn from two continuous populations. Then

$$D_i = x_{1i} - x_{2i}, \quad i = 1, 2, 3, \ldots, n$$

The differences D_i of the pairs of observations are computed and signs are recorded. The rest procedure is similar to the one developed earlier. The null hypothesis in this case is that the two populations have a common median, i.e., $M_{d1} = M_{d2}$. This means $M_{d1} - M_{d2} = 0$ or in other words, $M_{\text{diff}} = 0$.

Example 2

An income generating program was launched in countryside. Twenty farmers were selected randomly to study the result of the program. The average monthly incomes (in \$) before and after the program were recorded as follows:

Before

10,500	11,200	10,500	9600	7500	8200	12,800
10,250	11,800	10,100	12,000	8270	7800	11,200

After

10,200	10,250	9200	15,000	7600	5500	6250
11,400	12,250	12,200	17,000	9200	5600	6260
8900	8905	7825	15,000	13,250	12,500	
9000	8965	7800	16,000	13,000	12,600	

Test at 5 % significance level if there has been a change in the median income after the program.

The null and alternative hypotheses are formulated as follows:

$$H_0: M_{d1} - M_{d2} = 0$$
$$H_0: M_{d1} \neq M_{d2} \neq 0$$

This is a two-tail test. We have $\alpha = 0.05$; so, $\alpha/2 = 0.025$. From table $Z_{0.025} = 1.96$. The acceptance and rejection regions are shown in the following figure.

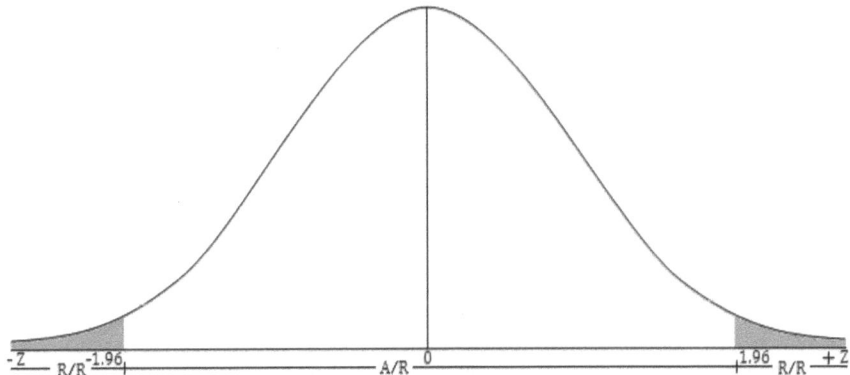

The paired observations are arranged in Table 9.2, differences are computed, and the signs are recorded.

Here $R+ = 6$ and $R- = 14$. Therefore, $R = 6$. The proportion $p = 6/20 = 0.30$. The test statistic is

$$
\begin{aligned}
Z_c &= \frac{p - 0.50}{0.50/\sqrt{n}} \\
&= \frac{0.30 - 0.50}{0.50/\sqrt{20}} \\
&= \frac{-0.20}{0.112} \\
&= -1.79 \\
[Z_c] &= 1.79
\end{aligned}
$$

This falls in the acceptance rejection. The null hypothesis is not rejected at 5 % significance level.

Conclusion: The two population medians are similar. There has not been any statistical change in the median income after the program.

Table 9.2 Income data before and after program

Observation	Before x_1	After x_2	Difference $D = x_1 - x_2$	Sign
1	10,500	10,250	+250	+
2	11,200	11,800	−600	−
3	10,500	10,100	+400	+
4	9600	12,000	−2400	−
5	7500	8270	−770	−
6	8200	7800	+400	+
7	12,800	11,200	+1600	+
8	10,200	11,400	−1200	−
9	10,250	12,250	−2000	−
10	9200	12,200	−3000	−
11	15,000	17,000	−2000	−
12	7600	9200	−1600	−
13	5500	5600	−100	−
14	6250	6260	−10	−
15	8900	9000	−100	−
16	8905	8965	−60	−
17	7825	7800	+25	+
18	15,000	16,000	−1000	−
19	13,250	13,000	+250	+
20	12,500	12,600	−100	−

Now suppose that the question is "has the median income increased after the program?" Note that R value (6) has come from $R+$ and the plus sign indicates that the income before is higher than the income after, since we have subtracted the second figure from the first. Therefore, median income after the program has increased would imply that the p value is smaller than 0.50 and consequently we would expect a negative test statistic. Therefore, the critical region (rejection region) would also lie on the left side of the distribution.

The hypotheses are

$$H_0: M_{d1} - M_{d2} = 0$$
$$H_a: M_{d1} - M_{d2} < 0$$

It is a one-tail test. Here $\alpha = 0.05$; $Z\alpha = 1.645$. The acceptance and rejection regions are shown in the following figure.

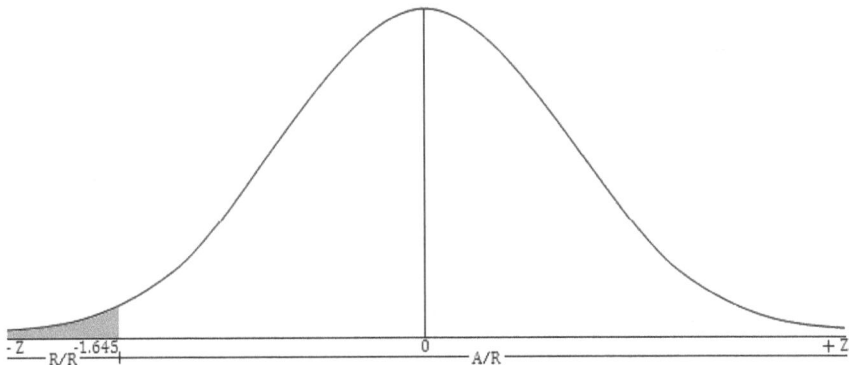

Z_c is calculated before $= -1.79$. This falls in the rejection region. The null hypothesis is, therefore, rejected at 0.05 level of significance.

Conclusion: The median income has increased after the program.

9.2 The Rank Test

Rank test is a nonparametric test and nonparametric test is characterized by being distribution free. No assumption is made regarding the distribution as already mentioned. In rank test the observations are ranked and on the basis of the ranking a statistic is calculated to test a hypothesis.

Rank tests are very efficient, and so are very popular. These are easy to understand and calculate. Rank tests have a crucial advantage for data which are originally ranked and not in numerical form. In such data, numerical calculations such as mean and standard deviation are not possible. If there are some outliers which are likely to create problems, then also the observations can be transformed into ranks and rank test can be performed.

9.2.1 The Wilcoxon Rank-Sum Test

Wilcoxon rank test is a typical and popular form of rank test. It is used to compare two population means. In order to perform the test, follow these steps:

(i) Make a combined listing of the scores of both the samples (n_1 and n_2) arranged in ascending order. The smaller sample is n_1 and the larger sample is n_2.

(ii) Rank the combined scores starting from the lowest score. The lowest score is rank 1, and the highest score is the highest rank.

(iii) Sum the ranks of both the samples separately for n_1 and n_2. Sum of the ranks of the smaller sample (n_1) is for testing. This is w. The sum of the ranks of the larger sample (n_2) is for getting an idea of the direction only.

(iv) Set the level of significance (say $\alpha = 0.05$) for testing. Formulate the hypothesis. The null hypothesis is of the form
H_0: The two population distributions are equal/identical/similar ($W_1 = W_2$). The alternative hypothesis (H_a) will depend on the type of question to be solved ($W_1 \neq W_2$, $W_1 > W_2$ or $W_1 < W_2$ as the case may be).

(v) Use the value of n_1, n_2, and w and from the table, and find the value of the probability.

(vi) If the probability thus found out is less than the set value of α, reject the null hypothesis (H_0). If it is a one-tail test, compare directly with α. If it is a two-tail test, compare with the value of $\alpha/2$.

(vii) Interpret the test result.

The operation of the test is demonstrated with the help of an example.

Example 3
The household annual incomes of two randomly selected samples from two groups of people in a city during a year were as follows (Table 9.3).
Our task is to test the following hypothesis (assumed $\alpha = 0.05$):

$H_0 =$ Incomes of the two groups are equal.

$H_a =$ People of group I are poorer than those of group II.

Obviously, it is a one-tail test. We rank the observations as shown in Table 9.4.
Two steps are important here. Identification of the smaller sample is one and ranking is the other. Ranking should start from the smallest score. In our example, group I is the smaller sample ($n_1 = 5$), and so rank of 3900 in n_1 is 1. The highest score is 25,000 (in n_2). So its rank is 11 ($n_1 + n_2 = 11$). The test statistic Wilcoxon rank sum w is the sum of the ranks of the smaller sample (n_1).

Table 9.3 Household annual income ($)

Group I	Group II
3900	7150
5000	8450
7500	9100
9750	11,000
17,400	13,500
	20,500

Table 9.4 Ranking of household incomes

Combined observations ordered		Combined rankings	
Group I	Group II	R_1	R_2
3900		1	
5000		2	
	7150		3
7500		4	
	8450		5
	9100		6
9750		7	
	11,000		8
	13,500		9
17,400		10	
	20,500		11

$n_1 = 5; n_2 = 6; w = 24$

In our example, $w = 24$, $n_1 = 5$, $n_2 = 6$. From the table, we note that the probability $p = 0.20$. This means that the probability of the two income sets to be equal is 0.20 or 20.00 %. This value is larger than our set value of α (0.05). So we do not reject the null hypothesis. This decision implies that the incomes of the two groups of people are identical/similar/equal.

Let us consider another example. Independent two random samples of men and women in Thailand gave the following incomes (Table 9.5).

Are the incomes of the groups—men and women—identical?

Solution

Let us assume $\alpha = 0.05$.

The null and the alternative hypotheses are formulated as follows:

H_0: Incomes of men and women are equal.

H_a: Incomes of men and women are different.

The alternative hypothesis tells us that it is a two-tail test. So we should compare the probability with $\alpha/2$ or $0.05/2$ or 0.025. The income observations arranged in order are as follows (Table 9.6).

Table 9.5 Income of men and women (Baht)

Men	Women
4700	3500
4000	2500
4900	3000
7800	1900
3490	4650
	4800
	1800
	1700
	8000

Table 9.6 Incomes of men and women (in order)

Men	Women	Combined rankings	
Sample 1(n_1)	Sample 2 (n_2)	R_1	R_2
	1700		1
	1800		2
	1900		3
	2500		4
	3000		5
3490		6	
	3500		7
4000		8	
	4650		9
4700		10	
4900	4800	12	11
7800		13	
	8000		14
		$w = 49$	(56)

Here, $n_1 = 5$, $n_2 = 9$, and $w = 49$. From table we note that value of p is between 0.05 and 0.10 (less than 0.10 and greater than 0.05). This is larger than $\alpha/2$ or 0.025. So we do not reject the null hypothesis. This implies that the incomes of the two groups are not different. Note that the table does not always provide the exact value of p. But the computer will do.

Now, with the same data sets consider the following question.

Question

Is the income of the women group more than that of the men group?

Here, the hypotheses should be formulated as follows:

H_0: Incomes of men and women are equal.

H_a: Income of the women group is more than that of the men group.

Here, obviously it is a one-tail test. So we should compare the p value with α (0.05) and not $\alpha/2$ (0.025).

From table for $n_1 = 5$, $n_2 = 9$, and $w = 49$, value of p is between 0.05 and 0.10 (less than 0.10 and greater than 0.05) as we have already observed. This p value is more than our set α value of 0.05. So we do not reject the null hypothesis. This implies that income of the women group is not more than that of the men group.

If the sample sizes of both the samples (n_1 and n_2) are larger than 10 ($n_1 > 10$ and $n_2 > 10$), we can use normal distribution (approximately) to test the hypothesis. The normal distribution has population mean $= \mu$ and standard deviation $= \sigma$. The calculation will be as follows:

$$\mu_1 = \frac{n_1(n_1 + n_2 + 1)}{2}$$

$$\sigma_1 = \sqrt{\frac{n_1 n_2(n_1 + n_2 + 1)}{12}}$$

$$z = \frac{w_1 - \mu_1}{\sigma_1}$$

This may be illustrated by an example. Let us suppose that in one study the researcher found $n_1 = 12$, $n_2 = 15$, and $w_1 = 155$. Test the hypothesis that the two populations are not equal.

Let $\alpha = 0.05$.

The hypotheses are formulated as follows:

H_0: The two population distributions are same.

H_a: The two population distributions are not same.

It is obviously a two-tail test. Therefore, $\alpha/2 = 0.05/2 = 0.025$ $z_{0.025} = 1.96$. Reject H_0 if $z_c > 1.96$ or $z_c < -1.96$.

$$\mu_1 = \frac{12(12 + 15 + 1)}{2} = 168$$

$$\sigma_1 = \sqrt{\frac{12 \times 15(12 + 15 + 1)}{12}} = \sqrt{\frac{5040}{12}} = \sqrt{420} = 20.494$$

$$z = \frac{155 - 168}{20.494} = \frac{-13}{20.494} = -0.634$$

This z value is larger than -1.96. It falls in the acceptance region of H_0. Therefore, the null hypothesis H_0 is not rejected. This implies that the two population distributions are same. These are not different.

9.2.2   The Spearman Rank Correlation

The Spearman rank correlation or simply Spearman correlation is a measure of the degree to which two numerical variables are monotonically related or associated. The Spearman correlation is used both as a measure of the degree of monotonic association and as a test statistic to test the hypothesis of monotonic association. Since variables that are linearly associated are necessarily also monotonically related, the spearman correlation may be used in situations in which the Pearson correlation (a measure of linear association) is used. The values and interpretations of the two correlations are essentially the same when the actual relation is linear.

The spearman correlation is defined by the following formula:

$$r_s = 1 - \frac{6 \sum d_i^2}{n(n^2 - 1)}$$

where
d_i difference of ranks of the ith pair of observation,
n no. of pairs of observations.

The values of the correlations may lie between -1 and $+1$ ($-1 \leq r_s \leq +1$). The sign indicates the kind of relationships. A positive value indicates agreement between the rankings in the same direction, i.e., high ranks in one series tend to go with high ranks in the second. A negative sign indicates the reverse.

The calculation of the spearman correlation is simple. Suppose, there are two variables x and y. First, reduce the x scores to ranks. Second, reduce the y scores to ranks. Third, calculate the difference (d_i) between the pairs of the scores. Finally, use the formula given above to calculate the spearman correlation (r_s).

For testing the hypothesis of association between two variables, the calculated correlation (r_s) is converted to t value and t test can be carried out as learnt earlier. The conversion may be done using the following formula:

$$t = r_s \sqrt{\frac{n-2}{1 - r_s^2}}$$

Based on the result of the t test, a decision is taken as either to reject or not to reject the null hypothesis. The null hypothesis would state that there is no association between the two variables (Table 9.7).

Table 9.7 Scores of X and Y

x	y	Rank x	Rank y	$d_i = $ Rank $x - $ Rank y	d_i^2
13	15	2	4	-2	4
14	12	3	2	1	1
15	13	4	3	1	1
16	19	5	6	-1	1
19	20	7	7	0	0
17	16	6	5	1	1
11	10	1	1	0	0
Total					8

Example 4

$$r_s = 1 - \frac{6 \sum d_i^2}{n(n^2 - 1)}$$
$$= 1 - \frac{6 * 8}{7(49 - 1)}$$
$$= 1 - \frac{48}{7 * 48}$$
$$= 0.857$$

H_0: $\rho_s = 0$; (ρ_s = population spearman correlation between the two populations)
H_a: $\rho_s > 0$
    Use $\alpha = 0.05$.
    Now,

$$t = r_s \sqrt{\frac{n - 2}{1 - r_s^2}}$$
$$= 0.857 \sqrt{\frac{7 - 2}{1 - (0.857)^2}}$$
$$= 0.857 \sqrt{\frac{5}{1 - 0.734}}$$
$$= 0.857 \sqrt{\frac{5}{0.266}}$$
$$= 0.857 * \sqrt{18.797}$$
$$= 0.857 * 4.336$$
$$= 3.716$$

From the t table, the critical value of $t = 1.943$ for α 0.05 and df 6 $(7 - 1 = 6)$. The calculated t value (3.716) is larger than the critical t value (1.943) and hence it falls on the rejection region. Therefore, the null hypothesis is rejected. This means that there is significant association between the two populations.

Regarding corrected rank order correlation, see Xycoon (2009).

9.3 Nonparametric Method in Analysis of Variance

Analysis of variance (one way) can be carried out by one of the nonparametric methods. The particular method is Kruskal–Wallis Test. The test is named after William Kruskal and W. Allen Wallis. It is a nonparametric test and it does not assume a normal population, unlike the analogous one-way analysis of variance. However, it assumes an identically shaped population.

The test involves ranking of all the data from all the groups together. The lowest value is assigned rank 1, next lowest value is assigned rank 2, and the highest value is assigned the rank of N (the total number of sample). If there is any tie, assign the average of the ranks they would have received had they not been tied.

After performing the transformation of the raw data to the ranks as stated above, the Kruska–Wallis test statistic K is calculated. The test statistic K is approximately followed the chi-square distribution. Then hypotheses are formulated as usual. The null hypothesis is "there is no difference between the mean ranks of the groups." This would mean that there is no difference between the means of the groups.

In calculating the test statistic K, different authors have suggested different formulae. But the one shown here seems to be easy and familiar. The procedure is as follows:

1. Calculate the sum of squares of the ranks of the data of the groups in the same way as was done in the case of one-way analysis of variance.
2. Next, calculate the test statistic K from the formula.

$$K = \frac{\text{SSr}}{N(N+1)/12}$$

3. Then compare this test K value with the critical chi-square from the table with appropriate α value and degrees of freedom. Reject null hypothesis if $K \geq \chi^2_{\alpha,\text{df}}$

Example 5

Incomes ($/month) of four groups of people are shown in the following table. Use Kruskal–Willis rank test to see if there is difference between the means of the incomes of four groups of people.

The raw data as well as the ranks of the combined data of all the four groups are shown in the table.

Raw data (group)				Rank data (group)				
1	2	3	4	1	2	3	4	Total
495	500	550	600	8.0	12.0	23.5	26.5	
490	510	545	625	5.5	15.0	22.0	33.0	
475	495	540	630	2.0	8.0	21.0	34.5	
498	515	550	622	10.0	16.5	23.5	32.0	
475	520	565	615	2.0	18.5	25.0	30.0	
489	528	600	660	4.0	20.0	26.5	37.0	
501	495	610	630	14.0	8.0	29.0	34.5	
500	475	620	640	12.0	2.0	29.0	36.0	
515	490	599		16.5	5.5	26.0		
520	500			18.5	12.0			
Total				92.5	117.5	225.5	263.5	703

Let SSr be sum of squares of ranks. Then

$$SSr = \frac{(92.5)^2}{10} + \frac{(117.5)^2}{10} + \frac{(225.5)^2}{9} + \frac{(263.5)^2}{8} - \frac{(703)^2}{37}$$
$$= 3425.64$$

Krustal–Wallis test statistic

$$K = \frac{SSr}{N(N+1)/12}$$
$$K = \frac{3425.64}{37(37+1)/12}$$
$$= \frac{3425.64}{117.167} = 29.237$$

Degree of Freedom (df) = no. of groups—1 = 4 − 1 = 3
Hypotheses are formulated as follows:

H_0: There is no difference between groups (all group rank means are same)

H_a: There is difference between groups (all group rank means are not same).

From the table we find $\chi^2_{0.05,3} = 7.81$.

The calculated chi-square value (29.237) being larger than the table value (7.81), the null hypothesis falls in the rejection region of H_0. Therefore, the null hypothesis is rejected. This means that the all rank means are not the same. This further means that the group means are different.

The same conclusion will be drawn if we run the test using the technique of analysis of variance.

Yet, another technique is available for testing the hypothesis involving rank data. This is rank transformation. The K statistic can be approximately transformed into the F test statistic in the following way:

$$F = \frac{K(g-1)}{(N-1-K)/(N-g)}$$

g the number of groups,
N total no. of observations,
F_r F statistic of the ranks.

In our example,

$$F_r = \frac{29.237(4-1)}{(37-1-29.237)/(37-4)}$$
$$= 428.07$$

Df for this F_r in this example are 3.33. From the F table, $F_{0.05,3.33} = 2.92$. Therefore, the calculated F_r value is larger than the table critical F value. Therefore, the null hypothesis is rejected.

A question may arise if both the techniques provide the same conclusion, then which technique is to be used. The answer to this question may be summarized in the following way.

(a) If the calculated test statistic value is marginal meaning that it is closer to the critical value (table value), then both the techniques may be used and the results may be reviewed as to which one is to be used depending on the exact situation of the problem.
(b) If the value of the calculated test statistic is too high as compared to the critical value (table), then any one test may be used. But if the nonparametric test is carried out, it may avoid the influence of the outliers.

Problems

9.1. Five photocopy machines of model A and eight photocopy machines of model B were randomly selected to determine the continuous service hours until they required rest. The following are the results:

Model A:	24,	20,	23,	22,	18			
Model B:	21,	19,	16,	25,	26,	17,	28,	30

Based on Wilcoxon rank test, calculate the probability of the service hours of the two machines being equal.

9.2. In a random sample of five courses the given grading and expected grading are as follows:

Given grading:	2.8	3.7	4.0	3.6	3.8
Expected grading:	2.6	2.9	3.3	3.2	3.1

Use Wilcoxon rank test to test the following:

(a) Is the expected grading less than the given grading?
(b) Is the given grading greater than the expected grading?
(c) Are the two grading equal? Use 5 % significance level.

9.3   Use the data given in problem 9.2 and answer the following:

(a) Calculate the Spearman rank correlation coefficient.
(b) Is there positive correlation between the two grading?
(c) Is the correlation between the two gradings significant?
In all the cases use 5 % significance level.

9.4   Ten students in AIT were selected at random. The purpose was to examine if there is a correlation between the undergraduate scores and first term scores of the students in AIT. The following were their scores.

Undergraduate scores	First term scores in AIT
56	60
70	75
50	60
65	64
90	92
49	45
63	59
69	69
71	81
81	85

(a) Calculate the rank correlation coefficient.
(b) Test at 5 % level of significance if the correlation coefficient is significant.

9.5   Distinguish between Spearman rank correlation coefficient and Pearson correlation coefficient.

9.6   The following numbers represent numbers of hours taken by two groups of people to complete a work. The two sample groups of people are comparable in performing the works.

Group 1:

15, 15, 12, 10, 11, 13, 18, 17, 15, 12, 18, 20, 22, 19, 15, 17, 20, 19, 18, 20.

14, 18, 15, 20, 16, 14, 19, 20, 15, 17, 14, 12, 10, 17, 19, 19, 18, 13, 19, 20.

Group 2:

Run the median test and conclude if the two works require different hours to complete the work.

9.7 Hours of study and grade points in respect of 8 students in a certain course are as follows:

Hours of study	Grade points
5.0	3.8
4.0	3.7
4.5	3.5
6.0	4.0
5.5	3.5
2.9	3.9
3.8	3.6
5.3	3.0

Test if there is relationship between hours of study and grade points, using rank correlation coefficient.

9.8 Two groups of students randomly drawn from a larger group were assigned two methods of instruction in a particular course. In a combined test the following were their scores:

Group 1: $(n = 12)$

180, 193, 142, 173, 155, 186, 192, 149, 169, 173, 182, 183

189, 148, 200, 190, 185, 177, 189, 156, 188, 171, 191, 299, 178, 189, 162.

Group 2: $(n = 15)$

Run the median test and conclude if there is any difference in the results of the two methods of instruction. (Determine the combined median. Count the frequencies below and above the combined median in each group. Prepare the r × c table. Run the chi-square test. Drop any observation falling exactly on the combined median).

9.9 A group of 10 students were selected at random and their scores on assignment and written examination were recorded as shown below:

Student	Assignment scores	Written exam scores
1	75	88
2	60	72
3	88	100
4	88	90
5	98	70
6	99	100
7	75	77
8	79	85
9	77	87
10	90	86

Compute the Spearman rank correlation coefficient and test if there is a relationship between the two types of tasks. Use 0.05 level of significance.

Answers

9.1. 0.262
9.2. (a) Yes; (b) Yes; (c) Yes
9.3. (a) 0.70; (b) No; (c) No
9.4. (a) 0.96 (b) Significant
9.5. Require similar time
9.6. Yes, there is a relationship
9.7. No difference
9.8. 0.327; no significant relationship

Reference

Xycoon: Statistics—Econometrics—Forecasting, Office of Research Development and Education, http://www.xycoon.com (30 Sept 2009)

Chapter 10
Correlation

Abstract The concept of correlation always refers to the linear relationship between two variables and uses their joint distribution. It is shown how to interpret the correlation coefficient. Through the use of examples, it is demonstrated how to formulate the hypotheses concerning correlation, interpretation, and draw conclusions.

Keywords Correlation · Linear relationship · Hypothesis formulation · Testing

We use correlation analysis to study the linear relationship between two variables. It is always advisable to make a plot of the variables and get an idea of the possibility of the linear relationship. For example, if two variables x and y are associated in the form of $x^2 + y^2 = a^2$, where "a" is a constant, the relationship between x and y is perfect. The relationship is circular and not linear (Fig. 10.1). But if we consider the concept of "correlation," which measures the linear relationship, the relationship would come out to be zero indicating no relationship. So it is important to remember that the concept of "correlation" refers always to linear relationship.

In correlation analysis we are concerned with the joint distribution of x and y. If there is a linear relationship between x and y, where x and y are random variables, the joint observations on pairs of x and y will tend to be clustered around a straight line. It is immaterial to think which set of observations represent x and which set of observations represent y.

In order to avoid the rigorous mathematical derivations, we want to state the computational formulae only. As usual, inferential statistics deal with inferences regarding the population correlation which is estimated from the sample correlation. Therefore, we set two sets of formulae, one for the population and the other for the sample.

Fig. 10.1 Circular correlation

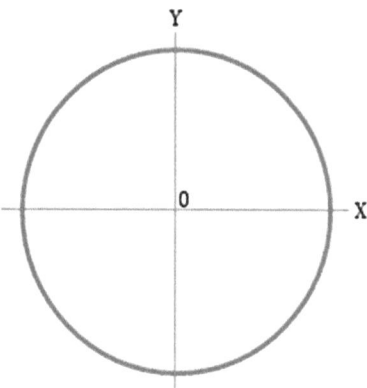

If X and Y are a pair of random variables with mean μ_x and μ_y and variances σ_x^2 and σ_y^2, then the population correlation coefficient ρ between X and Y is given as

$$\rho = \text{corr}(X, Y) = \frac{\text{Cov}(X, Y)}{\sigma_x \sigma_y}$$

$$\rho = \frac{E\{(X - \mu_x)(Y - \mu_y)\}}{\sqrt{\left[E\{(X - \mu_x)^2\}E\{(Y - \mu_y)^2\}\right]}}$$

The sample correlation coefficient is given as

$$r = \frac{\{(1/(n-1))\}\sum(x_i - \bar{x})(y_i - \bar{y})}{S_x S_y}$$

$$= \frac{\sum(x_i - \bar{x})(y_i - \bar{y})}{\sqrt{\left\{\sum(x_i - \bar{x})^2 \sum(y_i - \bar{y})^2\right\}}}$$

$$= \frac{\sum x_i y_i - n\bar{x}\bar{y}}{\sqrt{\{(\sum x_i^2 - n\bar{x}^2)(\sum y_i^2 - n\bar{y}^2)\}}}$$

Interpretation of the correlation coefficient is important. The correlation coefficient may lie between -1 and $+1$. In other words,

$$-1 \le \rho \le +1$$

A positive correlation means larger values of x are associated with larger values of y and smaller values of x are associated with smaller values of y. A negative correlation means larger values of x are associated with smaller values of y and smaller values of x are associated with larger values of y.

The interpretation of the correlation coefficient may be summarized as follows:

(a) a correlation coefficient of –1 implies perfect negative relationship (Fig. 10.2).
(b) a correlation coefficient of +1 implies perfect positive relationship (Fig. 10.3).
(c) a correlation coefficient of zero implies no linear relationship (Fig. 10.4).
(d) the larger the absolute value of the correlation coefficient, the stronger the linear relationship between the random variables.

Fig. 10.2 Perfect negative correlation

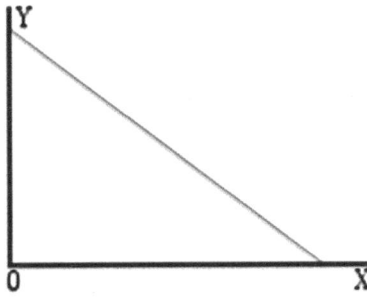

Fig. 10.3 Perfect positive correlation

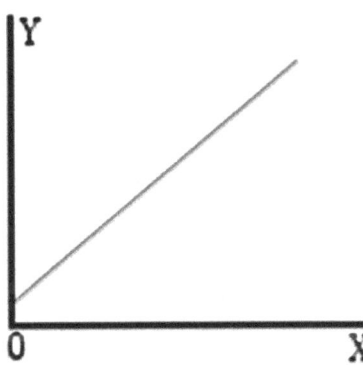

Fig. 10.4 No correlation

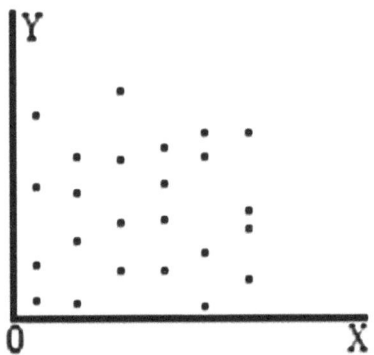

The sample correlation coefficient is quite useful. It measures the strength of linear relationship between two variables in a sample. But the sample correlation coefficient can also be used to test the hypothesis of no relationship between a pair of random variables in the population. To a researcher, this test is of great importance. In almost all practical cases, a value of r other than zero in absolute term will be obtained. What we want to study from the test is to see if the population correlation coefficient is significantly different from zero. The test can be accomplished by use of the t-test. It can be shown that the random variable

$$t = \frac{r\sqrt{(n-2)}}{\sqrt{(1-r^2)}}$$

has a student's t-distribution with $n-2$ degrees of freedom.

The test procedure can be carried out in the following manner:

(a) To test if the correlation coefficient is greater than zero.

The hypotheses are

$$H_0: \rho = 0$$
$$H_a: \rho > 0$$

$$t_c = \frac{r\sqrt{(n-2)}}{\sqrt{(1-r^2)}}$$

Reject H_0 if calculated $t > t_{\alpha, n-2}$.

(b) To test if the correlation coefficient is less than zero.

The hypotheses are

$$H_0: \rho = 0$$
$$H_a: \rho < 0$$

$$t_c = \frac{r\sqrt{(n-2)}}{\sqrt{(1-r^2)}}$$

Reject H_0 if calc. $t < -t_{\alpha, n-2}$.

(c) To test if there the correlation coefficient is other than zero.

The hypotheses are

$$H_0: \rho = 0$$
$$H_a: \rho \neq 0$$

$$t_c = \frac{r\sqrt{(n-2)}}{\sqrt{(1-r^2)}}$$

Reject H_0 if calc. $t < -t_{\alpha/2, n-2}$ or $t > t_{\alpha/2, n-2}$

Example 1

Students' scores in examination and assignment are assumed to be correlated. A random sample of 10 students had the following scores:
 Examination

81	62	74	78	93	69	72	83	90	84
76	71	69	76	87	62	80	75	92	79

Assignment

(a) Find the sample correlation coefficient.
(b) Test at 5 % significance level the hypothesis of no linear association between examination and assignment.

Solution:

(a) Here examination scores are taken as x scores and assignment scores are taken as y scores. From the data we calculate the following:

$$\sum x_i y_i = 60{,}862; \quad \bar{x} = 78.6; \quad \bar{y} = 76.7; n = 10$$
$$\sum x_i^2 = 62{,}604; \quad \sum y_i^2 = 59{,}497$$

$$r = \frac{\sum x_i y_i - n\bar{x}\bar{y}}{\sqrt{\{(\sum x_i^2 - n\bar{x}^2)(\sum y_i^2 - n\bar{y}^2)\}}}$$

$$= \frac{60{,}862 - 10 * 78.6 * 76.7}{\sqrt{\left\{62{,}604 - 10 * (78.6)^2\right\}\left\{59{,}497 - 10 * (76.7)^2\right\}}}$$

$$= \frac{60{,}862 - 60{,}286.2}{\sqrt{\{(62{,}604 - 61{,}779.6)(59{,}497 - 58{,}828.9)\}}}$$

$$= \frac{575.8}{\sqrt{(824.4 * 668.1)}}$$

$$= 0.776$$

(b)   The hypotheses are formulated as follows:

$$H_0: \rho = 0$$
$$H_a: \rho \neq 0$$

$$
\begin{aligned}
t_c &= \frac{r\sqrt{(n-2)}}{\sqrt{(1-r^2)}} \\
&= \frac{0.776\sqrt{(10-2)}}{\sqrt{\left\{1 - (0.778)^2\right\}}} \\
&= \frac{0.776 * 2.828}{0.631} \\
&= 3.478
\end{aligned}
$$

$\alpha = 0.05$; $\alpha/2 = 0.025$.

From table $t_{0.025,8} = 2.306$; $t_c > t_{0.025,8}$. Therefore, H_0 is rejected. This means that the population correlation coefficient between examination scores and assignment scores is not zero. This can also be stated that the population correlation coefficient between examination scores and assignment scores is significantly different from zero.

Example 2

In a market survey a random sample of eight textbooks sold during the year and the sale prices are shown.

Sales	122	186	292	157	254	352	147	111
Price ($)	292	305	297	313	308	299	278	270

(a)   Find the sample correlation between sales and price.

(b)   Test at 5 % significance level the hypothesis of no linear association between sales and price.

(c)   Test at 5 % significance level if the correlation between sales and price is greater than zero.

Solution:

(a)   Here sales and price are taken as x and y scores, respectively. From the data we can calculate the following:

$$\bar{x} = 202.6; \quad \bar{y} = 295.3; \quad \sum x_i y_i = 482{,}535; \quad n = 8$$
$$x_i^2 = 381{,}743; \quad y_i^2 = 698{,}916$$

$$r = \frac{\sum x_i y_i - n\bar{x}\bar{y}}{\sqrt{\{(\sum x_i^2 - n\bar{x}^2)(\sum y_i^2 - n\bar{y}^2)\}}}$$

$$= \frac{482{,}535 - 8 * 202.5 * 295.3}{\sqrt{\left[\{381{,}743 - 8 * (202.6)^2\}\{698{,}916 - 8 * (295.3)^2\}\right]}}$$

$$= \frac{3913}{\sqrt{\{(381{,}743 - 328{,}374)(698{,}916 - 697{,}617)\}}}$$

$$= \frac{3913}{\sqrt{(53{,}369 * 1299)}}$$

$$= 0.470$$

(b) The hypotheses are formulated as follows:

$$H_0\!: \rho = 0$$
$$H_a\!: \rho \neq 0$$

$$t_c = \frac{r\sqrt{(n-2)}}{\sqrt{(1-r^2)}}$$

$$= \frac{0.470\sqrt{(8-2)}}{\sqrt{\{1 - (0.470)^2\}}}$$

$$= \frac{0.470 * 2.449}{0.883}$$

$$= 1.304$$

$\alpha = 0.05$; $\alpha/2 = 0.025$.

From table $t_{0.025,6} = 2.447$

Calculated $t <$ table t. Therefore, H_0 is not rejected. This means that correlation between sales and price is not significantly different from zero. In other words, there is no significant relationship between sales and price.

(c) The hypotheses are formulated as follows:

$$H_0\!: \rho = 0$$
$$H_a\!: \rho > 0$$

Calculated $t = 1.304$ as worked out in (b) above. $\alpha = 0.05$. From table $t_{0.05,6} = 1.943$. Therefore, calculated $t <$ table t. The null hypothesis H_0 is, therefore, not rejected. This means that the correlation between sales and price is not greater than zero.

Example 3

Land cost varies inversely with the distance from the city center. In a sample survey in a city, the following data were recorded:

Distance (km)	1	2	4	8	15	20	25	30
Cost ($/m²)	100	80	70	60	50	40	30	25

(a) Calculate the sample correlation between distance and land cost.
(b) Test at 5 % significance level if the correlation between distance from the city center and the land cost is less than zero.

From the data we can calculate the following figures:

$$\bar{x} = 13.13; \quad \bar{y} = 56.88; \quad \sum x_i y_i = 4070$$
$$\sum x_i^2 = 2235; \quad \sum y_i^2 = 30{,}525$$

$$
r = \frac{\sum x_i y_i - n\bar{x}\bar{y}}{\sqrt{\{(\sum x_i^2 - n\bar{x}^2)(\sum y_i^2 - n\bar{y}^2)\}}}
$$
$$
= \frac{4070 - 8 * 13.13 * 56.88}{\sqrt{\left\{2235 - 8 * (13.13)^2\right\}\left\{30{,}525 - 8 * (56.88)^2\right\}}}
$$
$$
= \frac{-1904.68}{\sqrt{\{(855.82)(4642.32)\}}}
$$
$$
= \frac{-1904.68}{1993.24}
$$
$$
= -0.956
$$

(b) The hypotheses are formulated as follows:

$$H_0: \rho = 0$$
$$H_a: \rho < 0$$

$$t_c = \frac{r\sqrt{(n-2)}}{\sqrt{(1-r^2)}}$$

$$= \frac{-0.956\sqrt{(8-2)}}{\sqrt{\{1-(-0.956)^2\}}}$$

$$= \frac{-0.956 * 2.449}{0.293}$$

$$= -7.99$$

$\alpha = 0.05$. From table $t_{0.05,6} = -1.943$.

Calculated $t <$ table t. Therefore, the null hypothesis H_0 is rejected. This means that the correlation between distance from the city center and the land cost is less than zero.

Problems

10.1 A firm had the following profits and investment expenditures during the period 1982–1990.

Year	1982	1984	1986	1988	1990
Profit ($1000)	200	400	600	800	1000
Investment expenditure ($1000)	45	65	70	85	95

(a) Estimate the correlation coefficient between profit and investment expenditure.
(b) Test the significance of the correlation coefficient.

10.2 The following data pertain to selling prices and volume measured by the number of pages of new statistics books:

Price ($)	Number of pages
10	400
12	600
12	500
10	300
8	400
8	200

Is there any correlation between price and number of pages for the statistics books? Use 5 % level of significance.

10.3   Use the data in the following table, calculate the correlation coefficient and test the relationship between sales volume and promotional expenses with respect to a few products.

Product	Sales volume	Promotional expenses ($1000)
1	31	5
2	40	11
3	25	3
4	30	4
5	20	2
6	34	5

10.4   A researcher wanted to study if there was a relationship between GRE scores and GPA scores of students. A random sample of eight students shows the following scores:

GRE scores	GPA scores
480	2.70
490	2.90
510	3.30
510	2.90
530	3.10
550	3.00
610	3.20
640	3.70

Calculate the correlation coefficient between GRE and GPA scores and test at 0.02 level of significance if the relationship is significant.

10.5   A student in statistics class wants to study if there is any relationship between number of hours a student reads during the night preceding the examination day, and the score he obtains in the said examination. He selects a random sample of size 10 and finds the following results:

Number of hours read	Scores
5.0	70
6.0	72
4.0	55
4.5	50
6.5	75

(continued)

(continued)

Number of hours read	Scores
2.0	75
3.0	62
7.0	69
7.0	65
5.0	80

Conduct the correlation test at 0.05 level of significance and conclude if there is a significant correlation between hours of reading and scores.

10.6   In order to draw a demand curve, the unit prices and per capita consumptions of a commodity were recorded for several periods as follows:

Unit price	75.4	68.0	62.8	50.4	41.4	38.6
Per capita consumption	18.3	20.6	21.9	24.4	20.4	25.0

Name the techniques that may be used to test if there is a significant relationship between price and consumption. Do the test using any one. Use 5 % level of significance.

10.7   An economist is interested in the production cost of fertilizer in an industry. In four occasions he noted the following:

Production (000 tons)	Cost per ton ($)
3.0	40
2.4	50
2.6	55
5.5	30

Using the technique of linear correlation, test if there is a significant correlation between production and cost. Use 98 % confidence level.

10.8   In pond water, dissolved oxygen content and temperature vary with depth. The following data are taken from a research work conducted in an aquaculture pond in AIT (March 1993):

Depth (cm)	Dissolved oxygen (mg/L)	Temperature ($^\circ$C)
10	15	34.5
20	14	34.0
30	10	33.0
50	5	31.0
78	2	30.0

(a) Compute the correlation coefficients between depth and dissolved content as well as between depth and temperature.

(b) Test at 2 % level of significance if the correlation coefficients are significant.

Answers

10.1 $r_s = 0.986$; significant
10.2 $r_s = 0.79$; not significant
10.3 $r_s = 0.91$; significant
10.4 $r_s = 0.812$; significant
10.5 $r_s = 0.165$; not significant
10.6 $r_s = -0.70$; not significant
10.7   no significant relationship
10.8   (a) $r_s = -0.963$; -0.97; (b) both significant

Chapter 11
Simple Regression

Abstract A regression model is introduced. Basically, a simple regression model deals with one dependent variable and one independent variable. The technique is demonstrated to show how to formulate the hypotheses in a simple regression model and tested. Setting the confidence intervals of the parameters as well as checking the adequacy of a regression model is explained. It is demonstrated how to calculate the Coefficient of Determination and interpret it. With the help of examples it is shown how to transform data and convert a nonlinear model into a linear model thus making it easy to use the basic concept of a simple regression.

Keywords Simple regression · Dependent and independent variables · Parameter estimation · Coefficient of determination · Data transformation

In the correlation chapter we studied the relationship between two random variables x and y. Correlation shows the direction as well as the strength of relationship. In correlation the two variables are treated perfectly symmetrically. It is immaterial which variable is treated as x and which variable is treated as y. This means that we do not consider whether one variable is dependent on the other. In correlation, it is not evident how change in response in one variable takes place as a result of a unit change in the other variable.

Linear regression takes into account the dependency of one variable on one or more other variable(s). Also it exhibits the change in the response of the dependent variable as a result of a unit change in the independent variable(s). In other words, regression analysis serves the model building purpose. In this chapter we shall study the Simple Linear Regression Model. Multiple Regression Models will be dealt with in the next chapter.

11.1 Simple Linear Regression Model

In Simple Linear Regression only one independent variable is considered. To illustrate the concept, the yield and fertility of agricultural land may be cited as an example. The more the fertility of land, the more is the yield. Here yield and fertility

© Springer Science+Business Media Singapore 2016

A.Q. Miah, *Applied Statistics for Social and Management Sciences*,
DOI 10.1007/978-981-10-0401-8_11

are correlated. Yield depends on fertility and hence it is the dependent variable. Fertility controls the yield. So fertility is an independent variable.

In the Simple Regression Model we are interested in the relationship between the dependent variable (Y) and one independent variable (X). If the independent random variable X takes a specific value x_i and consequently the dependent variable Y takes the value y_i, the population regression model is expressed by the following:

$$Y_i = \alpha + \beta X_i + e_i$$

where α and β are constants (parameters) and e_i is a random variable with mean equal to zero. In fact, e_i is the error part which explains the variation in Y_i not explained by X_i. Our concern now is to estimate the coefficients α and β. The most commonly used method of estimating α and β is the Least Square Estimation. What we want to do is to find out a straight line through n pairs of observations (x_1, y_1), (x_2, y_2), (x_3, y_3), ... (x_n, y_n) in such a way that the sum of the squares of the deviations from the regression line is the minimum.

If a and b are the estimates of α and β, then the regression line would assume the form

$$y_i = a + bx_i + e_i$$

Therefore, the error part $e_i = y_i - (a + bx_i)$ and the square of the error part $e_i^2 = \{y_i - (a + bx_i)\}^2$. The principle in the Least Square Estimation is to minimize this square summed up to all pairs of the observations. Thus, our purpose is to minimize $\sum e_i^2$, i.e., $\sum \{y - (a + bx_i)\}^2$. Using this criterion and with the help of calculus, it can be shown that the resulting estimates a and b are as follows:

$$b = \frac{\sum (x_i - \bar{x})(y_i - \bar{y})}{\sum (x_i - \bar{x})^2}$$
$$= \frac{\sum x_i y_i - n\bar{x}\bar{y}}{\sum x_i^2 - n\bar{x}^2}$$
$$= \frac{S_{xy}}{S_{xx}}$$
$$a = \bar{y} - b\bar{x}$$

and the sample regression line of y on x is

$$y = a + bx$$

Example 1

An irrigation engineer wants to study the vertical temperature distribution in water reservoirs. In a reservoir he records the following observations:

Depth (x cm)	Temperature (y °C)
10	34.5
20	34.0
30	33.0
40	31.0
50	30.0
60	29.5
70	29.0
80	28.0
90	27.0
100	25.0

Find out the sample regression line showing the relationship between temperature and depth.

Solution:

From the data we can calculate the following quantities:

$$n = 10; \ \sum x_i = 550; \ \sum y_i = 301;$$
$$\bar{x} = 55; \ \bar{y} = 30.1$$
$$\sum x_i^2 = 38,500; \ \sum y_i^2 = 9145.5;$$
$$\sum x_i y_i = 15,725$$
$$S_{xy} = \sum x_i y_i - n\bar{x}\bar{y}$$
$$= 15,725 - 10 * 55 * 30.1$$
$$= -830$$
$$S_{xx} = \sum x_i^2 - n\bar{x}^2$$
$$= 38,500 - 10 * (55)^2$$
$$= 8250$$
$$b = \frac{S_{xy}}{S_{xx}}$$
$$= \frac{-830}{8250}$$
$$= -0.1006$$
$$a = \bar{y} - b\bar{x}$$
$$= 30.1 - (-0.1006) * 55$$
$$= 35.633$$

The regression line is

$$y = 35.633 - 0.1006\,x$$

11.2   Hypothesis Testing in Simple Linear Regression

Our interest in this section is to study the adequacy of the simple regression model. This we want to do by testing the hypothesis about the model parameters and constructing confidence intervals of the same. The important hypothesis is whether the estimated parameter is significant. This could be also said that we want to test if the estimated parameters are significantly different from zero. Thus, the hypotheses are

$$\beta = 0$$
$$\beta \neq 0$$

The test procedure is carried out in a specified manner. The sum of squares (S_{yy}) of the dependent variable, i.e., the sum of the squares of the deviations from the mean can be broken down into two parts—the regression sum of squares (SSR) and error sum of squares (SSE). This may be stated in the following way:

$$\text{SST} = \text{SSR} + \text{SSE}$$

where,

$$\begin{aligned}
\text{SST} &= S_{yy} \\
&= \sum (y_i - \bar{y})^2 \\
&= \sum y_i^2 - n\bar{y}^2 \\
\text{SSR} &= b * S_{xy} \\
&= b * \left\{ \sum x_i y_i - n\overline{xy} \right\} \\
\text{and } \text{SSE} &= \text{SST} - \text{SSR}
\end{aligned}$$

The SSR and SSE divided by their appropriate degrees of freedom will give respective Mean Squares. The degrees of freedom for SSR and SSE are 1 and $n - 2$ respectively. Therefore,

$$\text{MSR} = \frac{\text{SSR}}{1}$$
$$\text{MSE} = \frac{\text{SSE}}{n - 2}$$

Table 11.1 Analysis of variance

Source of variation	Sum of squares	Degrees of freedom	Mean squares	F_0
Regression	SSR = bS_{xy}	1	MSR	MSR/MSE
Error or residual	SSE =			
	$S_{yy} - bS_{xy}$	$n - 2$	MSE	
Total	S_{yy}	$n - 1$		

The test statistic for testing hypothesis formulated above is given by

$$F_0 = \frac{\text{MSR}}{\text{MSE}}$$

Reject H_0, if $F_0 > F_{\alpha,1,n-2}$. It is advisable to prepare the ANOVA (analysis of variance) table as shown (Table 11.1).

Example 2
For the temperature-depth problem, run a test for the significance of the coefficient. Use 5 % significance level.

Solution:
The hypotheses are formulated as follows:

$$\beta = 0$$
$$\beta \neq 0$$
$$S_{yy} = \sum (y_i - \bar{y})^2$$
$$= \sum y_i^2 - n\bar{y}^2$$
$$= 9145.5 - 10 * (30.1)^2$$
$$= 85.40$$
$$\text{SSR} = b * S_{xy}$$
$$= (-0.1006) * (-830)$$
$$= 83.50$$
$$\text{SSE} = S_{yy} - \text{SSR}$$
$$= 85.40 - 83.50$$
$$= 1.90$$

Now, $F_0 = 347.92$ and $F_{0.05,1,8} = 5.32$. Therefore, $F_0 > F_{0.05,8}$. The null hypothesis H_0 is rejected. This means that the coefficient is not equal to zero at 5 % level of significance (Table 11.2).

Table 11.2 Analysis of variance

Source of variation	Sum of squares	Degrees of freedom	Mean squares	F_0
Regression	83.50	1	83.50	347.92
Error or residual	1.90	$10 - 2 = 8$	0.24	
Total	85.40	$10 - 1 = 9$		

11.3 Confidence Intervals of Parameters

The confidence intervals for the parameters α and β are quite useful. With the help of these intervals, the quality of regression line can be judged. If α is the desired level of significance, then the confidence interval for α and β (parameters) are given by

$$a - t_{\alpha/2,n-2}\sqrt{\left[\text{MSE} * \left\{1/n + \bar{x}^2/S_{xx}\right\}\right]}$$

$$\leq \alpha \leq$$

$$a + t_{\alpha/2,n-2}\sqrt{\left[\text{MSE} * \left\{1/n + \bar{x}^2/S_{xx}\right\}\right]}$$

and

$$b - t_{\alpha/2,n-2}\sqrt{\left[\left\{\frac{\text{MSE}}{S_{xx}}\right\}\right]}$$

$$\leq \beta \leq$$

$$b + t_{\alpha/2,n-2}\sqrt{\left[\left\{\frac{\text{MSE}}{S_{xx}}\right\}\right]}$$

Example 3

For the temperature-depth problem, construct 95 % confidence intervals for the population parameters.

Solution:

$$\text{Here, } \alpha = 0.05 \quad \alpha/2 = 0.025 \quad n = 10$$

$$t_{0.025,8} = 2.306.$$

Confidence interval for α is given by

$$a - t_{\alpha/2,n-2}\sqrt{\left[\text{MSE}*\left\{1/n + \bar{x}^2/S_{xx}\right\}\right]}$$

$$\leq \alpha \leq$$

$$a + t_{\alpha/2,n-2}\sqrt{\left[\text{MSE}*\left\{1/n + \bar{x}^2/S_{xx}\right\}\right]}$$

$$\Rightarrow 35.633 - 2.306\sqrt{[0.24 * \{1/10 + (55)2/8250\}]}$$

$$\leq \alpha \leq$$

$$35.633 + 2.306\sqrt{[0.24 * \{1/10 + (55)2/8250\}]}$$

$$\Rightarrow 35.633 - 0.772 \leq \alpha \leq 35.633 + 0.772$$

$$\Rightarrow 34.862 \leq \alpha \leq 36.405$$

Confidence interval for β is given by

$$b - t_{\alpha/2, n-2} \sqrt{\left[\left\{\frac{MSE}{S_{xx}}\right\}\right]}$$
$$\leq \beta \leq$$
$$b + t_{\alpha/2, n-2} \sqrt{\left[\left\{\frac{MSE}{S_{xx}}\right\}\right]}$$
$$\Rightarrow b - t_{0.025, 8} \sqrt{\left[\left\{\frac{MSE}{S_{xx}}\right\}\right]}$$
$$\leq \beta \leq$$
$$\Rightarrow b + t_{0.025, 8} \sqrt{\left[\left\{\frac{MSE}{S_{xx}}\right\}\right]}$$
$$\Rightarrow -0.1006 - 2.306 \sqrt{\left[\left\{\frac{0.24}{8250}\right\}\right]}$$
$$\leq \beta \leq$$
$$-0.1006 + 2.306 \sqrt{\left[\left\{\frac{0.24}{8250}\right\}\right]}$$
$$\Rightarrow -0.1006 - 0.0124 \leq \beta \leq -0.1006 + 0.01214$$
$$\Rightarrow -0.1130 \leq \beta \leq -0.0882$$

11.4 Adequacy of the Regression Model

In fitting a regression model, several assumptions are made. In parameter estimation, it is assumed that the errors are uncorrelated random variables with mean zero and constant variance. In hypothesis testing, it is assumed that the errors are normally distributed. It is also assumed that the order of the model is correct (first-order polynomial, polynomial of degree two, etc.).

11.4.1 Residual Analysis to Test Uncorrelated Errors

The residuals are obtained by subtracting the predicted y observations based on regression line from the corresponding individual y observations. The residuals are plotted against the predicted y observations. By checking the plots it is possible to identify if the residuals are uncorrelated or not.

11.4.2   Residual Analysis to Test Normality

Normality of the residuals can also be checked by plotting of the residuals. The frequency histogram of the residuals is constructed and a normal distribution curve is superimposed on this. By comparing the frequency histogram and the normal curve, it can be checked if the residuals are normally distributed.

11.4.3   Lack-of-Fit Test

Regression models are fitted with the experimental data in which the true relationship is not previously known. Therefore, it is necessary to check whether the assumed model is correct. The check can be carried out with the help of the Lack-of-Fit Test.

The test is carried out by partitioning the SSE into two—sum of squares due to the pure error SSpe and sum of squares due to lack-of-fit SSlof. Thus,

$$SSE = SSpe + SSlof$$

First the sum of squares due to pure error SSpe is calculated and then sum of squares due to lack-of-fit SSlof is calculated from

$$SSlof = SSE - SSpe$$

For calculation of SSpe, there must be repeated observations on y for at least one level of x. The hypotheses to be tested are

H_0:   The model adequately fits the data
H_a: The model does not adequately fit the data

The technique is illustrated with the help of an example.

Example 4
Let the following be the pairs of observations from an experiment.

x:	2.0	2.0	3.0	4.4	4.4	5.1	5.1	5.1	5.8	6.0
y:	2.4	2.9	3.2	4.9	4.7	5.7	5.7	6.0	6.3	6.5

Solution:
In this example, there are two, two and three repeated observations on y for levels of x at 2.0, 4.4, and 5.1 respectively. We can calculate the following figures from the data.

$$\bar{x} = 4.29;\ \bar{y} = 4.83;\ \sum x_i y_i = 226.72$$

$$S_{xx} = \sum (x_i - \bar{x})^2 = 19.35$$

$$S_{yy} = \sum (y_i - \bar{y})^2 = 20.14$$

$$S_{xy} = \sum (x_i - \bar{x})(y_i - \bar{y})$$

$$= \frac{(\sum x_i)(\sum y_i)}{n}$$

$$= \sum x_i y_i - n\bar{x}\bar{y}$$

$$= 226.72 - 10 * 4.29 * 4.83$$

$$= 226.72 - 207.21$$

$$= 19.51$$

$$b = \frac{S_{xy}}{S_{xx}}$$

$$= \frac{19.51}{19.35}$$

$$= 1.0083$$

$$a = \bar{y} - b * \bar{x}$$

$$= 4.83 - 1.0083 * 4.29$$

$$= 0.51$$

The regression line is $y = 0.51 + 1.0083x$

$$SSR = b * S_{xx}$$

$$= 1.0083 * 19.35$$

$$= 19.51$$

The calculations for SSpe are shown in the following table (Table 11.3). Thereafter the ANOVA table is constructed as follows (Table 11.4): For Lack-of-Fit Test,

$$F_0 = \frac{MSlof}{MSpe}$$

$$= \frac{0.105}{0.053}$$

$$= 1.98$$

From the table

$F_{0.05,4,4} = 6.39$. Therefore, $F_0 < F_{0.05,4,4}$. Therefore, we do not reject H_0. This means that the model describes the data adequately.

Table 11.3 The SSpe

Level of x	y_i	$\bar{y}_i$	$(y_i - \bar{y})^2$	D.F
2.0	2.4, 2.9	2.65	0.125	$2 - 1 = 1$
4.4	4.9, 4.7	4.80	0.020	$2 - 1 = 1$
5.1	5.7, 5.7, 6.0	5.80	0.060	$3 - 1 = 2$
Totals		SSpe = 0.205		4

Table 11.4 Analysis of variance

Source of variation	Sum of squares	Degrees of freedom	Mean squares	F_0
Regression	19.51	1	19.51	246.96
Residual	0.63	8	0.079	1.98
(lack of fit)	0.42	4	0.105	
(pure error)	0.21	4	0.053	
Total	20.14	9		

11.5 Coefficient of Determination

The coefficient of determination is used to measure the adequacy of the regression model. It is defined by

$$R^2 = \frac{\text{SSR}}{\text{SST}} = 1 - \frac{\text{SSE}}{\text{SST}}$$

Its value lies between zero and one, i.e.,

$$0 \le R^2 \le 1$$

R^2 refers to the variability in the data explained by the regression model. In the previous example, R^2 = SSR/SST = 19.51/20.14 = 0.9687. This means 96.87 % of the variability in the data is accounted for by the regression model.

If more and more terms are added to the regression model, R^2 will generally increase. But this does not always mean that the model fit is better with the additional terms. The fit will be better if the mean square error is reduced. This implies larger values of F_0.

11.6   Data Transformation

Sometimes theory specifies a nonlinear model. It is possible to transform the nonlinear model to linear one and then the Least Square Technique can be applied to estimate the model parameters. As an example, the length–weight relationship of fish usually or always can adequately be represented by

$$w = al^b$$

where w = weight, l = length; and a and b are coefficients to be estimated. Taking logarithm of both sides we can get

$$\log w = \log a + b * \log(l),$$

and this is equivalent to

$$y = a' + bx$$

First the sample weights w_i are to be transformed into $\log w_i$ to be represented by y, and sample lengths l_i are to be transformed into $\log (l_i)$ to be represented by x. Then the estimated coefficient b of x is the same as the exponent b of l. From the regression, a' will be known. So the coefficient a can be computed from the known transformation $\log a = a'$.

Example 5
In an experiment dealing with length–weight relationship of a particular fish species, the following observations were recorded from a random sample of 4.

weight (wg):	0.13	1.8	13	90
Length (mm):	20	50	100	200

Determine the coefficients a and b of the theoretical model

$$w = al^b$$

Solution:
Here we need to estimate the model already specified. In this prescribed model we need logarithmic transformation. The calculations are shown hereafter (Table 11.5).

Table 11.5   Model estimates

w (g)	l (mm)	y (log w)	x (log l)
0.13	20	−0.886	1.301
1.80	50	0.255	1.699
13.00	100	1.114	2.000
90.00	200	1.954	2.301

From these data we can calculate the following figures:

$$\bar{x} = 1.825; \; \bar{y} = 0.6093; \; \sum x_i y_i = 6.005$$

$$S_{xy} = \sum x_i y_i - n\bar{x}\bar{y}$$
$$= 6.005 - 4 * 1.825 * 0.6093$$
$$= 6.005 - 4.448$$
$$= 1.557$$
$$S_{xx} = 0.548$$
$$b = \frac{S_{xy}}{S_{xx}}$$
$$= \frac{1.557}{0.548}$$
$$= 2.841$$
$$a' = y - b * \bar{x}$$
$$= 0.6093 - 2.841 * 1.825$$
$$= 0.6093 - 5.185$$
$$= -4.5757$$
$$\log a = a'$$
$$= -4.5757$$

Taking antilog we get $a = 2.6564 \times 10^{-5}$
Therefore, we get the estimated model

$$w = 2.6564 \times 10^{-5}(1)^{2.841}$$

For this model too hypothesis for significance of the coefficients can be tested, their confidence levels can be constructed, coefficient of determination can be computed, and lack of fit can be tested in the usual process.

The data transformation technique can be applied in a variety of problems. Calculation of population growth rate and forecasting are good examples. The multiplicative and exponential population forecasting model is $P_n = P_0 (1 + r)^n$, where P_0 is the population of the base year, P_n is the population at the end of n years, r is the growth rate, and n is the number of years. Let us now see how the simple regression can be used after data transformation and how the growth rate can be calculated. For generalization purpose, we shall use t for n indicating time in the above model.

$$P_t = P_0(1 + r)^t$$

Taking logarithm of both sides we get,

$$\log P_t = \log\{P_0(1+r)^t\}$$
$$= \log P_0 + \log(1+r)^t$$
$$= \log P_0 + t * \log(1+r)$$
$$= \log P_0 + \log(1+r) * t$$

Setting

$$Y = \log P_t,$$
$$A = \log P_0 \text{ and}$$
$$B = \log(1+r),$$

we get,

$$y = A + Bt$$

which is linear. Now suppose, in a regression analysis, based on several years population data, we have the following model equation:

$$Y = 2.36412 + 0.02952t$$
$$\text{Here,} \quad B = 0.02952$$
$$\Rightarrow \log(1+r) = 0.02952$$
$$\Rightarrow \quad 1 + r = \text{anti - } \log(0.02952)$$
$$= 1.0703$$
$$\Rightarrow \quad r = 1.0703 - 1$$
$$= 0.0703$$
$$= 7.03\%$$

This growth rate, if assumed to remain steady, can be used in the original model and forecast of population for some future period can be made.

11.7 Interpretation of Simple Regression Model

In practice, regression models are built based on several pairs of observations. Manual computations in such cases are extremely difficult and time-consuming. With the advent of computers, it is also not necessary to do the manual calculations. Software packages such as SPSS are used and easily the model building purpose is accomplished. What is the most important of the regression models is, therefore, the interpretation. Decisions are based on the interpretation of the regression models. As such it is desirable to deal with this aspect more specifically.

Problems

11.1   In a pilot study of a new fertilizer, four standard plots were selected at random. The following results were noted:

Fertilizer (lbs)	Yield (lbs)
1	70
2	70
3	80
4	100

(a)   Establish the sample regression model.

(b)   Predict the yield if 5 lbs of fertilizer is used.

11.2   While examining the relationship between water depth and dissolved oxygen concentration in a pond, an aquaculturist recorded the following the observations:

Depth (cm)	10	20	30	40	50	60	70	80	90	100
DO (mg/L)	15	14	13	12	10	8	5	4	2	2

(a)   Establish the sample regression.
(b)   Estimate the DO concentration at 105 cm depth
(c)   Test if the coefficient is significant (use 5 % level of significance).
(d)   Find the 95 % confidence interval of the constant term.
(e)   Find the 95 % confidence interval of the coefficient.
(f)   Find the coefficient of determination.

11.3   The demand (y) for a commodity and the corresponding price (x) are shown as follows:

Price (x)	45	40	35	32	30	25	24	23	22	20
Demand (y)	500	550	560	580	585	600	610	620	650	670

(a)   Based on the data estimate the demand function

$$Y = AX^b e^u (e \text{ is the error term})$$

(b) How much variability in demand is explained by the price of the commodity?

(c) Estimate the demand if the price is 60 and 15?

11.4 An investigator wants to study the percentage of urban population based on percentage of literacy. He knows that percentage of urban population of a country depends on its percentage of literacy. In a study of 50 country observations the simple linear regression analysis shows the following results.

Multiple R 0.75705
R square 0.57313
Standard Error 10.43147

Analysis of variance

Source of variation	Sum of squares	DF	Mean square	F	Sig. of F
Regression	7012.75950	1	7012.75950	64.44623	0.0000
Residual	5223.15170	48	108.81566		
Total	12,235.91120				

Variables in the equation

Variable	B	SE B	T	Sig T
% literacy	1.227722	0.152933	8.028	0.0000
(Constant)	−1.946212	3.103777	−0.627	0.5336

Dependent variable is percent urban population.

(a) Write down the sample regression equation.

(b) Write down the interpretations.

11.5 Available number of scientists and engineers in a country depends on its population size. A simple linear regression based on 24 country studies shows the following results.

Multiple R 0.74246
R square 0.55124
Standard Error 1523456.611

Analysis of variance

Source of variation	Sum of squares	DF	Mean square	F	Sig. of F
Regression	6272 (exp 10)	1	6272 (exp 10)	27.02396	0.0000
Residual	5106 (exp 10)	22	232 (exp 10)		
Total	11378 (exp 10)	23			

Variables in the equation

Variable	B	SE B	T	Sig T
Population	5365.866	1032.20	5.198	0.0000
(Constant)	454201.617	370257.74	1.227	0.2329

Dependent variable is available number of scientists and engineers. Population is in million.

(a) Write down the sample regression equation.

(b) Use this model and estimate the number of scientists and engineers of your country.

11.6   Population of Bangladesh figures during several census periods is shown.

Census year	Population
1901	28,972,786
1911	31,555,056
1921	33,254,096
1931	35,604,170
1941	41,997,297
1951	44,165,740
1961	55,222,663
1974	76,398,000
1981	89,912,000

Calculate the growth rate of population of Bangladesh during 1901–1981.

11.7   The Gross National Incomes per capita ($) for India and Pakistan for several years are shown in the following table:

Year	India	Pakistan
1978	270	260
1979	250	270
1980	250	280
1981	260	290
1982	270	290
1983	280	300
1984	280	310
1985	300	320
1986	300	330
1987	300	340

(a) Calculate the growth rates of India and Pakistan and compare these.

(b) Project the per capita gross national income of India and Pakistan during 1995.

11.8   The Gross Domestic Investment as percentage of gross domestic product for Thailand and Singapore for few years is shown in the following table:

Year	Singapore	Thailand
1972	41.4	21.7
1974	45.3	26.6
1976	40.8	24.0
1979	43.4	27.2
1981	46.3	26.3
1983	47.9	25.9
1985	42.5	24.0
1987	39.4	23.8

(a) Calculate the growth rates for Singapore and Thailand during the periods 1972–1987.

(b) Calculate the growth rates of Singapore and Thailand during 1972–1983 and compare these growth rates with those calculated for the period 1972–1987 in (a).

Answers

11.1 (a) $y = 55 + 10x$, i.e.; yield = 55 + 10 (fertilizer quantity)
        (b) 105
11.2 (a) DO = $17.467 - 0.1630$ (depth)
        (b) 0.352 mg/L; (c) significant
        (c) $16.695 \leq \alpha \leq 18.239$
        (d) $(-0.1837) \leq ? \leq (-0.1423)$
        (e) 0.9766
11.3 (a) $y = 1650.89x^{-0.3065}$
        (b) 92.16 %; (c) 471; 720
11.4 (a) Urban Population (%)
        $=-1.946212 + 1.227722$ (% literacy)
11.5 (a) No. of Scientists and Engineers
        $=454,201.617 + 5365.866$ (Population million)
11.6 1.39 %
11.7 (a) India = 2.08 %; Pakistan = 2.90 %; (b) India = 354; Pakistan = 427
11.8 (a) Singapore = -0.03 % (not good estimate);
        Pakistan = -0.22 % (not good estimate);
        (b) Singapore = 1.09 %; Pakistan = 1.26 %

Chapter 12
Multiple Regression

Abstract A multiple regression model deals with one dependent variable and two or more independent variables. A very important feature of a multiple regression model is interpretation. It is demonstrated and explained here how to use a categorical variable in a regression model by use of dummy variables. A useful feature of a regression model is prediction/forecasting. It is demonstrated through examples. The technique of how to transform nonlinear relationships into linear relationships is also explained, thus making it suitable for use of the basic concept of linear regression models in such cases.

Keywords Multiple regression model · Interpretation · Prediction/forecasting · Dummy variables · Other regression model

12.1 Multiple Regression Model

In Chap. 11 we studied the simple linear regression. In simple linear regression model one dependent variable and only one independent variable are involved. There we studied the influence of the independent variable on the dependent variable. In practice, there are several situations where more than one independent variable makes influences on the dependent variable. In such situations we need to study the simultaneous influences of all the independent variables on the dependent variables.

Let us take the example of household income. It may depend on educational level of the household head, length of service in a job, income of other members, income from enterprises, income from shares, and a few other variables. Multiple regression deals with such problems, i.e., problems dealing with the simultaneous influences of several (more than one) independent variables on a single dependent variable.

The general regression model is

$$Y = \beta_0 + \beta_1 X_1 + \beta_2 X_2 + \beta_3 X_3 + \cdots + \beta_k X_k + e$$

© Springer Science+Business Media Singapore 2016

A.Q. Miah, *Applied Statistics for Social and Management Sciences*,
DOI 10.1007/978-981-10-0401-8_12

If $(y_1, x_{11}, x_{21}, x_{31}, \ldots x_{k1})$, $(y_2, x_{12}, x_{22}, x_{32}, \ldots, x_{k2})$, $(y_3, x_{13}, x_{23}, x_{33}, \ldots, x_{k3})$, $\ldots$ are the sets of sample observations, a is an estimator of β_0 and $b_1, b_2, b_3, \ldots, b_k$ are the estimators of $\beta_1, \beta_2, \beta_3, \ldots, \beta_k$, respectively, then the model assumes the form

$$y_i = a + b_1 x_1 + b_2 x_2 + b_3 x_3 + \cdots + b_k x_k + e_i$$

Therefore, the error term is

$$e_i = \{y_i - (a + b_1 x_1 + b_2 x_2 + b_3 x_3 + \cdots + b_k x_k)\}$$

and the square of the error part is

$$e_i^2 = \{y_i - (a + b_1 x_1 + b_2 x_2 + b_3 x_3 + \cdots + b_k x_k)\}^2$$

The principle of least square estimation is to sum up all the squares of these error parts across all sets of observations and minimize these sums of squares with respect to the coefficients $a, b_1, b_2, b_3, \ldots, b_k$.

In the case of simple linear regression we used two simple expressions to calculate the values of a and b. But in multiple regression when several independent variables are involved, it becomes extremely difficult to calculate the values of a, $b_1, b_2, b_3, \ldots, b_k$ manually. Fortunately, computerization has made this task easy and we do not need to do the computation manually. For this reason, derivations of the computation formulae are avoided here. However, one important criterion that must be kept in mind is that the computed values of $a, b_1, b_2, b_3, \ldots, b_k$ are based on the least square estimation. Computers compute these coefficients using the matrix algebra.

In the computer outputs, the statistics for testing the significance of the regression such as

$$H_0: \beta_1 = \beta_2 = \beta_3 = \cdots = \beta_k = 0$$
$$H_a: \beta_j \neq 0 \text{ for at least one } j.$$

are available. Also available are the statistics to test the individual regression coefficients such as

$$H_0: \beta_j = 0$$
$$H_a: \beta_j \neq 0$$

Usually, standard errors of the individual coefficients are provided along with the desired confidence levels. The coefficient of determination R^2 can also be obtained from the computer output.

12.2   Interpretation

It is quite important to interpret the multiple regression analysis produced from a computer run. The basic idea of the meaning of the regression model/equation and the various coefficients along with their statistics is of utmost importance. We must see that we are able to use the regression equation judicially and guard against misinterpretation.

The R^2 as usual, will mean the proportion of the variability of the dependent variable explained by the set of the independent variables. The F-value tells us the adequacy of the regression model as a whole. The t-values and the corresponding significance levels indicate the statistics to test the null hypothesis of the individual coefficients (H_0: $a = 0$; $b_j = 0$). Let the sample regression equation be

$$y = a + b_1 x_1 + b_2 x_2 + b_3 x_3 + \cdots + b_k x_k$$

Here,

a intercept on the y-axis. If all x's assume the value of zero, y will be equal to a

b_1 magnitude of the change in y for a unit change in x_1. It is the rate of change of y with respect to x_1, keeping all other explanatory variables constant (partial differentiation)

b_2 magnitude of the change in y for a unit change in x_2. It is the rate of change of y with respect to x_2, keeping all other explanatory variables constant (partial differentiation)

b_k magnitude of the change in y for a unit change in x_k. It is the rate of change of y with respect to x_k, keeping all other explanatory variables constant (partial differentiation)

The change in y resulting from any amount of change in x_j (not unit change) is given by

$$\Delta y = b_j * \Delta x_j$$

If all the x's change, then the change in y is the sum of the changes in x's. This means

$$\Delta y = b_1 \Delta x_1 + b_2 \Delta x_2 + b_3 \Delta x_3 + \cdots + b_k \Delta x_k$$

12.3   Prediction

A regression model shows the relationship between the dependent variable and a set of independent variables. Once the model is established for a particular phenomenon, it can be used to predict the value of the dependent variable for other sets of values of the independent variables. However, when the regression model is used for some important forecasting purpose, the problems of multicollinearity, homoscedasticity/heteroscedasticity, and autocorrelation will have to be checked. These topics are not discussed here.

12.4   Use of Dummy Variables

Categorical data can also be used in regression analysis. In this case a dummy variable is to be created. A dummy variable may assume two values only—zero and one. Suppose we want to see the influence of sex characteristics in a regression analysis. Sex has two values, male and female. The task can be carried out creating two dummy variables. One dummy variable could be D_1 with values "1", if it is a case of male and "0" otherwise. The other dummy variable could be D_2 with values "1", if it is a case of female and "0" otherwise.

12.5   Other Regression Models

What we have studied above are linear models. These models are characterized by X's. No polynomial or fractional power of X's is involved. Also, no interaction terms such as $X_1 X_2$ are included. But models involving polynomial or fractional powers as well as with interaction terms are encountered. Regression models can be applied in some of these situations. One example of a polynomial model is yield of a crop, say paddy. Yield depends on fertilizer. The more the fertilizer used, the more the yield is. But this happens up to a certain limit only. Beyond that, if more fertilizer is used, the yield is reduced. This phenomenon may be represented by a model of polynomial degree two. Thus,

$$Y = a + b_1 X + b_2 X^2$$

In order to estimate the parameters, we need to define a set of new variables as

$$X_1 = X$$
$$X_2 = X^2$$

Therefore, we get the model as an ordinary multiple regression as

$$Y = a + b_1 X_1 + b_2 X_2$$

Let us have an illustration. An experiment with various quantities of fertilizer resulted in the following regression:

$$Y = 30 + 27.25X_1 + 4.16X_2$$

where Y is the yield, X_1 is the quantity of fertilizer, and X_2 is equal to square of X. If the fertilizer is represented by X, then the regression equation becomes

$$Y = 30 + 27.25X - 4.16X^2$$

It can be shown with the help of calculus that maximum yield results when $X = 3.28$. The yield increases with the increase of fertilizer so long as it does not exceed 3.28 units. If the quantity of fertilizer is increased beyond 3.28, the yield decreases.

In a similar way the following nonlinear models can be fitted using multiple regression:

$$Y = a + b_1X + b_2X^2 + b_3X^3 + \cdots + b_kX^k$$

$$\text{define as:}\quad X_1 = X$$
$$X_2 = X^2$$
$$X_3 = X^3$$
$$\cdots$$
$$X_k = X^k$$

so that

$$Y = a + b_1X_1 + b_2X_2 + b_3X_3 + \cdots + b_kX_k$$

Another form of a nonlinear model is

$$Y = a + b_1/X$$

defined as $X_1 = 1/X$
   so that

$$Y = a + b_1X_1$$

One important application of nonlinear model which is estimated using multiple regression technique is Cobb–Douglas production function of the type:

$$Y = aL^{b1}K^{b2}$$

where Y = output, L = labor input, and K = capital input. If we take logarithm of both sides, we can get

$$\ln Y = \ln a + b_1 \ln L + b_2 \ln K$$

$$\text{Here define} \quad \begin{aligned} Y_1 &= \ln Y \\ a_1 &= \ln a \\ X_1 &= \ln L \\ X_2 &= \ln K \end{aligned}$$

so that the transformed model is

$$Y_1 = a_1 + b_1 X_1 + b_2 X_2$$

Suppose in a sample survey of some industries the following estimated model is established:

$$\ln Y = 0.2852 + 0.8252 \ln L + 0.2319 \ln K$$

This can be written as

$$\ln Y = \ln 1.33 + 0.8252 \ln L + 0.2319 \ln K$$

which can easily be converted to the form

$$Y = 1.33 L^{0.8252} K^{0.2319}$$

So it has been possible to estimate the parameters of the Cobb–Douglas production function using least square regression.

There are certain situations where the effect of one independent variable (X_1) depends on the level of another independent variable (X_2). In this case, an interaction term ($X_1 X_2$) may be used in the model (think of the multicollinearity also).

Consider the following illustration:

In testing the efficiency of a chemical plant producing nitric acid from ammonia, the stack losses and several related variables were measured for 17 different days. Then the following regression equation was fitted by least squares (Wonnacott and Wonnacott 1990: 451):

$$Y = 1.4 + 0.07X_1 + 0.05X_2 + 0.0025X_1 X_2$$

where
Y stack loss (% of ammonia lost)
X_1 airflow, measured as deviation from the mean
X_2 temperature of cooling water, measured as deviation from the mean.

Example 1

Throughput, raw water consumption, and wastewater treatment demand for ten petrochemical industries are shown in the following table.

S. No.	Throughput (000 TPY)	Raw water consumption (m³/day)	Wastewater treatment demand (m³/day)
1	900	6500	852
2	700	4000	585
3	690	4250	600
4	800	4300	625
5	850	5000	750
6	250	1200	190
7	790	4200	636
8	125	650	85
9	1400	8500	1200
10	550	3100	400

Note 000TPY = thousand tons per year

Run a multiple regression analysis to see how wastewater treatment demand depends on raw water consumption and throughput. Write down the regression equation.

Solution

Analysis of variance

	Sum of squares	DF	Mean squares
Regression	916,871.20	2	458,435.60
Residual	4010.90	7	572.99
Total	920,882.10	9	
$F = 800.08$		Significance $F = 5.45 \times 10^{-9}$	

Variables in the equation

Variable	B	SE B	T	Sig T
Throughput	0.3860	0.1162	3.3210	0.01275
Raw water consumption	0.0803	0.01794	4.4800	0.002868
(Constant)	−15.0499	18.5034	−0.8134	0.4428

$R^2 = 0.9956$

In this example the dependent variable is wastewater treatment demand. The regression equation is

Wastewater treatment demand = −15.0499 + 0.3860 × throughput + 0.0803 × raw water consumption.

Units are same as in the input data.

We note that 99.56 % of the variability in the dependent variable (wastewater treatment demand) is explained by the two independent variables (throughput and raw water consumption).

Significance level of the coefficients of the two independent variables is low (less than 5 %, if we set the significance level in our analysis at 5 %). So these coefficients are significant. But the significance level of the constant term (intercept) is not significant (it is 44.28 % much higher than 5 %). So it is not good.

Since the significance level of the constant term is too high, we can force the regression equation to pass through zero by putting constant term = 0 in the regression analysis. If we do this, the output is as follows: Analysis of variance

	Sum of squares	DF	Mean squares
Regression	916,492.1	2	458,246.1
Residual	4390.0	8	548.7
Total	920,882.1	10	
$F = 835.08$		Significance $F = 4.7 \times 10^{-9}$

Variables in the equation

Variable	B	SE B	T	Sig T
Throughput	0.3419	0.1006	3.3976	0.009393
Raw water consumption	0.0848	0.0167	5.07056	0.000963

$R^2 = 0.9952$

In this case, significance level of the coefficients of both the independent variables is very low. This means that the coefficients are highly significant. In this case 99.52 % of the variability in the wastewater treatment demand is explained by the two independent variables, throughput and raw water consumption.

Problems

A few problems are set hereafter. These are taken from practical studies and are part of the computer outputs.

12.1 In order to predict the employment of developed countries, a multiple regression based on several country observations for a few years produced the following results.

     Multiple R 0.96233
     R^2 0.92609
     Standard error 0.04498

Analysis of variance

	Sum of squares	DF	Mean squares
Regression	1.77480	3	0.59160
Residual	0.14165	70	0.00202
Total	1.91645	73	
$F = 292.35604$		Significance $F = 0.0000$	

Variables in the equation

Variable	B	SE B	T	Sig T
Industry share of GNP	1.911553	0.11120	17.191	0.0000
Service share of GNP	0.982751	0.069089	15.760	0.0000
Manufacturing of GNP	−0.251779	0.069089	−3.644	0.0000
(Constant)	−0.701170	0.049748	−14.093	0.0000

Dependent variable is employment measured as percentage of population. Share is in percentage.

(a) Write down the regression model.
(b) Write down the interpretations.
(c) Using the model, predict the employment in Korea, given

   Industry share of GNP = 11.4 %
   Service share of GNP = 45.7 %
   Manufacturing share of GNP = 45.7 %

(d) If the industry share of GNP is increased by one percent, what change will take place in employment?
(e) If the service share of GNP is decreased by one percent, what change will take place in employment?
(f) If the manufacturing share of GNP is increased by one percent, what change will take place in employment?

12.2 In study of employment, a multiple regression based on several country observations spread over a few years yielded the following results.

Multiple R 0.99673
R^2 0.99347
Standard error   10.35127

Analysis of variance

	Sum of squares	DF	Mean squares
Regression	2,201,321.431	2	1,100,660.716
Residual	14,465.089	135	107.149
Total	2,215,786.52	137	
$F = 10272.263$		Significance $F = 0.0000$	

Variables in the equation

Variable	B	SE B	T	Sig T
Population (million)	0.472473	0.00332	142.119	0.0000
GNP/cap ($)	0.001192	0.00111	1.077	0.2832
(Constant)	−6.429729	1.067992	−6.020	0.0000

Dependent variable is employment (million).

(a) Write down the sample regression equation.

(b) Estimate the employment in your country using this model.

(c) What will be the change in employment if population of a country increases by 0.45 million and GNP/cap increases by $20?

12.3 A point of interest of an urban planner was to study the remittances sent by slum dwellers in cities to their village homes. Using least square multiple regression and based on several households of a number of slums in a city he developed a "remittance model." The following is a computer output.

Multiple R 0.58028
R^2 0.33672
Standard error   306.50434

Analysis of variance

	Sum of squares	DF	Mean squares
Regression	24.04 (exp 6)	7	3.43 (exp 6)
Residual	47.35 (exp 6)	504	0.09 (exp 6)
Total	71.39 (exp 6)	511	
$F = 36.55199$		Significance $F = 0.0000$	

Variables in the equation

Variable	B	SE B	T	Sig T
HH income (Taka)	0.060377	0.00838	7.203	0.0000
D_1 (dummy)	286.53479	32.87474	8.716	0.0000
HH member (no)	5.983047	7.38530	0.810	0.4182
Length of stay (years)	−4.619365	2.45172	−1.884	0.0601
School going children (no)	8.382230	20.73630	0.404	0.6862
D_2 (dummy)	95.891214	32.32095	2.967	0.0032
Housing consumption (Tk)	−0.221813	0.08963	−2.475	0.0137
(Constant)	−44.337606	49.40786	−0.897	0.3699

Dependent variable is the amount of money (Tk) sending to village home per month. Tk 33.00 = US $1.00. D_1 refers to family living at village. D_2 refers to availability of property at village.

(a) Write down the sample regression equation.
(b) Interpret the coefficients.
(c) Estimate the remittance to be sent by a slum dweller if

    (i) his HH income is Tk 2000,
    (ii) he has four persons in his family,
    (iii) he has been staying in the city for 9 years,
    (iv) he has three school going children,
    (v) his housing consumption is Tk 500,
    (vi) his family lives in the village, and
    (vii) he has no property in the village.

12.4 The following is the partial computer output of a multiple regression analysis:

Multiple R   0.9460
R^2 0.8949

Analysis of variance

	Sum of squares	DF	Mean squares
Regression	3085.78	2	1542.89
Residual	364.22	7	52.03
Total	3450.00	9	
$F = 36.55199$		Significance $F = 0.0000$	

Variables in the equation

Variable	B	SE B	T	Sig T
Price ($)	−7.19	2.55	−2.81	0.0260
Income ($)	0.014	0.01	1.28	0.2400
Constant	111.69	23.50	4.75	0.0021

Dependent variable is quantity demanded.

(a) Write down the sample regression model.
(b) Calculate the F-statistic.
(c) Predict the changes in quantity demanded, if price changes from $10 to $5 and income changes from $1200 to $1500.

Answers

12.1 (a) Employment(% Population) $= -0.701170 + 1.911553$(Industry share of GNP)
$$+ 0.982751(\text{Service share of GNP})$$
$$-0.251779(\text{Manufac./Industry share of GNP})$$

   (b) (i) $R^2 = 0.92609$; this means 92.61 variability of employment (%) is explained by the three cited independent variables.

   (ii) F value is high, significance of F is zero. This shows a very good fitting of the model (at least one of the coefficients is not zero).

   (iii) If industry share of GNP is increased by 1 %, employment (%) will increase by 1.911553 (decrease will cause decrease).

   (iv) If service share of GNP is increased by 1 %, employment (%) will increase by 0.982751 (decrease will cause decrease)

   (v) If manufacturing/industry share of GNP is increased by 1 %, employment (%) will be decrease by 0.251779 (decrease will cause increase)

   (c) 54.50; (d) 1.91 % increase; (e) 0.98 % decrease;
   (f) 0.25 % decrease

12.2 (a) Employment (million) $= -6.429729 + 0.472473$ (Population) + 0.001192 (GNP/cap).
   (c) Increase by 0.236 million
12.3 (c) Tk 259.55/month
12.4 (a) Quantity demanded $= 111.69 - 7.19$(price) + 0.014(income)
   (b) 29.65; (c) Increase by 40

Reference

Wonnacott, T.H., Wonnacott, R.J.: Introductory Statistics for Business and Economics. Wiley, Singapore (1990)

Chapter 13
Sampling Theory

Abstract For population inferences, a complete count of the population would be desirable. Time and resource usually do not permit this. So we go for sampling. For population inferences to be valid, sampling is to be done complying with the theoretical requirements. The theories are explained. Advantages of sampling and prior survey considerations are explained. Principal steps involved in the choice of sample size are mentioned. Commonly used sampling methods dealing with simple random sampling, systematic sampling, cluster sampling, and stratified random sampling are explained.

Keywords Sampling · Advantages · Types of sampling · Random · Systematic · Cluster · Stratified sampling

If we want to study a population, how can we do? This could be done by a complete enumeration or census of the aggregate. But this is not always feasible. Time and resource may not permit us to have the complete enumeration. So we go for sampling. We select a sample of size much smaller than the size of the population. We obtain data from the selected sample. We analyze these data and make inferences about the population. However, sampling has certain advantages.

13.1 Advantages of Sampling

One of the advantages of sampling is the reduction of cost. A complete enumeration would obviously necessitate a large amount of money. This can be reduced greatly by use of sampling. For designing and executing adequately the survey, relatively better resources are employed. Thus, cost per unit of observation is higher than that necessary for a complete count. But the total cost is much less in sampling than that in a complete count.

Another advantage of sampling is greater speed. In a sample survey, the number of observations is smaller than that in a complete count. So the data can be collected and processed quickly. This is a vital point in case the results are urgently required.

© Springer Science+Business Media Singapore 2016

A.Q. Miah, *Applied Statistics for Social and Management Sciences*,
DOI 10.1007/978-981-10-0401-8_13

Sampling allows collection of comprehensive data. Since the size of the sample is small, it may be investigated thoroughly. Also, sampling is flexible and adaptable. Thus, a variety of information may be gathered. In certain situations, specialized equipments and highly trained personnel need to be employed. In such situations a complete count becomes impossible. Either sampling is to be used or the idea of data collection is to be abandoned. In this sense, sampling has greater scope.

Sampling has an advantage over the complete count in terms of accuracy. In sampling qualified personnel can be employed and an intensive training can be imparted to them. Also, field work and data processing can be supervised and quality controlled. This can lead to greater accuracy of the results.

There are certain situations when sampling is the only recourse of data collection. In testing the crushing strength of bricks, we cannot crush all the bricks. In studying the harmful effects of insecticides in agricultural fields, we cannot take all the insecticides to the laboratory for testing.

Last but not the least advantage of sampling is that we can measure the reliability of the sample results.

13.2 Considerations Prior to Sample Survey

Surveys vary greatly depending on the type, context, and complexity. It is not possible to design a single format for all types of surveys. But there are certain considerations which help in designing any survey. These are summarized in the following paragraphs. A considerable coverage may be seen from Cochran's work (1984: 4–8).

A clear statement of the objectives is necessary. Otherwise, there may be a shortcoming in the data collection. Some vital data may be forgotten to collect and some collected data may be superfluous. Furthermore, it may deviate from the main path of the research in question.

The next point of consideration is target and sample population. For convenience, sampled population is studied and on the basis of this inferences are made regarding the target population. So the sampled population ought to coincide with the target population. Otherwise, inferences made from the sampled population may not be fruitfully relevant to the target population.

A coordination schema is of utmost importance. In its absence, some essential data may be omitted and some superfluous data may be collected. This can only be detected while analyzing the data and drawing conclusions in conformity with the objectives. Often there is a tendency to use a lengthy questionnaire incorporating unimportant questions. This lowers the quality of responses.

Based on the analysis of data collected from a sample, inferences are made regarding the population. Since the data are collected from a part of the population, there is a certain amount of error in the inferences. This refers to precision. To have a greater precision, a larger sample size would be needed, which in turn involves higher cost. Same degree of precision may also not be necessary in all research works. Thus, the degree of precision desired in a particular research work ought to be clearly spelled out.

Data when measured are normally either in qualitative or quantitative form. Quantitative data are easy to analyze, interpret, and are subject to more sophisticated tests. Therefore, quantitative data, as far as possible, should be collected. It should further be kept in mind that quantitative data can be converted to qualitative data later. But the reverse is not true. When measuring data, this point should be kept in mind.

Sampling unit is a very element in sample survey. Many-a-times major difficulties arise for confusion in units. If we are interested in studying the housing conditions of slum dwellers in a city, the unit may be a household. But if we want to study the employment opportunity based on level of education, the unit may be an individual. If we want to study agricultural crops, the unit may be a field, a farm or an area of known quantity such as an acre, a hectare. When the units are resolved, the frame will have to be prepared. The frame will consist of a list of all the units of the population under study. It should be a complete list and will not have an overlapping.

For the results to be useful to explain the population characteristics, an adequate sample size followed by a scientific method of sample selection is to be used. More detailed discussions will be made of sample selection later.

Often data are collected using standardized questionnaires. In constructing the questionnaire, it is not always possible to identify the field conditions. A pretesting is, therefore, useful. Pretesting almost in all cases results in improvement in the questionnaire.

Usually, the volume of the research is such that a number of enumerators or field assistants need to be employed. They must be trained for the methods of measurements to be employed. For quality control, an adequate supervision of the fieldwork should be ensured. While the field work continues, editing of the completed questionnaires is very useful. In this way errors can be detected and rectified. Also the erroneous errors may be deleted. For this purpose, a detailed plan of the fieldwork is to be made.

13.3 Considerations in Sampling

In course of planning a sample survey, we will reach at a stage when we must decide the size of the sample. This decision is important. The larger the sample size, the better is the precision and better is the result. But if a smaller size of sample serves the purpose in some situations, then taking a larger sample size will mean wastage of resources. Again, if the sample is too small, it diminishes the utility of the results. The decision, however, cannot be always made satisfactorily. Often we will not have enough information to be sure that the sample size we have selected is the best one.

Among other things, there are two major considerations which we must think of in determining the size of a sample. One is the precision. If the precision we desire is higher (smaller error) the sample size ought to be higher and vice versa. The other point of consideration is the cost. If the sample size is large, the cost involved is high. So if we want to have a higher degree of precision, a larger sample size becomes necessary which in turn requires larger amount of budget. So we cannot always go for larger sample size in view of budget constraint. If we go for a smaller sample size, we can manage with less cost. But in this case we are sacrificing higher degree of precision. Therefore, we need to make a balance between the degree of precision and cost in determining the sample size in any particular situation. The balance will aim at maximum precision obtainable with minimum cost.

The sampling cost needs some discussion. The cost of sampling includes the cost of designing the sample, constructing the frame, training the interviewers, collecting, compiling, and calculating the data, office expenses, other overhead expenses, etc. However, these costs can be divided into two categories, namely, fixed cost and variable cost. Thus, the cost function can be represented in a simple form

$$C = c_0 + c_1 * n$$

where C = total sampling cost, c_0 = fixed cost, c_1 = cost per unit of sample, and n = sample size. C_0 is the fixed cost and include office rent, fixed administrative costs, equipment costs, etc. This fixed cost does not depend on the size of the sample survey. The other part $c * n$ is the variable cost. If the sample size is large, $c * n$ will be large, but c_0 remains constant. It must be remembered that the cost function shown above is the simplest one. Other more complicated cost functions can be constructed depending on a particular situation. But it ought to be taken into consideration that the cost function should be an accurate and useable one.

13.4 Principal Steps Involved in the Choice of a Sample Size

Two major considerations in determining the sample size have been discussed in the previous subsection. Determining the sample size is really a big problem. The following principal steps will help us in the choice of a sample size.

(i) There must be a statement concerning what is expected of the sample. This statement may be in terms of the desired precision, or in terms of decision that is to be taken or action that is to be taken when sample results are known.

(ii) Some equation connecting n (sample size) with the desired precision of the sample must be found. The equation will vary with the content of the statement of precision and with the kind of sampling that is contemplated.

(iii) This equation will contain, as parameters, certain unknown properties of the population. These must be estimated in order to give specific results.

(iv) It may so happen that data are to be analyzed for certain major subdivisions of the population and the desired degree of precision may be different for each subdivision. In such cases, separate calculations are to be made for each of the subdivisions. The total sample size n is to be found out by addition.

(v) Usually, more than one item or characteristics is measured in a sample survey. Sometimes the number of items is large. If a desired degree of precision is prescribed for each item, the calculations may lead to a series of conflicting values of n, one for each item. These values must be reconciled by some methods.

(vi) The last step is the appraisal of the chosen value of n in order to see that it is consistent with the resources available to take the sample. This necessitates an estimation of the cost, labor, time, and materials needed to obtain the proposed size of the sample. It sometimes becomes apparent that n will have to be drastically reduced. A hard decision must be faced—whether to proceed with a much smaller sample size, thus reducing the precision, or to abandon the efforts until more resources become available.

13.5   Types of Commonly Used Sampling Methods

There are several considerations and specifications which lead to different classification of sampling. A detailed discussion on the classification and types of sampling is beyond the scope of this book. Here we shall limit our discussion on simple random sampling, systematic sampling, stratified sampling, and cluster sampling. These are the sampling techniques that we will most often use in our research work.

13.5.1   Simple Random Sampling

Simple random sampling is the most commonly used sampling method. In this method every unit of the population has the equal chance of being selected in the sample. There may be two ways of drawing the units, namely sampling with replacement and sampling without replacement. In the former case, the unit already drawn, has the equal chance, like other units, of being drawn again. In the latter case, the unit once drawn, no more qualifies for subsequent draws.

If there are N units in the population and a sample of n is to be drawn, then a particular unit has the probability of $1/N$ to be drawn. There may be $_NC_n$ number of possible samples. So the probability that a particular sample of size n will be drawn is $1/(_NC_n)$.

Drawing the selected units from the population may be accomplished in a systematic way. First, all the units of the population are numbered from 1 to N.

Then using either random numbers from random digits table or a suitable computer program, the actual numbers are drawn, unit by unit until the sample size of *n* is reached.

Let us now see how this technique may be applied in the selection of sampled households in one of the study areas, Kathal Bagan in Dhaka. There are 351 households. A sample size of 35 households is to be selected using random numbers.

No official list is available for the households of the slum in question. So one has to go along the lanes, by-lanes, and the passages inside the slum and identify the households. Then use the random table. In this case 351 is a three-digit figure. So we can use consecutive three columns of the random table. Either columns 1–3 or 2–4 from each group may be used. Let us use columns 1–3. Starting from the first group (column 1–4) we go down and pick up the following numbers (Table 13.1).

Table 13.1 Picked up random numbers

195	862	764	937	164	601
711	355	886	484	823	384
669	387	054	638	465	497
862	757	051	903	241	351
629	384	574	064	422	103
324	329	317	461	111	380
233	900	270	945	146	582
577	879	978	714	842	091
054	709	883	843	453	563
419	931	605	388	612	239
228	131	659	629	767	849
851	680	705	348	763	611
466	804	878	398	883	270
205	566	889	927	209	155
031	787	410	255	979	975
118	324	530	856	820	166
197	685	473	646	842	099
679	791	912	660	935	454
684	965	593	717	223	150
298	927	917	545	906	711
560	229	999	079	870	986
472	609	516	045	555	630
003	235	515	121	327	924
027	492	861	078	940	675
939	468	909	581	017	647

Note Started from first thousand, went down through second thousand, third thousand and until end of fourth thousand; then again started from first thousand and went down until end of second thousand to select tentatively 150 numbers

Table 13.2 Chosen random numbers

195	197	235	045	223	155
324	298	054	121	327	166
233	003	051	078	017	099
054	027	317	164	351	150
228	329	270	241	103	
205	131	064	111	091	
031	324	255	146	239	
118	229	079	209	270	

In our list we have no number above 351. So dropping the numbers above 351 the following numbers are picked up (Table 13.2).

We want to do the sampling without replacement. In the chosen numbers 054, 324, and 270 occur twice. Keep these numbers once only. Then order first 35 numbers. Ordering is necessary for convenience of interviewing, because first interviewing household number 195, then going to 324 and coming back to 233, and so forth causes inconvenience and wastage of time. When ordered the numbers are as follows (Table 13.3).

Keep the rest six numbers, i.e., 091, 239, 155, 166, 099, and 150 (without ordering) as reserved numbers to be used in case any household among the selected 35 is not, for some reason or other, available for interviewing. If one household from the chosen 35 is not available for interviewing select 091st household. In case another household is not available, select 239th household and so on.

It may be noted that in selecting 35 households (plus about 20 %), we had to use 150 random numbers from the table, resulting in a loss of about 3/4 of the random numbers. There are several ways/techniques of avoiding this wastage. For more details on this account, please, see, for example, Yamane's work (1967: 64–68).

One technique is demonstrated here. In our example we have 351 the highest number. We divide the chosen random numbers by 351 and use the reminders as the selected random numbers. This is illustrated in Table 13.4.

From the above list select the first 35 numbers. When these 35 numbers are ordered they stand as shown in Table 13.5.

Seven more numbers (about 20 %), i.e., 102, 215, 085, 334, 089, 263, and 225 may be kept as reserved. These should, however, not be ordered. This technique has obviously made efficient utilization of random numbers.

Table 13.3 Ordered chosen random numbers

003	054	118	197	233	317
017	064	121	205	235	324
027	078	131	209	241	327
031	079	146	223	255	329
045	103	164	228	270	351
051	111	195	229	298	

Table 13.4 Calculations for chosen random numbers

Random number	Calculation	Selected number	Random number	Calculation	Selected number
195	0 * 351 + 195	195	862	2 * 351 + 160	160x
711	2 * 351 + 009	009	355	1 * 351 + 004	004
669	1 * 351 + 318	318	387	1 * 351 + 036	036
862	2 * 351 + 160	160	757	2 * 351 + 055	055
629	1 * 351 + 278	278	384	1 * 351 + 033	033
324	0 * 351 + 324	324	329	0 * 351 + 329	329
233	0 * 351 + 233	233	900	2 * 351 + 198	198
577	1 * 351 + 226	226	879	2 * 351 + 177	177
054	0 * 351 + 054	054	709	2 * 351 + 007	007
419	1 * 351 + 068	068	931	2 * 351 + 229	229
228	0 * 351 + 228	228	131	0 * 351 + 131	131
851	2 * 351 + 149	149	680	1 * 351 + 329	329x
466	1 * 351 + 115	115	804	2 * 351 + 102	102
205	0 * 351 + 205	205	566	1 * 351 + 215	215
031	0 * 351 + 031	031	787	2 * 351 + 085	085
118	0 * 351 + 118	118	324	0 * 351 + 324	324x
197	0 * 351 + 197	197	685	1 * 351 + 334	334
679	1 * 351 + 328	328	791	2 * 351 + 089	089
684	1 * 351 + 333	333	965	2 * 351 + 263	263
298	0 * 351 + 298	298	927	2 * 351 + 225	225
560	1 * 351 + 209	209	229	0 * 351 + 229	229x
472	1 * 351 + 121	121	609	1 * 351 + 258	258
003	0 * 351 + 003	003	235	0 * 351 + 235	235
027	0 * 351 + 027	027	492	1 * 351 + 141	141
939	2 * 351 + 237	237	468	1 * 351 + 117	117

Note x skip since it occurred earlier

Table 13.5 Selected random numbers

003	033	115	177	226	298
004	036	118	195	228	318
007	054	121	197	229	324
009	055	131	198	233	328
027	068	149	205	237	329
031	089	160	209	209	

It may be noted here that numbers of households selected in this sample are different from those selected previously. This does not pose a problem at all. Number of all possible samples from 351 households taken 35 at a time is given by

$$_{351}C_{35} = \frac{351!}{35!(351-35)!}$$

$$= \frac{351!}{35!316!}$$

$$= 2.04 * 10^{48} \text{(approximately)}$$

The selected sample is one of those possible samples.

Nowadays the random numbers can be generated easily by the computer software.

13.5.2 Systematic Sampling

Another sampling technique is systematic sampling. When saving of time and efforts become important, systematic sampling technique is used. In certain situations it is more efficient than simple random sampling. It is easy in drawing a sample. If the population is spread more evenly then systematic sampling provides more representation of the population. However, one disadvantage of this technique is that the variance of estimators cannot be obtained from a single sample. Furthermore, if the population units are poorly arranged, this technique will produce a very inefficient sample.

If a sample size of n is to be drawn from a population N, then a sample fraction k is calculated ($k = n/N$). Suppose the sampling fraction is 0.04. This 4 % sampling fraction means one sample unit is to be chosen from every 25 units of the population. Then using simple random numbers, a number is chosen between 1 and 25. Suppose this number is 12. This will be the first unit in the sample. Subsequent units are chosen systematically adding 25 to each of the selected units. Thus, the selected units will be 12, 37, 62, 87 … etc. until the last unit chosen completes the sample size of n.

This sampling technique can be demonstrated by using a practical example. In Babupura, one of the study areas, total owner population was 161 HHs out of which 16 HHs were to be selected. Here the sampling fraction is 16/161 = 0.099. Let it be taken as 0.10 (10 %). This means one unit of sample is to be chosen from every 10 population units. Here 10 is a two-digit figure. Using last two columns of group 3, i.e., third thousand (we can chose any two columns from any group) of random numbers (Table A.13), the first 10 numbers are 77, 37, 67, 99, 33, 25, 94, 55, 82, 01. So 01 will be selected since it occurs first between 1 and 10. Therefore, the first chosen sample unit is 1st HH. The second sample unit will be 1 + 10, i.e., 11th HH. In this way the 16 sample HHs will be 1, 11, 21, 31, 41, 51, 61, 71, 81, 91, 101, 111, 121, 131, 141, and 151.

13.5.3   Cluster Sampling

In cluster sampling the population is divided into some clusters or groups. It is advantageous if the recording units are available in some suitable clusters. Out of several clusters, a few are selected using simple random sampling. From the selected clusters, individual recording units are again selected using simple random sampling.

Suppose, there are n clusters $(M_1, M_2, M_3, \ldots M_n)$ in a population. From these, m clusters are selected using simple random sampling. From each of the selected m clusters individual sample units n_i $(i = 1, 2, 3, 4, \ldots)$ are again selected using simple random technique.

The use of cluster sampling can be demonstrated with the help of an example. Suppose, there are nine slums under study (slum$_1$, slum$_2$, slum$_3$, ... slum$_9$). Three slums are to be selected. Using the random table suppose, the chosen three random numbers are 2, 5, and 8. Populations of these three slums are 161, 1131, and 252 respectively. It is decided that subsample sizes of 16, 113, and 25 are to be chosen from these slums respectively. This can be accomplished by using simple random technique. The entire process is illustrated schematically in Chart 13.1. The calculations are as follows:

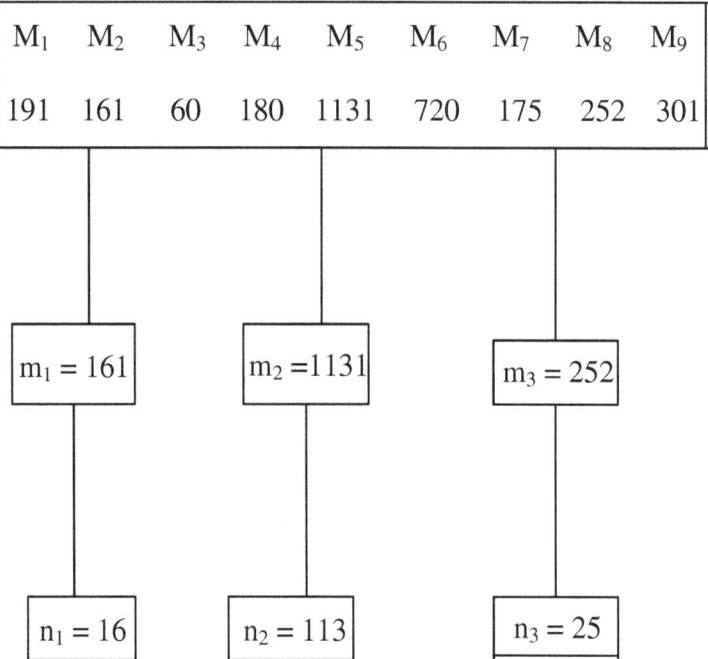

Chart 13.1 Example of cluster sampling

Population size

$$N = 161 + 119 + 60 + 180 + 1131 + 720 + 75 + 252 + 301$$
$$= 3099 \text{ HHs.}$$

Sample size

$$n = 16 + 113 + 25$$
$$= 154$$

$$\text{Sampling fraction} = 154/3099$$
$$= 0.0496$$
$$= 4.96\,\%$$

13.5.4 Stratified Random Sampling

In this sampling technique the population is divided into a number of nonoverlapping subpopulations based on certain criteria. Each subpopulation is known as stratum. From each of these strata, subsamples are chosen by simple random sampling. The master sample size is the sum of all subsamples drawn from all strata. This process is known as stratified random sampling.

Stratified sampling differs from cluster sampling. In cluster sampling a few clusters are chosen from several clusters and subsamples are drawn from these chosen clusters only. In stratified sampling subsamples are drawn from all the strata.

Stratified random sampling has some advantages. If the population can be stratified into homogeneous strata, increased precision, as compared to that in simple random sampling, can be obtained. Sometimes, it may be desirable to collect information concerning individual strata and make explicit the difference between the strata. In this sampling it may also be easier to collect information for either physical or administrative reasons.

Suppose, a population is divided into L number of strata ($L_1, L_2, L_3, \ldots$) with respective subpopulations of $N_1, N_2, N_3, \ldots$ Subsamples are drawn (using simple random technique) from each of these subpopulations. Let these subsamples be n_1, $n_2, n_3, \ldots$ The master sample size is obtained by $n = n_1 + n_2 + n_3 + \ldots$ The population size is given by $N = N_1 + N_2 + N_3 + N_4 \ldots$ Thus, the sample fraction is n/N.

The use of stratified sampling can be illustrated with the help of an example. In Wahab Colony, one large slum in Dhaka, it was desired to study the difference in various attributes between house owners and house renters who appeared to be two distinct groups in the said slum. The slum having 1732 households, was divided into two strata, namely house owners (720 households) and house renters (1012 households). Thereafter, using simple random sampling, 72 households were chosen from house owners and 101 households were chosen from 1012 renter households. Since simple random sampling was applied in each of the two strata, the technique here employed was a stratified random sampling. The sampling fractions in the house owners, house renters, and overall were 0.10, 0.0998, and 0.0998 respectively.

Chapter 14
Determination of Sample Size

Abstract Basic theoretical basis for determining required sample sizes using various theories are set. Examples are used in each case.

Keywords Basic principle · Sampling · Continuous data · Proportion · Allocation of subsamples

The larger the sample size, the better is the estimation. But always larger sample sizes cannot be used in view of time and budget constraints. Therefore, the practical way of calculating the sample size is to do it based on certain precision and reliability. There are several situations and in each case a different technique is employed to calculate the sample sizes. These are briefly discussed hereafter.

14.1 Basic Principle

Confidence interval, for example, for mean is given by

$$\bar{x} - z_{\alpha/2} * \frac{\sigma}{\sqrt{n}} \leq \mu \leq \bar{x} + z_{\alpha/2} * \frac{\sigma}{\sqrt{n}}$$

Thus, the length of confidence interval is $2 * z_{\alpha/2} * \sigma/\sqrt{n}$. The precision is 1/2 of the length of the confidence interval, i.e., $z_{\alpha/2} * \sigma/\sqrt{n}$.

$$\text{Precision} = z_{\alpha/2} * \sigma/\sqrt{n}$$

or,

$$\text{Precision} = (\text{reliability}) * (\text{standard error})$$

This is the basic relationship between precision, reliability, and standard error.

© Springer Science+Business Media Singapore 2016
A.Q. Miah, *Applied Statistics for Social and Management Sciences*,
DOI 10.1007/978-981-10-0401-8_14

14.2 Sample Size in the Case of Random Sampling (Continuous Data)

We have,

$$\text{Precision} = (\text{reliability}) * (\text{standard error})$$

or,

$$d = z_{\alpha/2} * \sigma/\sqrt{n}$$

If σ is not known, it may be estimated by sample standard deviation s. Therefore, $d = z_{\alpha/2} * s/\sqrt{n}$

When sampling is without replacement, the fpc is necessary and the relationship becomes

$$d = z_{\alpha/2} * \frac{s}{\sqrt{n}} * \sqrt{\left\{\frac{N-n}{N}\right\}}$$

Solving for n you get

$$n = \frac{N(zs)^2}{Nd^2 + (zs)^2}$$

Example 1

A sample of families is to be selected without replacement to estimate the average weekly expenditures on food. The precision is to be within +$4 with 95 % confidence level. It is known from past surveys that $s = \$12$. The total number of families is 2000. How large the sample must be?

$$n = \frac{N(zs)^2}{Nd^2 + (zs)^2}$$

here, $N = 2000$; $s = 12$, $z = 1.96$, $d = 4$

$$
\begin{aligned}
n &= \frac{2000(1.96 \times 12)^2}{2000(4)^2 + (1.96 \times 12)^2} \\
&= \frac{1,106,380.8}{32,000 + 553.19} \\
&= \frac{1,106,380.8}{32,553.19} \\
&= 34
\end{aligned}
$$

Sometimes the precision is given in relative term or for simplicity it may be desired to express the precision in terms of percentage of the mean. In that case the formula is to be modified.

Let d_0 be the desired precision in relative term and d be the precision in absolute term. We have the coefficient of variation $c = s/\bar{x}$. Now,

$$c = \frac{s}{\bar{x}}$$

Therefore, $s = \bar{x} * c$.

Again,

$$d_0(\text{precision in fraction}) = \frac{d}{\bar{x}}$$

Therefore, $d = \bar{x} * d_0$

Now,

$$
\begin{aligned}
n &= \frac{N(Z * s)^2}{N(d)^2 + (Zs)^2} \\
&= \frac{NZ^2(s)^2}{Nd^2 + (Z)^2(s)^2} \\
&= \frac{NZ^2(\bar{x} * c)^2}{N(d)^2 + (Z)^2(\bar{x} * c)^2} \\
&= \frac{NZ^2(\bar{x} * c)^2}{N(\bar{x} * d_0)^2 + (Z)^2(\bar{x} * c)^2} \\
&= \frac{NZ^2(\bar{x})^2(c)^2}{N(\bar{x})^2 d_0 + (Z)^2(\bar{x})^2(c)^2} \\
&= \frac{(\bar{x})^2(NZ^2 c)^2}{(\bar{x})^2 N d_0^2 + (\bar{x})^2(Zc)^2} \\
&= \frac{(\bar{x})^2\left\{N(Z * c)^2\right\}}{(\bar{x})^2\left\{N d_0^2 + (Zc)^2\right\}} \\
&= \frac{N(Z * c)^2}{N d_0^2 + (Zc)^2}
\end{aligned}
$$

Example 2

The quantity of cold drink consumed by students in AIT in a day is to be estimated. The error is to be within +5 % of the mean. A preliminary survey showed $s = 0.3$ L and $\bar{x} = 1.20\,L$. The total number of students in AIT is 2000. The estimate is to be made with 95 % confidence level. How large the sample size is to be?

Here,

N 2000

d 0.05

c 0.3/1.20 = 0.25

z 1.96 (for 95 % confidence level)

$$n = \frac{2000(1.96 \times 0.25)^2}{2000(0.05)^2 + (1.96 \times 0.25)^2}$$
$$= \frac{480.2}{5 + 0.24}$$
$$= \frac{480.2}{5.24}$$
$$= 91.6$$
$$= 92$$

14.3   Sample Size in Case of Simple Random Sampling (Proportion)

In case of proportion (sampling without replacement)

$$\text{Standard error} = \sqrt{\left\{ \frac{pq}{n-1} * \frac{N-n}{N} \right\}};$$

For practical purposes $n - 1$ may be replaced by n. Therefore,

$$\text{Standard error} = \sqrt{\left\{ \frac{pq}{n} * \frac{N-n}{N} \right\}}$$

From the basic relationship

$$\text{Precision} = (\text{reliability}) * (\text{standard error})$$

we get,

$$d = z * \sqrt{\left\{ \frac{pq}{n} * \frac{N-n}{N} \right\}}$$

Therefore,

$$d^2 = z^2 * \frac{pq}{n} * \frac{N-n}{N}$$

Solving for n we can get,

$$n = \frac{Nz^2pq}{Nd^2 + z^2pq}$$

Note that maximum variance (standard error) occurs when $p = 0.5$.

In this case $q = 1 - 0.5 = 0.5$.

Therefore, $pq = 0.5 \times 0.5 = 0.25$

Example 3

You are interested in estimating the proportion of AIT students who smoke. A preliminary random survey showed that out of 25 students 15 smoke.

(a) How large the sample must be if you want to estimate with 95 % confidence level and 5 % precision?
(b) If the preliminary survey is not available, how can you determine the sample size? Currently, there are 820 students in AIT.

(a) here, $p = 15/25 = 0.6$;
so $q = 1 - 0.6 = 0.4; z = 1.96$
We have,

$$n = \frac{Nz^2pq}{Nd^2 + z^2pq}$$

$$= \frac{820(1.96)^2(0.6 \times 0.4)}{820(0.05)^2 + (1.96)^2(0.6 \times 0.4)}$$

$$= \frac{756.03}{2.05 + 0.922}$$

$$= \frac{756.03}{2.972}$$

$$= 255$$

(b) If the preliminary survey is not available, the sample size can be determined using maximum variance condition.
Here, $p = 0.5$, $q = 0.5$

$$n = \frac{Nz^2pq}{Nd^2 + z^2pq}$$
$$= \frac{820(1.96)^2(0.5 \times 0.5)}{820(0.05)^2 + (1.96)^2(0.5 \times 0.5)}$$
$$= \frac{787.53}{2.05 + 0.96}$$
$$= \frac{787.53}{3.012}$$
$$= 262$$

14.4 Sample Size in the Case of Stratified Sampling (Means)

Several cases of determining the sample sizes in case of stratified sampling procedure are outlined in brief in the following sections.

14.4.1 Allocation of Equal Subsamples to All Strata

There are four ways of allocating subsamples to different strata, namely, equal subsamples, proportional, optimum, and Neyman allocations. The simple one is to allocate equal subsamples to each stratum. In proportional allocation, subsamples in each stratum are allocated according to proportion of the respective subpopulations. In optimum allocation, the varying sampling costs are taken into account. This implies that more units are to be chosen from the stratum where cost is less. Neyman allocation is applicable where sampling costs in different strata do not vary greatly. It is assumed that the sampling costs are equal in all strata.

Based on these principles, the methods of determining the sample sizes and allocating the subsamples to different strata are outlined hereafter.

The basic relationship is

$$\text{Precision} = (\text{reliability}) * (\text{standard error})$$

Therefore,

$$\text{standard error} = \frac{\text{precision}}{\text{reliability}}$$

or,

$$(\text{standard error})^2 = \frac{(\text{precision})^2}{(\text{reliability})^2}$$

or,

$$\text{variance} = \frac{d^2}{z^2}$$

This variance is desirable variance and may be denoted by D^2.
Therefore,

$$D^2 = \frac{d^2}{z^2}$$

value of d is in absolute figure.
   But for stratified sampling

$$\text{variance} = \frac{L}{N^2} \sum \frac{N_h^2 S_h^2}{n} - \frac{1}{N^2} \sum N_h S_h^2$$

where,
N_h population size of h stratum,
S_h^2 population variance of h stratum, (if not known, use sample s^2)
N total population size,
n sample size,
L no. of strata.

   Therefore,

$$D^2 = \frac{L}{N^2} \sum \frac{N_h^2 S_h^2}{n} - \frac{1}{N^2} \sum N_h S_h^2$$

Solving for n you can get,

$$n = \frac{L \sum N_h^2 S_h^2}{N^2 D^2 + \sum N_h S_h^2}$$

14.4.2   Proportional Allocation

Based on the same principle, it can be shown that

$$n = \frac{L \sum N_h S_h^2}{N^2 D^2 + \sum N_h S_h^2}$$

14.4.3   Optimum Allocation

Deduction according to the principle used above, the sample size comes out to be

$$n = \frac{\left(\sum N_h S_h \sqrt{C_h}\right)\left(\sum N_h S_h / \sqrt{C_h}\right)}{N^2 D^2 + \sum N_h S_h^2}$$

Neyman allocation
In this case the sample size works out to be

$$n = \frac{\left(\sum N_h S_h\right)^2}{N^2 D^2 + \sum N_h S_h^2}$$

Example 4
We want to study the average number of visitors per day in a city shops. There are 2000 shops in the locality which can be stratified into three categories namely, small (1200), medium (600), and large (200). Their standard deviations are 25, 35 and 55, respectively, known from past records. The costs of selecting sampling units are $c_1 = \$1$, $c_2 = \$2$ and $c_3 = \$3$ respectively.

(a) Calculate the sample sizes using various methods of stratified sampling. Use 95 % confidence level and a precision of ±3 customers. Calculations are summarized in the following tables:

Calculations for determination of sample size

Strata	N_h	S_h	S_h^2
1	1200	25	625
2	600	35	1225
3	200	55	3025
Total	2000		

Strata	$N_h * S_h$	$N_h * S_h^2$	$N_h^2 * S_h^2$
1	30,000	750,000	900,000,000
2	21,000	735,000	441,000,000
3	11,000	605,000	121,000,000
Total	62,000	2,090,000	1,462,000,000

Strata	$\sqrt{C_h}$	$N_h * S_h * \sqrt{C_h}$	$N_h * S_h / \sqrt{C_h}$
1	1.000	30,000	30,000
2	1.414	29,694	14,851
3	1.732	19,052	6351
Total		78,746	51,202

95 % confidence level. Therefore, $z = 1.96$

$$D = d/z = 3/1.96 = 1.53$$

Case 1 *Equal size sample*

$$n = \frac{L \sum N_h^2 S_h^2}{N^2 D^2 + \sum N_h S_h^2}$$

$$= \frac{3(1,462,000,000)}{(200)^2 (1.53)^2 + 2,090,000}$$

$$= 382.9$$

$$= 383$$

Sample in each stratum is

$$n_h = n/L = 383/3 = 128$$

Case 2 *Proportional allocation*

$$n = \frac{N \sum N_h S_h^2}{N^2 D^2 + \sum N_h S_h^2}$$

$$= \frac{2000(2,090,000)}{(2000)^2 (1.53)^2 + 2,090,000}$$

$$= 365$$

$$n_h = (N_h/N) * n$$
$$n_1 = (1200/2000) * 365 = 219$$
$$n_2 = (600/2000) * 365 = 110$$
$$n_3 = (200/2000) * 365 = 36$$
$$\text{Total} = \qquad\qquad\qquad 365$$

Case 3 *Optimum allocation*

$$n = \frac{\left(\sum N_h S_h \sqrt{C_h}\right)\left(\sum N_h S_h / \sqrt{C_h}\right)}{N^2 D^2 + \sum N_h S_h^2}$$

$$= \frac{(78{,}746)(51{,}202)}{(2000)^2 (1.53)^2 + 2{,}090{,}000}$$

$$= 352$$

$$n_h = \frac{N_h * S_h / \sqrt{C_h}}{\sum N_h * S_h * / \sqrt{C_h}} * n$$

$$n_1 = (30{,}000/51{,}202) * 229 = 206$$
$$n_2 = (14{,}851/51{,}202) * 229 = 102$$
$$n_3 = (6351/51{,}202) * 229 \quad = \quad 44$$
$$\text{Total} = \qquad\qquad\qquad\qquad\qquad 352$$

Case 4 *Neyman allocation*
Here all C_h are equal.

$$n = \frac{\left(\sum N_h S_h\right)^2}{N^2 D^2 + \sum N_h S_h^2}$$

$$= \frac{(62{,}000)^2}{(2000)^2 (1.53)^2 + 2{,}090{,}000}$$

$$= 336$$

$$n_h = \frac{N_h * S_h}{\sum N_h S_h} * n$$

$$n_1 = (30{,}000/62{,}000) \times 336 = 162$$
$$n_2 = (21{,}000/62{,}000) \times 336 = 114$$
$$n_3 = (11{,}000/62{,}000) \times 336 = \quad 60$$
$$\text{Total} = \qquad\qquad\qquad\qquad\qquad 336$$

Notes: Precision may be expressed

(a) in terms of standard error (D), which equals to d/z; d is in absolute figure.
(b) in terms of absolute figure of d. Then you need to calculate D from $D = d/z$.
(c) in terms of percentage. In this case you need to know the mean. Calculate d in absolute figure from $d = (\bar{x} * \text{percentage})/100$. Then calculate D from $D = d/z$.

14.5   Sample Size in the Case of Stratified Sampling (Proportion)

Sample allocations are summarized first hereafter.

Case 1 *Proportional allocation*

$$n_h = \frac{N_1}{N} * n; \text{ similar for means.}$$

Case 2 *Optimum allocation*

$$n_h = \frac{N_h \sqrt{(P_h * Q_h / \sqrt{C_h})}}{\sum N_h \sqrt{(P_h * Q_h / \sqrt{C_h})}} * n$$

P_h and Q_h refer to stratum population. If unknown, use stratum sample p_h and q_h from preliminary survey.

Case 3 *Neyman allocation*

$$n_h = \frac{N_h \sqrt{(P_h * Q_h)}}{\sum N_h \sqrt{(P_h * Q_h)}} * n$$

Note that in cases of 2 and 3, the formulas are obtained from those for means, replacing S_h by $\sqrt{(P_h * Q_h)}$.

Given the sample allocation procedure, the determination of sample sizes are outlined below.

Case 1 *Proportional allocation*

$$n = \frac{N \sum N_h P_h Q_h}{N^2 D^2 + \sum N_h P_h * Q_h}$$

$D = d/z$; d is in percentage points.

Case 2 *Optimum allocation*

$$n = \frac{\left\{ \left(\sum N_h \sqrt{(C_h P_h * Q_h)} \right) \right\} * \left\{ \left(\sum N_h \sqrt{(P_h * Q_h)} / \sqrt{C_h} \right) \right\}}{N^2 D^2 + \sum N_h P_h * Q_h}$$

Case 3 *Neyman allocation*

$$n = \frac{\left\{N_h \sqrt{(P_h * Q_h)}\right\}^2}{N^2 D^2 + \sum N_h P_h * Q_h}$$

Note that the above three formulas are obtained from those for means replacing S_h by $\sqrt{(P_h * Q_h)}$.

Example 5

We are interested in estimating the proportion of families in 3 slums, remitting money to village homes. The numbers of families in the slums are 4000, 6000, and 10,000. A preliminary survey showed that 12, 15, and 25 % families in those slums remit money. Sampling costs per unit in those slums are $1, $2, and $3 respectively. We need to estimate with ±3 % precision and 95 % confidence level. Find the sample sizes using various methods of stratified sampling.

Calculations are summarized in the following tables:

Calculations for determination of sample size

Slums	Families (N_h)	p_h	$N_h * p_h$
1	4000	0.12	480
2	6000	0.15	900
3	10,000	0.25	2500
Total	20,000		3880
Slums	$N_h * p_h * q_h$	$N_h * \sqrt{(p_h * q_h)}$	$C_h * \sqrt{(C_h p_h q_h)}$
1	422.40	1300	0.32
2	765.50	2142	0.50
3	1875.00	4330	0.75
Total	3062.40	7772	
Slums	$N_h * \sqrt{(C_h p_h q_h)}$	$\sqrt{(p_h * q_h/C_h)}$	$N_h * \sqrt{(p_h * q_h/C_h)}$
1	1300	0.32	1280
2	3030	0.25	1500
3	7500	0.25	2500
Total	11,830	0.82	5280

Case 1 *Proportional allocation*

$$D = 0.03/1.96 = 0.015$$

$$n = \frac{N \sum N_h P_h Q_h}{N^2 D^2 + \sum N_h P_h * Q_h}$$

$$= \frac{20,000 * 3062.40}{(20,000 * 0.015)^2 + 3062.40}$$

$$= 658$$

$$n_h = \frac{N_1}{N} * n$$

$$n_1 = \frac{4000}{20,000} * 658 = 132$$

$$n_2 = \frac{6000}{20,000} * 658 = 197$$

$$n_3 = \frac{10,000}{20,000} * 658 = 329$$

$$\overline{\text{Total} = \qquad\qquad 658}$$

Case 2 *Optimum allocation*

$$n = \frac{\left\{\left(\sum N_h \sqrt{(C_h P_h * Q_h)}\right)\right\}\left\{\left(\sum N_h \sqrt{(P_h * Q_h/C_h)}\right)\right\}}{N^2 D^2 + \sum N_h P_h * Q_h}$$

$$= \frac{11,830 * 5280}{(20,000 * 0.015)^2 + 3062.40}$$

$$= 671$$

$$n_h = \frac{N_h \sqrt{(P_h * Q_h/C_h)}}{\sum N_h \sqrt{(P_h * Q_h/C_h)}} * n$$

$$n_1 = \frac{1280}{5280} * 671 = 163$$

$$n_2 = \frac{1500}{5280} * 671 = 191$$

$$n_3 = \frac{2500}{5280} * 671 = 317$$

$$\overline{\text{Total} = \qquad\qquad 671}$$

Case 3 *Neyman allocation*

$$n = \frac{\left\{\left(\sum N_h \sqrt{(P_h * Q_h)}\right)\right\}^2}{N^2 D^2 + \sum N_h P_h * Q_h}$$

$$= \frac{(7772)^2}{(20,000 * 0.015)^2 + 3062.40}$$

$$= 649$$

$$n_h = \frac{N_h \sqrt{(P_h * Q_h)}}{\sum N_h \sqrt{(P_h * Q_h)}} * n$$

$$n_1 = \frac{1300}{7772} * 649 = 109$$

$$n_2 = \frac{2142}{7772} * 649 = 179$$

$$n_3 = \frac{4330}{7772} * 649 = 361$$

$$\text{Total} = \underline{\hspace{4cm}} \quad 649$$

14.6   Simple Cluster Sampling

The mean size of the secondary sampling units may be calculated using the following relationship, based on optimum allocation principle.

$$n = \sqrt{\left\{\frac{c_1}{c_2} * \frac{s_2^2}{s_1^2}\right\}}$$

where,

n mean size of the secondary sampling units

c_1 per unit cost of selecting primary units (clusters)

c_2 per unit cost of selecting secondary units (units of observations)

s_1^2 within-cluster variance

s_2^2 between-cluster variance.

If the intra-cluster correlation coefficient (ρ) is known, the following formula can be used:

$$n = \sqrt{\left\{\frac{c_1}{c_2} * \frac{1-\rho}{\rho}\right\}}$$

Cost function is given by

$$c = c_0 + c_1 * m + c_2 * mn$$

where,

c total cost

c_0 fixed cost

m no. of clusters

n, c_1 and c_2 as above.

Example 6

In a sample survey the cost of traveling to a slum is $50 and that of interviewing each household (secondary sampling units) in a slum is $5. The fixed cost is $400 and a budget of $2000. The intra-cluster correlation coefficient is estimated to be 0.15. Determine the number of slums and the average number of households to be interviewed.

Here

$$c_1 = \$50, \ c_2 = \$5, \ c_0 = \$400,$$
$$c = \$2000, \ \rho = 0.15$$

$$n = \sqrt{\left\{ \frac{c_1}{c_2} * \frac{1-\rho}{\rho} \right\}}$$

$$= \sqrt{\left(\frac{50}{5} * \frac{1-0.15}{0.15} \right)}$$

$$= 7.5$$
$$\cong 8$$

Select 8 units from each selected slum.
Now,

$$c = c_0 + c_1 * m + c_2 mn$$

or,

$$2000 = 400 + 50 * m + 5 * m * 8$$

or,

$$90m = 2000 - 400$$
$$= 1600$$
$$m = 1600/90$$
$$= 18$$

Select 18 slums from all slums of the city.

Problems

14.1   You are given the following information:

Population size   = 15,000
Desired confidence level = 15,000
Desired precision   = ±3 %
Desired CV   = 55 %

Total budget   = $3000
Fixed cost   = $500
Variable cost   = $2/unit
Calculate the sample size

14.2 Suppose you are interested in estimating the mean household income of the people of your city, by using simple random sampling.
      You want to be 95 % confident and have a precision of ± 4 %. Aim at 35 % as the coefficient of variation.
      How large the sample size is to be? (If the number of households of your city is not readily available, compute it assuming the hh size to be 6).

14.3 You want to study the proportion of farmers who are aware of environmental pollution. The study area has a farmer population of 10,500. Assuming the use of simple random sampling technique, calculate the sample size based on

      (a) expected rate of occurrence,
      (b) maximum variance.
           The precision is to be ±3 %. Assume 5 % level of significance.

14.4 You are given the following:

      Population size   = 15,000
      Desired confidence level   = 95 %
      Desired precision   = ±3 %
      (i) Desired CV   = 55 %
          (for continuous variables)
      (ii) Expected rate of occurrence   = 0.30
          (for bivariate variables)
      (iii) Budget

      Total amount   = $2500
      Fixed cost   = $480
      Variable cost   = $2/unit

      What should be the sample size?

14.5 Us the data in problem 14.4 (i) except population size.

      (a) Calculate the sample sizes for population sizes of 1000, 5000, 10,000, 20,000, and 40,000. Draw a graph.
      (b) Draw a graph of sample size with the population size.
      (c) What is your comment?

14.6 Repeat problem 14.5 using data of 14.4 (ii).
14.7 The manufacturer of an industrial plant wishes to estimate the mean daily yield of a chemical during a year. A preliminary survey showed the mean daily yield to be 871 tons with a standard deviation of 51 tons.

How many days the manufacturer should observe the yield, if the precision is desired to be ±2 %?

14.8   In a preliminary random sample of 100 students selected from 5000, it is found that 40 students own cars.
How large a sample must be selected to have a precision of ±5 % with a confidence level of 95 %?

14.9   In the eve of an election a sample survey shows that 83 out of 155 intend to vote for a particular candidate. What sample size should be taken to convince the candidate that he has a 99 % chance of being elected? The estimate should have precision of (a) ±3 % and (b) ±2 %.

14.10   A manufacturer of refrigerator claims that there is a 99 % chance of the temperature in the refrigerator to be within ±0.1 °C. A customer research group wants to check this. A preliminary survey shows that the mean temperature is 2.5 °C with a standard deviation of 0.25 °C. What should be the sample size to conduct the survey, if the total number of users of the same refrigerator in the city is 2835?

14.11   In a certain population you want to do cluster sampling based on optimum allocation principle. Your total budget is $2000. The overhead cost is $200. The estimated cost per unit of observation is $2.5. The estimated cost of traveling to a cluster and other related work is $30. A preliminary survey shows that the between cluster variance is 12 times of the within cluster variance. How large the sample size should be?

14.12   In a student population of 4000 you want to estimate the proportion of students who smoke. In a preliminary survey of 50 students 30 were found smoking.
How large a sample must be if you want to estimate with ±5 % precision at 99 % confidence level?

14.13   The Student Union of AIT wants to petition the Dean of Students Affairs to increase the bursary on the ground that the cost of living has increased significantly. The Students Union wants to substantiate it by estimating the current food expenditure of students through a simple random sample survey.
The estimate is to have a precision of ±3 and 99 % confidence level. From a preliminary survey it has been found that the mean monthly food expenditure is Baht 3125 with a standard deviation of Baht 640. How large a sample size must be? Currently there are 920 students in AIT.
How large the sample size must be if the standard deviation would be Baht 128?

14.14   You are required to conduct a survey on birth rate. It has been decided that a simple cluster sampling will be used for the purpose.
The following information are available to you:

(i)   Total budget for the survey is $22,000.
(ii) The overhead cost is estimated to be $2000.
(iii) Enumerator cost per month is $400.

(iv) The enumerator has to spend 2.5 days, on average, in contacting a cluster and for other preliminary works.

(v) On average, the enumerator can interview 40 persons per day.

(vi) The intra-cluster correlation coefficient is estimated to be 0.001. Calculate, based on optimum allocation principle.

(a) The average number of persons from each cluster.

(b) The number of clusters.

(c) Total sample size.

14.15 It is desired to estimate the mean of a normally distributed population with an error of less than 0.50 with a probability of 0.90. It is known that the variance of the population is 4. How large the sample size is to be in order to achieve the accuracy stated above? The population size is 12,000.

14.16 A company wishes to estimate the proportion of the people who prefer their brand of soft drink. It wishes to keep the error within 2 %, with a risk of 0.0456. How large a sample must be taken?

14.17 You have a sample of size 935. The approximate fixed and variable costs are Baht 15 and Baht 33 respectively. The fixed cost is converted to per unit of sample. What is the total sampling cost?

14.18 You wish to determine the sample size for estimating the farmers' income. The organization sponsoring the research can spend a maximum sum of Baht 75,000. The fixed and variable costs (all converted to per unit basis) are Baht 25 and Baht 43 respectively. What sample size should you choose?

14.19 In a population of families the household income is normally distributed with a standard deviation of $1200. How large a sample is to be to determine the mean income if it is desired that the probability of the sampling error of more than $55 be less than 5 %?

14.20 You wish to estimate the proportion of defectives in large production lot within 0.05 of the true proportion, with 90 % confidence. How large a sample must be?

Answers

14.1 1189

14.3 (b) 969

14.4 1010

14.5 (a) 564; 1026; 1144; 1213; 1251

   (c) There is no straight line relationship between sample size and population size. Rate of increase of sample size diminishes with the increase of population size.

14.6 473; 760; 823; 858; 877

14.7 31 days

14.8 343

14.9 (a) 1549; (b) 2920 (assuming $N = 10,000$)

14.10   34
14.11   360
14.12   550
14.13   199
14.14   (a) 316; (b) 145; (c) 45,820
14.15   43
14.16 $n = N/(0.0004\ N + 1)$ for maximum variance
14.17   Baht 44,880
14.18   Max. 1147
14.19   1829
14.20   271

Chapter 15
Index Numbers

Abstract Techniques are developed and it is shown how scales are constructed for the qualitative data and a wide variety of analytical tools are used to analyze these. Techniques in constructing several indices are shown. These include priority, satisfaction, agreement, performance, price, laspeyres, paasche price, fisher's ideal price, quantity, total cost, cost of living, and standard of living indices. The technique for calculating rate of inflation is shown. The Rule of 70 is explained.

Keywords Index numbers · Priority · Satisfaction · Agreement · Performance · Price · Laspeyres · Paasche price · Fisher's ideal price · Quantity · Total cost · Cost of living · Standard of living indices · Rate of inflation · Rule of 70

Sometimes, it is possible to rank the qualitative data. If that can be done, a wide variety of analytical techniques can be applied to analyze them. When some sort of rankings is possible, the intended practice is to construct some scales and on the basis of those scales, indexes can be constructed. These indexes then will serve a useful device to analyze and infer.

Scales have the notion of continuous measurements. For example, in a continuous scale we can measure 2, 3 cm, etc. A measurement between 2 and 3 cm can have any value, say 2.35 cm. There is no jump from 2 to 3 cm. The idea of constructing a continuous scale for qualitative data is similar. Values of the scales are conveniently selected and it is assumed that the measurement between any two values in the scale is continuous. Although this gives an approximation, it has meaningful interpretation in practical applications. An example may serve the purpose of explaining the concept.

The employers of AIT alumni were in a survey requested to rate the emphasis laid on the components of a post-graduate program of studies at master level. The following scale was used.

© Springer Science+Business Media Singapore 2016
A.Q. Miah, *Applied Statistics for Social and Management Sciences*,
DOI 10.1007/978-981-10-0401-8_15

Scale:

Emphasis	No	Low	Normal	High
Scale value	0	1	2	3

Clearly, there is some meaningful ranking in the scale. Normal emphasis with value 2 is at a higher level than low emphasis with value 1. Although the respondents quote only 0, 1, 2, or 3, yet the scale may be assumed to be continuous. The employers' responses on "individual research work" were, on average, rated 2.39. This is easily interpretable. The overall emphasis laid by the employers on individual research work was between normal emphasis and high emphasis, almost in the middle of the two scale values. This type of scales also serves as a basis by which several items can be compared. For example, the average emphasis laid by the employers on "field work" was rated at 1.98. Thus, it can easily be concluded that the employers lay more emphasis on individual research work than on field trip.

It is to be kept in mind that statistical tests on the indexes constructed based on such scales can be applied easily. Thus, the index values can be of sufficient importance in inferential statistics. In the sections hereafter, several scale and index construction procedures are outlined.

15.1 Priority

Often we want to study the priorities attached by the target group on certain components of a development program. Different respondents will cite different priorities for different components. Our task is to summarize them. How to do them? Construction of an index is a good answer. Suppose we want to record up to the fourth priority. The scale may be constructed as follows:

Scale:

Priority	First	Second	Third	Fourth	No
Scale value	1.00	0.75	0.50	0.25	0.00

The purpose here is to keep the index value between 0 and 1 for convenience and easy interpretation. There are four gaps, i.e., 0–0.25, 0.25–0.50, 0.50–0.75, and 0.75–1.00. So 1 was divided by four and the scale step was computed as 0.25. Suppose, now that we want to measure up to the fifth priority. What should be the scale? Obviously, there are five gaps now. So the scale step can be 1/5, i.e., 0.20 and the scale can be constructed as follows.

Scale:

Priority	First	Second	Third	Fourth	Fifth	No
Scale value	1.00	0.80	0.60	0.40	0.20	0.00

Once the scale is decided, computation of the index is simple. In this case the index is computed by the following formula:

$$I = \frac{\sum s_i f_i}{N}$$

where
I priority index such that $0 \leq I \leq 1$,
s_i scale value at ith priority,
f_i frequency of ith priority,
N total no. of observations $= \sum f_i$.

Example 1

In a slum dwellers study, the dwellers were asked to mention their priorities (up to fifth priority) of upgrading against seven components. The responses are shown in Table 15.1. Construct a suitable scale, and calculate the priority indexes against each of the components and compare.

Table 15.1 Priority for upgrading

Priority	Frequency of responses against the components of the upgrading for						
	Hsng	Wspl	Eltc	Gspl	Envt	Road	Drng
1	121	162	118	196	39	90	13
2	79	92	125	201	89	78	66
3	120	43	31	150	144	119	126
4	163	33	12	32	209	115	163
5	127	90	23	15	153	138	169
0	113	298	408	124	99	191	193
Total	723	718	717	718	733	731	730

Notes Hsng = Housing; Wspl = Water supply; Eltc = Electricity; Gspl = Gas supply; Envt = Environment; Drng = Drainage
Priorities beyond fifth were treated as no priority

The scale values are set according to the number of priorities as follows:
Scale:

Priority	First	Second	Third	Fourth	Fifth	No
Scale value	1.00	0.80	0.60	0.40	0.20	0.00

The priority indexes are calculated hereafter.
Housing

$$I = \frac{1.0 * 121 + 0.8 * 79 + 0.6 * 120 + 0.4 * 163 + 0.2 * 127 + 0.0 * 113}{723}$$
$$= \frac{346.8}{723}$$
$$= 0.48$$

Water supply

$$I = \frac{1.0 * 162 + 0.8 * 92 + 0.6 * 43 + 0.4 * 33 + 0.2 * 90 + 0.0 * 298}{718}$$
$$= \frac{292.6}{718}$$
$$= 0.41$$

Electricity

$$I = \frac{1.0 * 118 + 0.8 * 125 + 0.6 * 31 + 0.4 * 12 + 0.2 * 23 + 0.0 * 408}{717}$$
$$= \frac{246.0}{717}$$
$$= 0.34$$

Gas supply

$$I = \frac{1.0 * 196 + 0.8 * 201 + 0.6 * 150 + 0.4 * 32 + 0.2 * 15 + 0.0 * 124}{718} I$$
$$= \frac{462.6}{718}$$
$$= 0.64$$

Table 15.2 Priority indexes of components of upgrading program

Components for upgrading	Priority index	Rankings in order of priority
Housing	0.48	II
Water supply	0.41	IV
Electricity	0.34	VI
Gas supply	0.64	I
Environment	0.42	III
Road	0.41	IV
Drainage	0.33	VII

Environment

$$I = \frac{1.0 * 39 + 0.8 * 89 + 0.6 * 144 + 0.4 * 209 + 0.2 * 153 + 0.0 * 99}{733}$$
$$= \frac{310.8}{733}$$
$$= 0.42$$

Road

$$I = \frac{1.0 * 90 + 0.8 * 78 + 0.6 * 119 + 0.4 * 115 + 0.2 * 138 + 0.0 * 191}{731}$$
$$= \frac{297.4}{731}$$
$$= 0.41$$

Drainage

$$I = \frac{1.0 * 13 + 0.8 * 66 + 0.6 * 126 + 0.4 * 163 + 0.2 * 169 + 0.0 * 193}{730}$$
$$= \frac{240.4}{730}$$
$$= 0.33$$

The indexes are summarized in Table 15.2. Rankings are also shown to have the comparison of various upgrading components in order of priority.

The residents attach the highest priority to gas supply followed by housing improvement. The lowest priority is attached to improvement of drainage.

15.2 Satisfaction

Sometimes we are interested in studying the satisfaction of the target group against an attribute or a set of attributes. Satisfaction with job, satisfaction with the housing condition, satisfaction with supply of agricultural inputs are some of the numerous examples where satisfaction measurements become relevant.

Three different scales can be suggested for measuring satisfaction depending on the type/value of responses recorded. If the responses are in two values only namely, "satisfied" and "not satisfied," the scale can be constructed as follows:

Dissatisfied	Satisfied
−1.00	1.00

The index can be computed using the following simple computational formula:

$$I = \frac{f_s - f_d}{N} I$$

where
I satisfaction index such that $-1 \leq I \leq +1$,
f_s frequency of responses indicating satisfaction,
f_d frequency of responses indicating dissatisfaction,
N total no. of observations $= \sum f_i$.

Example 2
In a sample survey in a village, the farmers were asked whether they were satisfied with supply of seeds, fertilizer, and pesticides. The responses were as follows: (Table 15.3).

Table 15.3 Farmers' satisfaction with supply of inputs

Supply of	Frequency of responses		Total
	Satisfied	Not satisfied	
Seeds	120	180	300
Fertilizer	98	189	287
Pesticides	162	120	282
Over all	380	489	869

Compute the satisfaction indexes and compare them.

Computation:

For seeds:

$$I = \frac{f_s - f_d}{N}$$
$$= \frac{120 - 180}{300}$$
$$= \frac{-60}{300}$$
$$= -0.20$$

For Fertilizer:

$$I = \frac{f_s - f_d}{N}$$
$$= \frac{98 - 189}{287}$$
$$= \frac{-91}{287}$$
$$= -0.32$$

For Pesticides:

$$I = \frac{f_s - f_d}{N}$$
$$= \frac{162 - 120}{282}$$
$$= \frac{42}{282}$$
$$= 0.15$$

For overall:

$$I = \frac{f_s - f_d}{N}$$
$$= \frac{380 - 489}{869}$$
$$= \frac{-109}{860}$$
$$= -0.13$$

The indexes are summarized in Table 15.4. Rankings are also shown to have the comparison of satisfaction level against input supply.

Table 15.4 Indexes of satisfaction for supply of agricultural inputs

Supply of	Index value	Rankings in order of satisfaction
Seeds	−0.20	II
Fertilizer	−0.32	III
Pesticides	0.15	I

The farmers are satisfied to some extent with the supply of pesticides. They are dissatisfied with the supply of seeds. They are dissatisfied to the greatest extent with the supply of fertilizer. Overall, the input supply is not satisfactory.

If the responses are recorded at three values, namely, "satisfied," "neither satisfied nor dissatisfied," and "dissatisfied," then the scale can be constructed as follows:

Satisfaction	Dissatisfied	Neutral	Satisfied
Scale value	−1.0	0.0	+1.0

The computational formula for calculation of satisfaction index stands as follows:

$$I = \frac{1.0 * f_s + 0.0 * f_0 - 1.0 * f_d}{N}$$

where

I satisfaction index such that $-1 \leq I \leq +1$,
f_s frequency of responses indicating satisfaction,
f_0 frequency of responses indicating neutral,
f_d frequency of responses indicating dissatisfaction,
N total no. of observations $= \Sigma f_i = f_s + f_0 + f_d$.

Example 3
Satisfaction levels of residents in a city with different aspects were enquired. The responses recorded on a three-point scale are shown in Table 15.5. Calculate the satisfaction index and rank them.

Housing

$$I = \frac{1.0 * 102 + 0.0 * 521 - 1.0 * 108}{731}$$
$$= \frac{-6.0}{731}$$
$$= -0.008$$

Table 15.5 Satisfaction with housing condition

Aspects	Frequency of responses indicating			Total
	Satisfaction	Neutral	Dissatisfaction	
Housing	102	521	108	731
Water supply	372	144	20	536
Electricity	431	36	17	484
Gas supply	125	13	18	156
Environment	6	339	384	729
Road	178	269	259	706
Drainage	16	298	259	573

Water supply

$$I = \frac{1.0 * 372 + 0.0 * 144 - 1.0 * 20}{536}$$
$$= \frac{352.0}{536}$$
$$= 0.657$$

Electricity

$$I = \frac{1.0 * 431 + 0.0 * 36 - 1.0 * 17}{484}$$
$$= \frac{414.0}{484}$$
$$= 0.855$$

Gas supply

$$I = \frac{1.0 * 125 + 0.0 * 13 - 1.0 * 18}{156}$$
$$= \frac{107.0}{156}$$
$$= 0.686$$

Environment

$$I = \frac{1.0 * 6 + 0.0 * 339 - 1.0 * 384}{729}$$
$$= \frac{-378.0}{729}$$
$$= -0.519$$

Table 15.6 Satisfaction index of residents with different housing aspects	Housing aspect	Index	Rankings in order of satisfaction
	Housing	−0.008	IV
	Water supply	0.66	III
	Electricity	0.86	I
	Gas supply	0.69	II
	Environment	−0.519	VII
	Road	−0.115	V
	Drainage	−0.424	VI

Road

$$I = \frac{1.0 * 178 + 0.0 * 269 - 1.0 * 259}{706}$$
$$= \frac{-81.0}{706}$$
$$= -0.115$$

Drainage

$$I = \frac{1.0 * 16 + 0.0 * 298 - 1.0 * 259}{573}$$
$$= \frac{-243.0}{573}$$
$$= -0.424$$

Satisfaction levels are summarized in Table 15.6. The rankings are also shown.

15.3 Agreement

In certain situations the investigators want to see if the respondents agree to a certain proposition or not. The expected responses could be a bivariate one "agree" or "do not agree." But from such responses, the degree of agreement or disagreement cannot be measured. The following one gives a better measurement.

Scale:

Agreement	Strongly disagree	Disagree	Neutral	Agree	Strongly agree
Scale value	−2.0	−1.0	0.0	+1.0	+2.0

This five-point scale allows recording positive as well as negative responses indicating agreement and disagreement, respectively.

Table 15.7 Farmers' agreement on satisfactory extension services

Agreement	Responses		
	Current year	Previous year	Year before previous year
Strongly agree	165	140	90
Agree	120	100	85
Neutral	35	30	25
Disagree	80	120	140
Strongly disagree	60	75	122
Total	460	465	462

Note Figures are frequencies of responses

Example 4

In a countryside sample survey the proposition that was put to the farmers was "extension services were satisfactory during the current year, the previous year and the year preceding the previous year." The farmers were asked to respond whether they agreed to it, according to the five-point scale. The responses were as follows (Table 15.7):

Calculate the indexes and interpret the results based on the indexes.

Computations:

Current year

$$I = \frac{2.0 * 165 + 1.0 * 120 + 0.0 * 35 - 1.0 * 80 - 2.0 * 60}{460}$$
$$= \frac{250.0}{460}$$
$$= 0.543$$

Previous year

$$I = \frac{2.0 * 140 + 1.0 * 100 + 0.0 * 30 - 1.0 * 120 - 2.0 * 75}{465}$$
$$= \frac{110.0}{465}$$
$$= 0.237$$

Year before previous year

$$I = \frac{2.0 * 90 + 1.0 * 85 + 0.0 * 25 - 1.0 * 140 - 2.0 * 122}{462}$$
$$= \frac{-119.0}{462}$$
$$= -0.258$$

Table 15.8 Summary of indexes

Reference year	Index value
Current	+0.543
Previous	+0.237
Year before previous year	−0.258

The index values are summarized in Table 15.8 hereafter.

Positive index implies agreement meaning extension services were satisfactory. Higher index value implies greater agreement on satisfactory service. Negative index value implies disagreement and greater negative value suggests greater disagreement on satisfactory service. The index value may be as follows:

$$-2 \leq I \leq +2$$

In the present example during the year before the previous year the index value is negative. This means that the service in general was not satisfactory. During the previous year and the current year the index values are positive. This implies that the services on the average are satisfactory. Furthermore, the index values have a consistent and steady increase from −0.258 to +0.304. This suggests that extension services have been performing well gradually over time.

15.4   Performance

We can construct a scale to measure the degree of performance. We could do it in the way "agreement" scale was constructed. But that could not be suitable. Agreement scale ranges from −2 to +2. But the performance scale should not read minus. There is, in fact, no sense in saying "minus performance." However, the five-point scale can be constructed in this case too. The scale should be as follows.

Scale:

Performance	Poor				Excellent
Scale value	0	1	2	3	4

The higher the scale value, the better is the performance. However, the greatest value can be 4 according to the scale. Therefore, the index value may lie between 0 and 4, i.e.,

$$0 \leq 1 \leq 4$$

Example 5

In a sample survey the employers of AIT alumni were requested to indicate on a five-point scale the performance of the AIT graduates in certain qualities. The responses are shown in Table 15.9. Calculate the performance indexes.

Solution:

The scale is constructed as shown hereafter. The higher the scale value, the better is the performance.

Scale:

Performance	Poor				Excellent
Scale value	0	1	2	3	4

Calculations of indexes are demonstrated hereafter.

General Technical Ability

$$I = \frac{0*3+1*6+2*26+3*39+4*22}{96}$$

$$= \frac{263}{96}$$

$$= 2.74$$

Special Technical Ability

$$I = \frac{0*4+1*4+2*18+3*42+4*20}{88}$$

$$= \frac{246}{88}$$

$$= 2.80$$

Calculations of indexes for other qualities are left to the readers. The indexes thus computed are shown in Table 15.10.

Table 15.9 Performance of AIT graduates

Qualities	Values					Total
	0	1	2	3	4	
General technical ability	3	6	26	39	22	96
Special technical ability	4	4	18	42	20	88
Planning ability	3	11	25	21	22	82
Implementation ability	3	11	16	33	15	78
Problem solving	5	9	21	39	26	100
Organizing work	4	5	26	32	27	94
Coordination	5	7	19	33	25	89
Teach/Train ability	2	12	16	33	26	89

Note Figures are frequencies of responses

Table 15.10 Index of performance of AIT graduates

Qualities	Indexes
General technical ability	2.74
Special technical ability	2.80
Planning ability	2.59
Implementation ability	2.59
Problem solving	2.72
Organizing work	2.78
Coordination	2.74
Teach/Train ability	2.78

15.5   Price Index

Price index helps us to study the movement of prices of an item or a group of items over certain period of time. Price index is a relative measure and is based on a base period. For a single item or commodity, the index is calculated as

$$\text{Price index} = \frac{P_t * 100}{P_0}$$

where
P_0 price of the item in base period,
P_t price of the item in period under study.

In order to avoid decimal point or fraction, it is customarily multiplied by 100. Some authors use the term "price relative" instead of price index when one item is involved.

Let us use one example to calculate the price index. Suppose the prices of a food item during 1980 and 1985 were \$2.25 and \$3.00, respectively. What is the price index of the item in 1985?

Here our base period is 1980. Therefore, the price index is calculated as follows:

$$\text{Price index} = \frac{3.00 * 100}{2.25}$$
$$= 133$$

This indicates that if the price in the base year (1980) is 100, the price in 1985 would have been 133 implying 33 % increase over the base period.

In practice we are more concerned with the price movement of several items together rather than one item only. The following subsections will illustrate the use and techniques in this regard.

15.5.1 Laspeyres Price Index

When we calculated the price index for a single item, we used only the unit prices over two periods of time. No weightage was used for the quantity consumed during the periods. In Laspeyres price index, we need to introduce the quantities too, thus, introducing the total cost concept.

Laspeyres price index is the total cost of purchase of a basket of goods traded at the base period at the prices in the period under study, expressed as a percentage of the total cost of purchase of the same quantities (base period quantities) at the base period price. Thus,

$$\text{Laspeyres price index} = \frac{\sum P_t Q_0 * 100}{\sum P_0 Q_0}$$

where
P_0 price in the base period,
Q_0 quantity in the base period,
P_t price in the current period.

Here $P_t Q_0$ is the total cost of purchase of the base period quantity at the current period price. $P_0 Q_0$ is the total cost of purchase of the base period quantity at the base period price. It may be noted here that in this case only the quantity traded during the base period is considered.

Example 6
The unit prices of five commodities and the quantities sold during two periods of time are shown. Calculate the Laspeyres price index.
    Solution:
The calculations are summarized hereafter in Table 15.12.

$$\text{Laspeyres price index} = \frac{\sum P_t Q_0 * 100}{\sum P_0 Q_0}$$
$$= \frac{1525 * 100}{1365}$$
$$= 112$$

15.5.2 Paasche Price Index

It may be noted that in Laspeyres price index only the quantity traded during the base period (Q_0) was considered. It is easy and suitable if the quantities traded during the latter period is difficult to obtain. If quantities during both the periods can be obtained, then Paasche price index may be used. It is the total cost of purchase of the quantities in the period under study at the same period price, expressed as a percentage of the total cost of purchase of the quantities in the latter period at the base period price. Thus,

$$\text{Paasche price index} = \frac{\sum P_t Q_t * 100}{\sum P_0 Q_0}$$

Using the example in Sect. 15.5.1 above, we can calculate the Paasche price index as follows:

$$\text{Paasche price index} = \frac{\sum P_t Q_t * 100}{\sum P_0 Q_t}$$
$$= \frac{1648 * 100}{1240}$$
$$= 133$$

15.5.3 Fisher's Ideal Price Index

We have noted that because of use of quantities in different periods, Laspeyres and Paasche price indexes differ. Fisher's ideal index is a solution to this problem. Fisher's ideal price index is the geometric mean of Laspeyres price index and Paasche price index. Thus,

Fisher's ideal price index = $\sqrt{(\text{Laspeyres p.} * \text{Paasche p. index})}$.

Let us use the same data as in Sect. 15.5.1 to calculate the Fisher's ideal price index:

$$\text{Fisher's ideal price index} = \sqrt{(112 * 133)}$$
$$= 122$$

15.5.4 Quantity Index

So far we have studied the price changes with the help of price indexes. In other words, it measures the change in "cost of living." There is another set of indexes. These measure the increase or decrease of quantities indicating change in "standard of living."

The following set of three formulae will be used to calculate the quantity indexes:

$$\text{Laspeyres quantity index} = \frac{\sum P_t Q_0 * 100}{\sum P_0 Q_0}$$

$$\text{Paasche quantity index} = \frac{\sum Q_t P_t * 100}{\sum Q_0 P_t}$$

Fisher's ideal quantity index = $\sqrt{}$(Laspeyres q. index * Paasche q. index).

To demonstrate their uses we may use the same data as given in Tables 15.11 and 15.12.

$$\text{Laspeyres quantity index} = \frac{\sum Q_t P_0 * 100}{\sum Q_0 P_0}$$

$$= \frac{1240 * 100}{1354}$$

$$= 91$$

$$\text{Paasche quantity index} = \frac{\sum Q_t P_t * 100}{\sum Q_0 P_t}$$

$$= \frac{1648 * 100}{1525}$$

$$= 108$$

$$\text{Fisher's ideal quantity index} = \sqrt{}(91 * 108)$$

$$= 99$$

Table 15.11 Prices and quantities traded

Commodities	1985		1990	
	Q_0	P_0	Q_t	P_t
Rice (kg)	500	0.40	500	0.50
Meat (kg)	100	1.25	80	1.50
Gasoline (l)	500	0.40	600	0.23
Shirt (no)	20	10.00	30	12.00
Color TV (no)	2	320.00	2	360.00

Note Prices are in $

Table 15.12 Total cost summary

Commodities	Cost ($)			
	$P_0 Q_0$	$P_t Q_0$	$P_t Q_t$	$P_0 Q_t$
Rice	200	250	250	200
Meat	125	150	120	100
Gasoline	200	165	198	240
Shirt	200	240	360	300
Color TV	640	720	720	640
Total	1354	1525	1648	1240

15.5.5   Total Cost Index

We have studied the price indexes (cost of living index) and quantity indexes (standard of living index). Here we want to study the total cost of index which indicates the combined effect of price change and quantity change. Specifically, we want to study how much the total cost changes as a result of changes in both price and quantity. The computational formula is as follows:

$$\text{Total cost index} = \frac{\sum P_t Q_t * 100}{\sum P_0 Q_0}$$

Using the data of Tables 15.11 and 15.12 we can calculate the total cost index as follows:

$$\begin{aligned}
\text{Total cost index} &= \frac{\sum P_t Q_t * 100}{\sum P_0 Q_0} \\
&= \frac{1648 * 100}{1354} \\
&= 121
\end{aligned}$$

Using Fisher's ideal indexes we can derive another relationship. Let us examine the following:

$$\begin{aligned}
&(\text{price index}) * (\text{quantity index}) \\
&= \sqrt{\{(\text{Laspeyres p. index} * \text{Paasche p. index})} \\
&\quad * (\text{Laspeyres q. index} * \text{Paasche q. index})\} \\
&= \sqrt{\left\{ \frac{\sum P_t Q_0}{\sum P_0 Q_0} * \frac{\sum P_t Q_t}{\sum P_0 Q_t} * \frac{\sum Q_t P_0}{\sum Q_0 P_0} * \frac{\sum Q_t P_t}{\sum Q_0 P_t} \right\}} \\
&= \sqrt{\left\{ \frac{\sum P_t Q_0}{\sum P_0 Q_0} * \frac{\sum P_t Q_t}{\sum P_0 Q_t} * \frac{\sum Q_t P_0}{\sum Q_0 P_0} * \frac{\sum Q_t P_t}{\sum Q_0 P_t} \right\}} \\
&= \sqrt{\left\{ \frac{(\sum P_t Q_t)^2}{\sum (P_0 Q_0)^2} \right\}} \\
&= \frac{\sum P_t Q_t}{\sum P_0 Q_0} \\
&= \text{Total cost index}
\end{aligned}$$

This is known as the factor reversal test. We can check it in our previous example:

$$(\text{Fisher's price index}) * (\text{Fisher's quantity index}) = (1.22) * (0.99)$$
$$= 1.21$$
$$= \text{Total cost index.}$$

15.5.6   Cost of Living/Standard of Living Index

There are three cost of living indexes. These are price indexes namely, Laspeyres price index, Paasche price index, and Fisher's ideal price index. There is not much superiority of one index over others, but Laspeyres price index is used more widely. However, Fishers' ideal price index would be preferred, if there is not much difficulty in collecting data and computing them.

Similar to the cost of living indexes, there are three standard of living indexes namely, Laspeyres quantity index, Paasche quantity index, and Fisher's ideal quantity index.

Table 15.13 Quantities and prices of goods and services

Commodities	1985		1990	
	Q_0	$P_0(\$)$	Q_t	$P_t(\$)$
Rice (kg)	2.00	0.40	2.00	0.50
Meat (kg)	0.50	1.20	0.55	1.60
Fish (kg)	1.00	0.50	1.25	1.00
Bread (kg)	1.00	0.60	1.00	1.00
Vegetable (kg)	1.00	0.40	1.25	0.80
Fruit (kg)	0.50	0.60	1.00	1.25
Spice (kg)	0.70	4.00	0.80	6.00
Cold drink (no)	7.00	0.12	10.00	0.24
Tea/coffee (cup)	3.00	0.12	4.00	0.20
Milk (l)	3.00	0.40	3.50	0.60
Detergent (kg)	0.50	0.60	0.80	0.80
Soap (no)	0.60	0.80	1.00	1.25
Gas (cyl)	0.05	5.00	0.07	6.60
Electricity (kwh)	2.00	0.07	3.00	0.10
Gasoline (l)	2.00	0.50	3.00	0.36
Shirt (no)	0.25	6.00	0.35	12.00
Pant (no)	0.25	14.00	0.30	20.00
Shoe (pair)	0.20	10.00	0.30	21.00
Academic (ls)	1.00	3.00	1.00	5.00
House rent (no)	1.00	20.00	1.00	30.00
Education (no)	0.40	100.00	0.40	150.00
Color TV (no)	0.01	400.00	0.01	600.00
Refrigerator (no)	0.01	300.00	0.01	500.00
Telephone (no)	1.00	6.00	1.00	8.00

A major work in computing the indexes is to collect data. Usually, a basket of goods and services that a man consumes during a certain period of time, say one week/one month, is considered. Appropriate sample sizes of the consumers are taken. Their quantities of consumption of goods and services are recorded. Unit prices of those goods and services are also noted. This is done during any part of the base year. During a latter year the same process is repeated. From these two sets of data, computations of indexes are done.

Example 7

In a sample survey of consumers in a city during 1985 and 1990, the average quantities of goods and services consumed by a consumer per week along with the average unit prices of those goods and services are shown in Table 15.13. Calculate the cost of living index, standard of living index and carry out the factor reversal test.

Cost components and their calculations are summarized in Table 15.14.

Table 15.14 Cost components ($)

Commodities	1985		1990	
	$P_0 Q_0$	$P_t Q_t$	$Q_0 P_t$	$Q_t P_0$
Rice (kg)	0.80	1.00	1.00	0.80
Meat (kg)	0.60	0.88	0.80	0.66
Fish (kg)	0.50	1.25	1.00	0.63
Bread (kg)	0.60	1.00	1.00	0.60
Vegetable (kg)	0.40	1.00	0.80	0.50
Fruit (kg)	0.30	1.25	0.63	0.60
Spice (kg)	2.80	4.80	4.20	3.20
Cold drink (no)	0.84	2.40	1.68	1.20
Tea/coffee (cup)	0.36	0.80	0.60	0.48
Milk (l)	1.20	2.40	1.80	1.40
Detergent (kg)	0.30	0.64	0.40	0.48
Soap (no)	0.48	1.25	0.75	0.80
Gas (cyl)	0.25	0.46	0.33	0.35
Electricity (kwh)	0.14	0.30	0.20	0.21
Gasoline (l)	1.00	1.08	0.72	1.50
Shirt (no)	1.50	4.20	3.00	2.10
Pant (no)	3.50	6.00	5.00	4.20
Shoe (pair)	2.00	6.30	4.20	3.00
Academic (ls)	3.00	5.00	5.00	3.00
House rent (no)	20.00	30.00	30.00	20.00
Education (no)	40.00	60.00	60.00	40.00
Color TV (no)	4.00	6.00	6.00	04.00
Refrigerator (no)	3.00	5.00	5.00	03.00
Telephone (no)	6.00	8.00	8.00	6.00
Total	93.57	150.71	142.11	98.71

Cost of Living Index:

$$\text{Laspeyres price index} = \frac{\sum P_t Q_0 * 100}{\sum P_0 Q_0}$$

$$= \frac{142.11 * 100}{93.57}$$

$$= 152$$

$$\text{Paasche price index} = \frac{\sum P_t Q_t * 100}{P_0 Q_t}$$

$$= \frac{150.71 * 100}{98.71}$$

$$= 153$$

$$\text{Fisher's ideal price index} = \sqrt{(152 * 153)}$$

$$= 152.5$$

Standard of Living Index:

$$\text{Laspeyres quantity index} = \frac{\sum Q_t P_0 * 100}{\sum Q_0 P_0}$$

$$= \frac{98.71 * 100}{93.57}$$

$$= 105$$

$$\text{Paasche quantity index} = \frac{\sum Q_t P_t * 100}{\sum Q_0 P_t}$$

$$= \frac{150.71 * 100}{142.11}$$

$$= 106$$

$$\text{Fisher's ideal quantity index} = \sqrt{(105 * 106)}$$

$$= 105.5$$

Factor reversal test:

$$\text{Total cost index} = \frac{\sum P_t Q_t * 100}{\sum P_0 Q_0}$$

$$= \frac{150.71 * 100}{93.57}$$

$$= 161$$

Using Fisher's ideal index both for cost of living and standard of living, we can get

$$(\text{Fisher's ideal price index}) * (\text{Fisher's ideal quantity index}) = (1.525) * (1.055)$$
$$= 1.61$$
$$= \text{total cost index}.$$

15.5.7 Inflation

The price index shows the movement of prices. Associated within this is the term "Inflation." Inflation means general rising level of prices. However, it does not mean that all prices are rising. In fact over a specific period of time some prices may rise, some prices may remain constant, and some prices may even fall. If the prices fall, we call it deflation.

Inflation is measured by price index numbers. As such, it would be useful to understand what is the meaning of price index. Suppose in 1990 the consumer price index was 125 with 1985 as the base year. This means the price in the base year 1985 was taken to be 100. Relative to this, the price level rose to 125 in 1990. This means an increase of price by 25 % over the base year price. If a bundle of commodities would cost $100 in 1985, it costs $125 in 1990.

Like price index, inflation also involves setting up a base year and measuring the rate. The rate of inflation is calculated for any given year by subtracting the previous year's price index from the current year's price index and dividing this difference by the previous year's price index. Usually, it is expressed in percentage. To illustrate how inflation is measured, suppose that the consumer price indexes were 255 in 1989 and 265 in 1990 relative to certain base year. Then

$$\text{Inflation rate} = \frac{265 - 255}{255} * 100$$
$$= 3.92\%$$

If the price indexes for two consecutive years are not available, then the inflation rate should be calculated using the formula:

$$r = \text{anti } Ln\left\{\frac{LnP_2 - LnP_1}{n}\right\} - 1$$

where
r annual inflation rate (multiply r by 100 to get it in percentage),
P_2 price index in the latter period,
P_1 price index in the former period,
n number of years between the two periods.

Suppose the price indexes in 1986 and 1991 were 260 and 320, respectively. Then the rate of inflation is calculated as follows:

$$r = \text{anti Ln} \left\{ \frac{\text{Ln}P_2 - \text{Ln}P_1}{n} \right\} - 1$$

$$r = \text{anti Ln} \left\{ \frac{\text{Ln}320 - \text{Ln}260}{5} \right\} - 1$$

$$= \text{anti Ln} \left\{ \frac{5.768 - 5.561}{5} \right\} - 1$$

$$= \text{anti Ln}(0.0414) - 1$$

$$= 1.0423 - 1$$

$$= 0.0423$$

$$= 4.23\%$$

The same formula may be rearranged and used to calculate the number of years after which the price level will be double. The above formula, if rearranged, stands as follows:

$$(1 + r)^n = P_2 / P_1$$

In the present context $P_2 = 2P_1$. Therefore,

$$(1 + r)^n = 2P_1 / P_1 = 2$$

Taking logarithm of both sides we can get

$$n \ln(1 + r) = \ln 2 = 0.693$$

Therefore,

$$n = \frac{0.693}{\ln(1 + r)}$$

$$= \frac{0.693}{r + r^2/2 + r^3/3 + \cdots}$$

$$= \frac{0.70}{r} \text{ (approximately)}$$

$$= \frac{0.70 * 100}{r * 100}$$

$$= \frac{70}{r} \text{ (in \%)}$$

This means dividing 70 by the annual inflation rate we get the number of years after which the price level will be double the present price level. Thus, if the inflation rate is 3.5, the number of years after which the price level will be double is 70/3.5 = 20 years. Similarly, an inflation rate of 8 % will double the price level in 70/8, i.e., about 9 years. This is known as the so-called RULE OF 70.

Problems

15.1 In a sample survey for food sufficiency in a rural community, 61 households reported that the food they produced was sufficient, while 108 households reported that it was insufficient. Construct a suitable scale, and calculate the sufficiency index and comment.

15.2 The unit prices of three commodities and the quantities sold during 1988 and 1991 are as follows:

Commodities	1988		1991	
	Q_0	P_0	Q_t	P_t
Rice (kg)	400	$0.40	400	$0.50
Meat (kg)	80	$1.25	64	$1.50
Gasoline (l)	400	$0.40	480	$0.33

(a) Calculate the Laspeyres, Paasche, and Fisher's Price and Quantity indexes.

15.3 Three items of food in a restaurant were checked during 12 months of a year. The average prices (in $/kg) for the period are shown in the following table.

Table: Prices ($) of three items of food

Month	Rice	Fish	Chicken
January	0.32	2.40	1.41
February	0.33	2.38	1.19
March	0.35	2.00	1.20
April	0.38	2.61	1.40
May	0.39	2.60	1.40
June	0.40	2.76	1.60
July	0.42	2.78	1.44
August	0.45	2.75	1.58
September	0.48	2.80	1.60
October	0.48	3.10	1.61
November	0.50	3.11	1.90
December	0.51	3.15	2.21

The following table shows the quantities of the three items of food sold during the same 12 months.

Table: Quantities (kg) of three items of food ordered

Month	Rice	Fish	Chicken
January	180	247	355
February	170	234	370
March	175	264	387
April	160	222	336
May	168	204	385
June	150	188	345
July	140	188	325
August	135	263	300
September	179	240	428
October	185	247	439
November	187	257	475
December	190	282	510

    Take January as the base period and calculate Laspeyres, Paasche, and Fisher's Price and Quantity Indexes.

15.4 For data in problems 15.2 and 15.3 run the factor reversal test.

15.5 The consumer price index (CPI) rose 40 % in 1980 and again 80 % in 1990. How much did the CPI rise over the entire period? Suppose, this overall rise of CPI was steady at 7.6 % per year. What was the base year? Calculate the indexes for all the three years, taking 1980 as the base year.

15.6 The price index was 190 last year and 210 this year.

(a) Calculate the rate of inflation
(b) What is the RULE OF 70?
(c) How long will it take for the price level to be double?
(d) How long will it take for the price level to be double if the annual inflation rates are 4, 7.5, and 10 %?

15.7 The unit prices of three commodities and the quantities sold during 1988 and 1991 are shown below:

Commodities	1988		1991	
	Q_0	P_0	Q_t	P_t
Rice (kg)	400	$0.40	400	$0.50
Meat (kg)	80	$1.25	64	$1.50
Gasoline (l)	400	$0.40	480	$0.33

Use the data and show that
(Fisher's ideal price index) * (Fishers' ideal quantity index) = Total cost
index.

Answers

15.1 $-1 \leq I \leq +1$; -0.28
15.2   LPI = 107.6; PPI = 105.2; LQI = 79.0; PQI = 100.5
15.3   The indexes are summarized in the following table.

Month	LPI	PPI	LQI	PQI
January	100.00	100.00	100.00	100.00
February	92.94	92.58	98.84	98.46
March	85.41	85.29	107.32	107.18
April	105.13	104.99	91.90	91.78
May	105.07	104.48	94.37	93.84
June	114.83	104.48	85.63	85.56
July	110.60	109.97	82.91	82.41
August	114.78	114.63	95.34	95.22
September	116.94	116.65	107.45	107.18
October	123.69	122.84	110.43	109.66
November	133.16	133.34	116.98	117.13
December	143.74	145.00	126.56	127.67

15.5   152 %; 1977; CPI of 1977 = 71; CPI of 1980 = 100;
        CPI of 1990 = 180
15.6 (a)   10.535;
    (b)   if 70 is divided by the inflation rate (in %), a figure is obtained. This
          figure represents the number of years when the price index will be double
          the base year price. This is known as the Rule of 70.
    (c)   6.65 years; (d) 17.5 years; 9.33 years; 7 years

Chapter 16
Analysis of Financial Data

Abstract The chapter shows the techniques for analysis of financial data which usually entail time series data. Examples are provided to show how to analyze project investment data and to get financial terms such as NPVC, NPVB, and BCR. Technique for statistical estimation of IRR is demonstrated. Qualitative and quantitative assessment of risk is shown. It is shown how to do trend analysis using time series data. The popular autoregressive model is developed. The technique for forecasting with and without the use of a model is demonstrated. Examples are used.

Keywords Financial data · NPVC · NPVB · BCR IRR · Statistical estimation of IRR · Autoregressive model · Risk assessment

16.1 Financial Terms

There are some terms that will be necessary to study the analysis of financial data. These terms are explained hereafter.

Time Value of Money: The concept says that the value of money changes over time. Suppose we have $100 and we deposit in the bank. The interest rate is 5 %. After one year, the value of $100 will be equal to $100 + interest for one year = $100 + $100 * 0.05 = $100 + $5 = $105. This is the future value.

Present Value, Future Value and Discounting Rate: In the example above, the present value is $100 and the future value is $105. If we know the amount of money we want to invest, we can calculate the future value using the interest rate (this rate is discounting rate). From the future value we can calculate the present value. The relation between present value and future value is

$$P_n = P_0(1+r)^n$$

where,
P_0 present value
P_n future value at the end of nth period

© Springer Science+Business Media Singapore 2016

A.Q. Miah, *Applied Statistics for Social and Management Sciences*,
DOI 10.1007/978-981-10-0401-8_16

n number of period (years)
r discounting rate

This r is a fractional number and not percentage. For example, if the discounting rate is 5 %, $r = 0.05$. Here n and r are to match. For example, if yearly rate is counted, n will be in years and r will also be per year. If quarterly rate is counted, n will be in quarter and r will also be per quarter.

Net Present Value: In financial appraisal or feasibility, there are several figures for income or expenditure. All of these are converted to their respective present values. All these present values are added. The figure we get is the net present value.

IRR: This is internal rate of return. The IRR of an investment is the discount rate at which the following happens.

Net present value of costs (negative cash flows of the investment)—net present of benefits (positive cash flows of the investment) = 0.

It is used to evaluate the desirability of the investment. It is the break-even interest rate.

Benefit Cost Ratio (BCR): It is a ratio showing net present value of benefit divided by net value of cost as shown hereafter.

$$BCR = \frac{Net\,present\,value\,of\,benefit}{Net\,present\,value\,of\,cost}$$

Break-Even Point: The break-even point of an investment is the time at which the present value of cost equals the present value of benefits.

ROI (Return on Investment also ROR Rate of Return): The formula for calculating ROI is given below:

$$ROI = \frac{Gain\,from\,investment - Cost\,of\,investment}{Cost\,of\,investment}$$

ROE (Return on Equity) = Defined by the following:

$$ROE = \frac{Net\,income}{Shareholder\,Equity}$$

Cash Flow: Cash flow is the movement of cash in and out of the investment project.

Depreciation: Depreciation is the apportionment of cost of capital during the life expectancy of the building/structure/asset. There are different ways of calculating depreciation. Of these, the straight-line method is popular and widely used. Suppose there is a building constructed with capital cost of $50,000 and the life expectancy is assumed to be 40 years. According to the straight-line method, the depreciation rate will be 50,000/40 = $1250/year. This means that the building will depreciate at the rate of $1250 per year.

16.2   Calculation of Net Present Values

Refer to Sect. 3.1.1 (Geometric Mean) wherein it was shown by deduction that

$$P_n = P_0^*(1+r)^n$$

where,

P_n future population
P_0 present population
R growth rate
n number of years

The same concept can be applied for calculating the future value of money if we know the present value of money and the interest rate. Conversely, we can calculate the present value of money if its future value and the interest rate are known. It is the basic concept in financial appraisal of a project. Look at the following example.

The estimated breakdown of estimated cost and benefits of an industrial estate in Shanghai are shown in the following table. The cost and benefits are in current prices. Using the Excel spread sheet, the present values have been calculated using an interest rate known here as 'discount rate' of 10 % per annum.

Each future value has been converted to respective present values using the discount rate of 10 % per annum. Calculation proceeds as follows:

Example 1
Amounts are in thousand US$

Year	Estimated cost	Estimated benefit	Present value of cost (PVC)	Present value of benefit (PVB)
1995	–	–	–	–
1996	32,312	–	26,704	–
1997	40,604	35,171	30,506	26,424
1998	42,634	40,604	29,120	27,733
1999	3534	42,541	2194	26,415
2000	92,638	86,455	52,292	48,802
2001	34,198	92,016	17,549	47,219
2002	38,639	96,477	18,025	45,007
2003	83,276	61,323	35,317	26,007
2004	48,534	63,684	18,712	24,553
2005	2950	66,046	1034	23,149
2006	72,950	68,360	23,244	21,782
2007	57,932	70,721	16,781	20,485
2008	150,183	72,309	39,548	19,041
2009	306,320	302,657	73,331	72,454

(continued)

(continued)

Year	Estimated cost	Estimated benefit	Present value of cost (PVC)	Present value of benefit (PVB)
2010	165,577	312,130	36,034	67,929
2011	173,869	321,616	34,399	63,630
Total	1,346,150	1,732,110	454,790	560,629

In 2000, the estimated cost is US\$92,638 thousand dollars. This is expenditure to the project. Number of years beginning from 1995 to 2000 is $n = 6$, $r = 0.10$ (discount rate). So the present value is given by $PVC = 92{,}638/(1 + 0.10)^6 = 52{,}292$. The estimated benefit $= 86{,}455$. So the $PVB = 86{,}455/(1 + 0.10)^6 = 48{,}802$. Calculation for other years is similar.

The NPVC (net present value of cost) = US\$454,790,000
The NPVB (net present value of benefit) = US\$560,629,000

We may notice that in this example, the benefit is more than the cost. So the project is profitable. Furthermore, the BCR (benefit cost ratio) = NPVB/NPVC = 560,629/454,790 = 1.23. This means that based on the NPV, the project is expected to provide a profit of 23 %.

16.3 Project Investment Data

Before making a decision whether to invest in a particular project, a thorough analysis is done. One criterion is to examine whether the IRR is above or equal to a pre-set IRR. Application of this concept is demonstrated with the help of an example. The following example is a real-world example. This is the example of an industrial estate project in China. Here the analysis is made considering the project life of 16 years (1995–2011). Project life could be considered longer than this. But one thing is to be kept in mind that the maintenance cost increases over time and may offset the IRR. Furthermore, the investor may not like to wait for long time to get his return.

Project cost includes capital cost of construction but does not include the cost of land. Land cost usually goes up over time and may distort the rate of return. Revenue comes from the sale/or lease of the developed industrial estate.

The analysis is made using Excel. Discount rate used is 10 %. Normally the discount rate is the short time lending rate that the Federal Bank charges to other banks. If discount rate is taken to be different, the calculated IRR will not be affected. Discount rate is only a reference rate to start with.

Example 2
Amounts are in US$ million (rounding)

1	Construction schedule	Total	1995	1996	1997	1998	1999	2000
2	Pre-investment	4.50	4.50					
3	Project cost	1346		32.31	40.60	42.63	3.53	92.64
4	O & M cost	80.88			1.14	1.69	1.77	6.91
5	Depreciation	34.13				2.44	2.44	2.44
6	Revenue	1732			35.17	40.60	42.54	86.46
7	Taxes	72.05						
8	ROI							
	Inflow	1651			34.03	38.92	40.77	79.55
	Outflow	1346		32.31	40.60	42.63	3.53	92.64
	IRR	32 %	NPVB	533.64	NPVC	454.79	BCR	1.17
9	ROE							
	Inflow	1545			34.03	36.48	38.33	77.11
	Outflow	1346		32.31	40.60	42.63	3.53	92/64
	IRR	25 %	NPVB	500.44	NPVC	454.79	BCR	1.10

Amounts are in US$ million (rounding)

1	Construction schedule	Total	2001	2002	2003	2004	2005	2006
2	Pre-investment	4.50	4.50					
3	Project cost	1346	34.20	38.64	83.28	48.53	2.95	72.95
4	O & M cost	80.88	7.15	3.44	2.53	2.57	2.68	2.88
5	Depreciation	34.13	2.44	2.44	2.44	2.44	2.44	2.44
6	Revenue	1732	92.02	96.48	61.32	63.68	66.05	68.36
7	Taxes	72.05	3.80	5.71		2.26	10.88	
8	ROI							
	Inflow	1651	84.86	93.04	58.80	61.11	63.37	65.48
	Outflow	1346	34.20	38.64	83.28	48.53	2.95	72.95
	IRR							
9	ROE							
	Inflow	1545	78.63	84.89	56.36	56.41	50.06	63.05
	Outflow	1346	34.20	38.64	83.28	48.53	2.95	57.93
	IRR							

Amounts are in US$ million (rounding)

1	Construction schedule	Total	2007	2008	2009	2010	2011
2	Pre-investment	4.50					
3	Project cost	1346	57.93	150.18	306.32	165.58	173.87
4	O & M cost	80.88	2.89	2.98	12.49	14.61	15.16
5	Depreciation	34.13	2.44	2.44	2.44	2.44	2.44
6	Revenue	1732	70.72	72.31	302.66	312.13	321.62
7	Taxes	72.05	1.78			23.75	23.87
8	ROI						
	Inflow	1651	67.83	69.33	290.16	297.52	306.46
	Outflow	1346	57.93	150.18	306.32	165.58	173.87
	IRR						
9	ROE						
	Inflow	1545	63.61	66.90	287.73	271.33	280.15
	Outflow	1346	57.93	150.18	306.32	165.58	173.87
	IRR						

The analysis above gives the following results:
ROI
IRR = 32 %; BCR = 1.17
ROE
IRR = 25 %; BCR = 1.10
In both the cases of ROI and ROE, the IRR is attractive. The BCR is more than 1.00, which shows benefit is more than the cost measured at present value, which also shows that the investment is attractive.

The IRR shows the rate of return the investor is expected to get. It does not, however, show the absolute value (amount) the investor will get during the period. The NPV does not show the rate of return but shows the absolute value (amount) the investor is expected get. While the IRRs are comparable for different projects of different sizes, the NPVs are not.

16.4 Risk and Statistical Estimation of IRR

16.4.1 Risk

Risks are future problems in a project that can be avoided or mitigated. Risks are not the current problems that must immediately be resolved.

Risks involve probability of occurring at a certain time in future in a project.

In many projects requiring financial investment, there remain some uncertainties. The uncertainties create risks to the investor. So calculating the risks and their financial burden on the part of the investor are important. We shall demonstrate here how to assess the risks. Suppose that a building is going to be constructed.

Risk may be measured qualitatively and quantitatively. Qualitative assessment of risk does not give a good idea. For risks may be stated to be low risk, medium risk, and high risk. From these we do not get a good idea of the risk. Quantitative assessment involves statistical methods. Mathematically it can be stated in the following way:

Risk = (probability of occurring of the event) × (expected value of the event)

Example 3
As another illustration, suppose you are going to construct a building. It will cost $850,000. If the earthquake effect is taken into consideration, the design will be stronger, but the building costs an additional amount of $7000. The probability of earthquake occurring is 0.005.

Do you want to consider earthquake effect in the design and construct accordingly?

Calculate the risk involved and then decide.

If the earthquake occurs then the estimated relevant costs are

Cost of rebuilding the building	$850,000
Damage to other properties	$400,000
Injury to people residing	$700,000
Fine by government agency for not complying earthquake regulations	$50,000
Total	$2,000,000

$$R = 0.005 \times 2,000,000$$
$$= \$10,000 \text{ (loss)}$$

So, if earthquake is not taken into account, risk is $10,000.

If earthquake is taken into account the loss is $7000.

Note: Probability values lie between 0.00 and 1.00. In any case, probability cannot be negative. Also in no case probability can be greater than 1.00. Furthermore, probability 0.00 means that the event cannot occur and probability 1.00 means that the event is certain to occur.

16.4.2   *Statistical Estimation of IRR*

IRR is an important tool for making a project investment decision. The IRR is calculated based on costs and benefits spread over a certain period of time and by converting to present values. If any cost or benefit is changed any time, the IRR may change. So there is a risk in using this tool (IRR). For this reason an interval (range) estimation of IRR is useful. This technique may be demonstrated with the help of an example. We shall use here the example provided in the previous section.

In this example, the cost is changed by 2.5, 5, 7.5, and 10 % upward and downward. The resulting IRRs and NCFs (net cash flow) are as follows:

Example 4

Cash flow is in NPV

IRR (%)	NCF (net cash flow in 000 $)
44	439.694
40	406.040
38	372.387
35	338.733
32	305.079
29	271.425
26	237.772
23	204.118
20	170.464

By linear regression, we can get the following model ($R^2 = 0.999$; $\alpha < 0.05$).

$$IRR = 5.146 + 0.088\ NCF$$

The interpretation is simple. If net cash flow in present value is by $200 (actually $200,000 since the units are in thousand $), the IRR will be as follows:

$$IRR = 5.146 + 0.088\ NCF$$
$$IRR = 5.146 + 0.088 \times 200 = 5.146 + 17.6 = 22.70 \rightarrow 22.75\,\%$$

This model is a particular case for a particular cash flow. It cannot be used for all cash flow systems. The above analysis is to be made for the particular project. For any project investment, different situation will be considered that may affect the cash flow. Then in each of the situation, the resulting IRR and NCF are to be recorded. Thereafter, run the regression using SPSS for Windows. The interval estimation of the IRR will be made. The IRR may be assumed to follow normal distribution. Note that in this example, t-value is used not the z-value. This is done because the sample size is small. The technique is an example to show how to analyze any case. In this case we may provide the interval estimation (range) of IRR associated with its probability, instead of a single value. The interval estimate of IRR is given by the following:

$$\bar{x} - t_{\alpha/2,n-1} * \frac{s}{\sqrt{n}} \leq \mu \leq \bar{x} + t_{\alpha/2,n-1} * \frac{s}{\sqrt{n}}$$

Here,

$\bar{x}$ sample mean = 31.8889

t-value at $\alpha/2$ 2.306 ($\alpha = 0.05$, level of significance)

s	standard deviation of sample = 2.2949
n	sample size = 9
μ	IRR being estimated

Putting these values we get

$$30.00 \le \mu \le 33.40$$

This means that the IRR will lie between 30.00 and 33.40 and the probability of this is 95 %. When we say that its probability is 95 %, it means the risk is 100.00–95.00 = 5 $\rightarrow$ 5.00 %.

Risk is of more concern to any investor. In the above example, the interval estimation of IRR with 95 % confidence level is 30.00–33.40 %. This is quite attractive. However, in analyzing the risk further, the investor may want to know something more. Let us suppose he may want to see 'what is the probability that the IRR may fall below 20 %'.

This may be calculated as follows:

For t distribution we know

$$t = \frac{\bar{x} - \mu}{s/\sqrt{n}}$$

Note that μ is the mean of $\bar{x}$. Therefore, from original data we can get

$$t = \frac{x - \bar{x}}{s/\sqrt{n}}$$

Our task is to calculate the value of α and $p(\alpha)$ for $x \le 20$. First we need to calculate the t value. In our example, $x = 20$, $\bar{x} = 31.889$, $s = 2.2949$, and $n = 9$. Putting these values we get $t = -16.90$. From t table the corresponding α value is <0.001 (df is $10 - 1 = 9$). This means that the probability of IRR falling below 20 is less than 0.001 or 0.10 %. In other words, the chance of the IRR falling below 20 is extremely low. With this information, the investor can take a better decision.

16.5 Time Series Financial Data

A time series is a sequence of data points, measured typically at successive points in time spaced at uniform time intervals. Two points are important here. The data points must be measured at successive intervals of time and the time intervals must be same. Index of a stock exchange is an example of a time series data.

The purpose of analyzing time series data is generally: (i) extracting meaningful statistics and other characteristics of the data, and (ii) modeling and forecasting future values.

16.5.1 Trend Analysis

Trend can better be understood by plotting the records. Since the data are time series, the x-axis will always be time. The other set of records will be along the y-axis. If the series shows many ups and downs, trend can be analyzed by some treatment to the data. One way is to smooth the data by use of the technique of moving average. The plot with the smoothened data may make it easier to have an idea of the trend.

The moving average may be 2-period, 3-period, 4-period, 5-period, etc. The table hereafter shows the 3-period moving averages of index values of Stock Exchange of Thailand (SET) data for March 2013. The calculation may be explained by an example. The first average of 3-period moving average = (1st record + 2nd record + 3rd record)/3 and it will be placed along the 3rd record. In the following table, this value is 1543.28. The next moving average = (2nd record + 3rd record + 4th record)/3. In the table it is 1549.79 and it is placed along the 4th record.

Example 5
Moving Average of SET Index Data

Date	Index	Moving average	Date	Index	Moving average
1	1539.80		18	1591.65	1592.19
4	1540.72		19	1568.25	1586.01
5	1549.31	1543.28	20	1543.67	1567.86
6	1559.35	1549.79	21	1529.52	1547.15
7	1560.98	1556.55	22	1478.97	1517.39
8	1566.92	1562.42	25	1523.95	1510.81
11	1577.64	1568.51	26	1544.03	1515.65
12	1576.68	1573.75	27	1560.87	1542.95
13	1578.70	1577.67	28	1544.57	1549.82
14	1586.79	1580.72	29	1561.06	1555.50
15	1598.13	1587.87			

Note Dates refer to March 2013. Moving average is 3-period moving average

Another treatment to the data is exponential smoothening. This technique provides more weight to the current and records those that are near the current record. However, there is no fixed rule how much weight is to be given and where.

For analysis of the trend, Fig. 16.1 showing the Stock Exchange of Thailand index during March 2013 is plotted and may be used. There is no clear trend of the index. However, during the first half of the month, the trend is rising of the index. Thereafter, the index in general starts dropping and then again rising.

Further, we can draw a moving average line. Figure 16.2 shows the moving average (3 period moving average) line (thin line). This also does not show

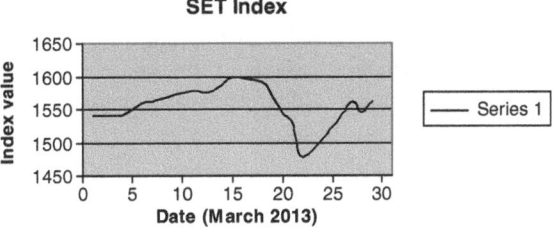

Fig. 16.1   Plot of Stock Exchange of Thailand Index During March 2013

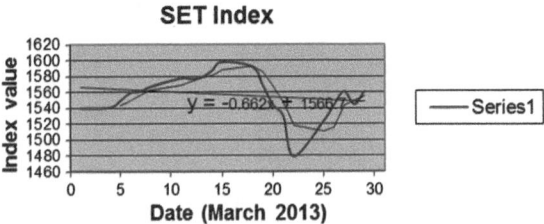

Fig. 16.2   SET Index Trend during March 2013

anything clear. However, the regression line (shown in straight line) shows a clear line with regression equation. The straight line shows a clear trend. The trend is 'the index is decreasing'. The negative coefficient of x also indicates the trend that the index is decreasing.

16.6   Models for Time Series Financial Data

There are a few models for the time series data such as autoregressive (AR) models, integrated (I) models, moving average models, and combinations of these models such as autoregressive moving average (ARMA) model and autoregressive integrated moving average ARIMA models, etc. Here we shall discuss the autoregressive model. It is quite popular with the users.

16.6.1   Autoregressive Model

Generally, the autoregressive model is given by

$$Y_t = b_0 + b_1 Y_{t-1} + b_2 Y_{t-2} + b_3 Y_{t-3} + b_4 Y_{t-4} + b_5 Y_{t-5} + \cdots + e_t$$

where,

Y_t	dependent variable at the time t
Y_{t-i}	independent variable at the time $t - i$ ($i = 1, 2, 3, \ldots$)
b_0, b_i ($i = 1, 2, 3 \ldots$)	regression coefficients; b_0 is a constant or intercept
e_i	error term (residual)

The coefficients are to be estimated.

Analysis shown here used SPSS for Windows. For using the SPSS for Windows programme, the following sequence is provided for guidance to those who are not quite conversant with SPSS for Windows.

Enter the data exactly in the same sequence as in the model. Then follow this:

SPSS	$\rightarrow$Analysis $\rightarrow$ Time Series $\rightarrow$ Variables
	$\rightarrow$ Dependable variable (highlight and press arrow)
	$\rightarrow$ Independent variables (highlight and press arrow)
	$\rightarrow$ Method $\rightarrow$ Prais–Winsten
	$\rightarrow$ Constant term (may include, later if found not significant, omit it)
	$\rightarrow$ Option $\rightarrow$ create variables $\rightarrow$ do not include
	$\rightarrow$ Predict cases $\rightarrow$ predict from estimation period through last case $\rightarrow$ Continue
	$\rightarrow$ Option $\rightarrow$ Display $\rightarrow$ Initial and final parameters without iteration summary or initial and final parameters with iteration details or final parameters only (if you want all details, any one; otherwise last one only preferably)
	$\rightarrow$ Continue $\rightarrow$ OK

The SPSS for Windows programme will run and provide results.

(a) Model Fitting

Important indicators for checking the model fitting: One is R^2. If R^2 value is high (say 0.90 or more), the model is ok. If the significance of any parameters including the constant term is less than 0.05 ($\alpha \leq 0.05$), the parameter is ok. If it is greater than 0.05 ($\alpha \geq 0.05$), the corresponding variable should be omitted. In practice maintaining this precisely may not be workable. If the significance level is not far above 0.05, it may be accepted. In such a case the error in the model may slightly be higher; but the model is workable. Do not remove all the variables at a time. Remove first the one whose α value is far above 0.05. Then check the result. Next remove the variable whose α value is far this time. Continue this until the α value is found acceptable. Then write down the model with the estimated parameters.

(b) Checking Autocorrelation (Durbin–Watson Statistic)

In Autoregressive model, the autocorrelation between the errors is an important item to check. In running the model our assumption is that the errors are independent or in other words, there is no correlation between the errors (residuals). The

Durbin–Watson d statistic is an important tool for checking this correlation. The statistic is given by

$$d = \frac{\sum_{t=2}^{n}(e_i - e_{i-1})^2}{\sum_{i=1}^{n} e_1^2}$$

Interpretation of Durbin–Watson d statistic:

$0 < d < dl$: there is positive autocorrelation
$dl < d < du$: inconclusive
$du < d < 2 + dl$: there is positive autocorrelation
$2 + dl < d < 4$: there is negative autocorrelation
d = Durbin–Watson d statistics obtained from computer run
dl = lower limit
du = upper limit

Durbin–Watson d statistics table provides the two values dl and du.

If the Durbin–Watson table is not readily available, the following simplified interpretation may be used:

$d = 0$: in this case, $\rho = 1$; it indicates positive correlation
$d = 2$: in this case, $\rho = 0$; it indicates no correlation
$d = 4$: in this case, $\rho = -1$; it indicates negative correlation

Correlation here means the correlation between errors or residuals.

In general the assessment of correlation between errors (residuals) can also be made from the following general criteria:

1. If d is closer to 2, there is no correlation or we can say the errors are independent.
2. If d is closer to 0 or 4 (two extreme values), there is correlation or we can say the errors are not independent.

Example 6
Consumer Price Index (CPI) Autoregression Data

Year	Quarter	Y_t	Y_{t-1}	Y_{t-2}	Y_{t-3}	Y_{t-4}	Y_{t-5}	Y_{t-6}	Y_{t-7}	Y_{t-8}	Y_{t-9}	Y_{t-10}
2001	1	226.7	220.2	216.7	211.1	211.1	202.4	198.3	190.7	185.2	181.7	177.1
2001	2	227.7	221.3	216.7	212.2	211.7	203.5	198.7	191.8	186.2	183.1	177.8
2001	3	229.4	223.5	217.6	212.7	213.5	205.4	199.8	193.3	187.4	184.2	178.8
2001	4	230.1	224.9	218	213.2	214.8	206.7	201.5	194.6	188	183.8	179.8
2001	5	229.8	226	218.2	213.9	216.6	207.9	202.5	194.4	189.1	183.5	179.8
2001	6	229.5	225.7	218	215.7	218.8	208.4	202.9	194.5	189.7	183.7	179.9
2001	7	229.1	225.9	218	215.4	220	208.3	203.5	195.4	189.4	183.9	180.1
2001	8	230.4	226.5	218.3	215.8	219.1	207.9	203.9	196.4	189.5	184.6	180.7

(continued)

(continued)

Year	Quarter	Y_t	Y_{t-1}	Y_{t-2}	Y_{t-3}	Y_{t-4}	Y_{t-5}	Y_{t-6}	Y_{t-7}	Y_{t-8}	Y_{t-9}	Y_{t-10}
2001	9	231.4	226.9	218.4	216	218.8	208.5	202.9	198.8	189.9	185.2	181
2001	10	231.3	226.4	218.7	216.2	216.6	208.9	201.8	199.2	190.9	185	181.3
2001	11	230.2	226.2	218.8	216.3	212.4	210.2	201.5	197.6	191	184.5	181.3
2001	12	229.6	225.7	219.2	215.9	210.2	210	201.8	196.8	190.3	184.3	180.9
Yt	2012											

Data Source Raw Data from Federal Reserve Bank of Dallas: re-arranged by author. Index 2005 = 100

Results (computer output) of the autoregressive model using SPSS for Windows are summarized here.

Model Description:

Variable	Y_t	(2012); dependent variables
Regressors	Y_{t-1}	(2011)
	Y_{t-2}	(2010)
	Y_{t-3}	(2009)
	Y_{t-4}	(2008)
	Y_{t-5}	(2007)
	Y_{t-7}	(2005)
	Y_{t-8}	(2004)
	Y_{t-9}	(2003)

Y_{t-6} was not significant, so it was removed. All others were included in the model.

Summary of the computer output (result) is provided hereafter.

Result Description

All the variables were entered in the SPSS run. The dependent variable, as always the case, was Y_t.

Estimation of autocorrelation coefficient	
Rho	0

Prais–Winsten estimates	
Multiple R	0.99998464
R-squared	0.99996928
Standard error	12.148562
Durbin–Watson	0.95602977

Analysis of Variance:

	DF	Sum of squares	Mean square
Regression	2	48046536.1	24023268.1
Residuals	10	1475.9	147.6

Variables in the Equation:

	B	SEB	BETA	T	Sig T
Y_{t-1}	−6.870162	1.0387128	−0.0252689	−6.61411	0.0000592
Y_{t-9}	10.474562	0.0391419	1.0223705	267.60478	0.0000000

(c) Interpretation of the result

Based on the output as shown above, the autoregressive model may now be written as follows:

$$Y_t = -6.70162Y_{t-1} + 10.474562\, Y_{t-9}$$

(a) The constant term is not significant, so it does not appear in the model.
(b) The coefficients of Y_{t-1} (−6.870162) and Y_{t-9} (10.474562) are significant ($\alpha \leq 0.05$), so only these two independent variables appear in the model.
(c) Value of R^2 is 0.99996928 which means that more than 99.99 % variation in the dependent variable (Y_t) is explained by these two independent variables (Y_{t-1} and Y_{t-9}). Thus, the model is a good fit.
(d) Durbin–Watson d statistic is 0.95602977 which is closer to one. This means that there is some autocorrelation between errors.
(e) Other things being constant, if CPI in the year Y_{t-1} increases by 1, the CPI of the current year would be decreased by 6.70162. If CPI in the year Y_{t-9} increases by one, the current year CPI would be increased by 10.474562.

Example 7
Personal consumption expenditure (PCE) inflation rate data for Dallas from 2001 to 2012 are shown in the following table. Establish an autoregressive model and interpret the results.

	Y_t	Y_{t-1}	Y_{t-2}	Y_{t-3}	Y_{t-4}	Y_{t-5}	Y_{t-6}	Y_{t-7}	Y_{t-8}	Y_{t-9}	Y_{t-10}	Y_{t-11}
Month	2012	2011	2010	2009	2008	2007	2006	2005	2004	2003	2002	2001
January	2.08	0.98	1.31	2.42	2.56	2.88	2.47	2.41	1.87	2.08	2.24	2.47
February	2.02	1.12	1.19	2.40	2.45	3.03	2.37	2.45	1.92	2.04	2.23	2.48
March	2.02	1.27	1.10	2.28	2.52	2.97	2.39	2.48	1.92	2.03	2.24	2.44

(continued)

(continued)

	Y_t	Y_{t-1}	Y_{t-2}	Y_{t-3}	Y_{t-4}	Y_{t-5}	Y_{t-6}	Y_{t-7}	Y_{t-8}	Y_{t-9}	Y_{t-10}	Y_{t-11}
April	1.94	1.43	1.00	2.22	2.56	2.86	2.49	2.39	2.13	1.88	2.21	2.53
May	1.89	1.55	0.95	2.09	2.63	2.70	2.63	2.32	2.25	1.85	2.16	2.55
June	1.87	1.62	0.89	1.97	2.72	2.60	2.78	2.23	2.38	1.79	2.05	2.64
July	1.82	1.71	0.93	1.68	2.85	2.54	2.83	2.21	2.39	1.80	2.07	2.59
August	1.74	1.78	0.95	1.67	2.79	2.46	2.91	2.24	2.36	1.83	2.06	2.62
Sept	1.75	1.81	0.96	1.62	2.65	2.56	2.77	2.43	2.30	1.79	2.13	2.50
October	1.72	1.91	0.83	1.65	2.54	2.62	2.71	2.52	2.32	1.77	2.13	2.47
Nov	1.67	1.97	0.85	1.54	2.47	2.76	2.63	2.55	2.34	1.75	2.12	2.44
Dec	1.58	2.06	0.88	1.50	2.35	2.70	2.76	2.53	2.35	1.79	2.12	2.36

Data Source Raw Data from Federal Reserve Bank of Dallas: re-arranged by author
The Output from the computer run is shown hereafter

Model Description:

Variable	Y_t	Dependent variable
Regressors	Y_{t-1}	
	Y_{t-3}	
	Y_{t-4}	
	Y_{t-7}	

Final Parameters:

Estimate of autocorrelation coefficient	
Rho	0

Prais–Winsten estimates	
Multiple R	0.99997356
R-squared	0.99994712
Standard error	0.01645571
Durbin-Watson	2.4941414

Analysis of Variance:

	DF	Sum of squares	Mean square
Regression	4	40.963834	10.240958
Residuals	8	0.002166	0.000271

Variables in the Equation:

	B	SEB	BETA	T	Sig T
Y_{t-1}	-0.1852705	0.04832493	-0.16384964	-3.833849	0.00499057
Y_{t-3}	0.28084766	0.04515018	0.29621570	6.220299	0.00025376
Y_{t-4}	0.37198087	0.03132903	0.52234736	11.873361	0.00000232
Y_{t-7}	0.37198087	0.04499271	0.34424727	5.891641	0.00036517

The model is therefore,

$$Y_t = -0.18527050Y_{t-1} + 0.28084766Y_{t-3} + 0.37198087Y_{t-4} + 0.26508091Y_{t-7}$$

Interpretation

The dependent variable is Y_t;

1. The R^2 value is 0.99994712. This is close to 1.00. Therefore, the model has a good fitting;
2. Coefficients of the independent variables Y_{t-1}, Y_{t-3}, Y_{t-4}, and Y_{t-7} are significant ($\alpha \le 0.05$). Therefore, only these variables are in the model.
3. Durbin–Watson statistic $d = 2.4941414$. This is closer to 2. So practically there is no autocorrelation between the errors.

The coefficients of Y_{t-1}, Y_{t-3} and Y_{t-4} and Y_{t-7} are the quantity increase or decrease in the value of Y_t for one unit change in the respective variables. By partial differentiation, it can also be shown that these are the rates of changes in Y_t with respect to the respective variables.

16.7 Forecasting

16.7.1 Forecasting Without a Model

Forecasting is the process of making prediction on some future values of an event based on past or present values of the event. This refers to time series dimension. Forecasting for some period based on time series data can be done in a simple way using some trend. However, this simple way of forecasting has some limitation. Let us use an example to show the technique.

CPI for first quarter and second quarter of USA during 2012 were as follows:

Quarter I	227.9066
Quarter II	224.7790

Between Quarter I & II

Per Cent Change $= 100 \times (227.9066 - 224.7790)/227.9066 = 1.3723$

If this is accepted, then CPI for the whole year will be $227.9066 * (1 + 0.013723) = 231.03$. Actually the average CPI of the whole year was 229.594.

Based on the principle we can also calculate the returns on financial investment. Returns may be expressed in two ways—simple return (usually denoted by R) and log return (usually denoted by r). The calculation is as follows:

Let us suppose that in a time series data price of an asset was \$227.9066 (P_t) at a time t and \$224.7790 (P_{t-1}) at a time $t - 1$. The simple return is defined by

$$R = \frac{P_t - P_{t-1}}{P_{t-1}}$$

$\rightarrow$

$$R = \frac{227.9066 - 224.7790}{224.7790} = \frac{3.1276}{224.7790} = 0.0139$$

In percentage form we can say that the simple return is 1.39 %. The other form is log return (r) to be calculated as

$$r = \text{Log} \frac{P_t}{P_{t-1}} = \text{Log} P_t - P_{t-1}$$

Another expression,

$$\text{Log}(1 + R) = \text{Log}\left(1 + \frac{P_t - P_{t-1}}{P_{t-1}}\right) = \frac{P_{t-1} + P_t - P_{t-1}}{P_{t-1}} = \frac{P_t}{P_{t-1}} = r$$

So,

$$r = \text{Log} 227.9066 - \text{Log} 224.7790 = 2.357757 - 2.351756 = 0.006001$$

In percentage form we can say that the log return (r) is 0.060 %.

Note that the simple return (R) and log return (r) do not produce the same figures. These are two items and so the figures are different. Usually in financial investment log return is used.

16.7.2   Forecasting Using a Model

Forecasting is the most important application of model. In the previous sections only exploratory analyses have been done. This section will deal with the forecasting. There are different ways to forecast. We shall discuss the ones which are easy to apply. We shall use the inflation rate example. The data structure is as follows:

Example 8

Y_t	Y_{t-1}	Y_{t-2}	Y_{t-3}	Y_{t-4}	Y_{t-5}	Y_{t-6}	Y_{t-7}	Y_{t-8}	Y_{t-9}	Y_{t-10}	Y_{t-11}
2012	2011	2010	2009	2008	2007	2006	2005	2004	2003	2002	2001

The model is

$$Y_t = -0.18527050\, Y_{t-1} + 0.28084766 Y_{t-3} + 0.37198087 Y_{t-4} + 0.26508091 Y_{t-7}$$

If we want to forecast the inflation rate in 2013, the time lag will be as follows:

Variable	Year	Average inflation rate
Y_t	2013	
Y_{t-1}	2012	1.81
Y_{t-3}	2010	0.88
Y_{t-4}	2009	1.92
Y_{t-7}	2006	2.65

Putting these values we get

$$Y_t = -0.18527050 * 1.84 + 0.28084766 * 0.88$$
$$+ 0.37198087 \times 1.92 + 0.26508091 * 2.65$$

Or,

$$Y_t = -0.3408 + 0.2471 + 0.7142 + 0.7024 = 1.3229$$

So the inflation rate in 2013 is likely to be 1.3229.

If you want to estimate the inflation rate in 2014, you will have to estimate it for 2013 and then 2014.

16.8 Seasonal Variation

Different authors have explained some techniques to calculate the seasonal variation in time series data. Some of those techniques are too technical and theoretical. These are complicated and in practical field these are difficult to apply. In this section we shall explain how easily the seasonal variation can be calculated by using dummy variables.

Suppose we want to estimate the seasonal variation occurring due to change in quarters of the year. In such a case, appropriate dummy variables may be introduced. These dummy variables may be taken to be D_1 (1st quarter), D_2 (2nd quarter), D_3 (3rd quarter), and D_4 (4th quarter). Values of these dummy variables are as follows:

D_1 1, if the record falls in first quarter;
    0, otherwise
D_2 1, if the record falls in second quarter;
    0, otherwise
D_3 1, if the record falls in third quarter;
    0, otherwise
D_4 1, if the record falls in fourth quarter;
    0, otherwise

This may be explained further with the help of an example.

Example 9

Consumer price indexes for USA for 2008–2012 are shown in the following table. The example shows how to estimate the seasonal variation by use of dummy variables.

The data structure is shown in the following table.

Year	2012	2011	2010	2009	2008				
Variable	Y_t	Y_{t-1}	Y_{t-2}	Y_{t-3}	Y_{t-4}	D_1	D_2	D_3	D_4
Jan	226.7	220.2	216.7	211.1	211.1	1	0	0	0
Feb	227.7	221.3	216.7	212.2	211.7	1	0	0	0
Mar	229.4	223.5	217.6	212.7	213.5	1	0	0	0
Apr	230.1	224.9	218.0	213.2	214.8	0	1	0	0
May	229.8	226.0	218.2	213.9	216.6	0	1	0	0
June	229.5	225.7	218.0	215.7	218.8	0	1	0	0
July	229.1	225.9	218.0	215.4	220.0	0	0	1	0
August	230.4	226.5	218.3	215.8	219.1	0	0	1	0
Sept	231.4	226.9	218.4	216.0	218.8	0	0	1	0
Oct	231.3	226.4	218.7	216.2	216.6	0	0	0	1
Nov	230.2	226.2	218.8	216.3	212.4	0	0	0	1
Dec	229.6	225.7	219.2	215.9	210.2	0	0	0	1

Data Source Raw data from Consumer Price Index (CPI-U), Department of Labor, Bureau of Labour Statistics, Washington D.C., 20212; rest arranged by author

Summary of the output appears below. The output is obtained from the run of SPSS Windows (time series data; autoregressive model).

Model Description:

All the variables were entered into the run.

Final Parameters:

Estimate of autocorrelation coefficient	
Rho	0

Prais–Winsten estimates	
Multiple R	0.99999355
R-squared	0.99998709
Adjusted R-squared	0.99998064
Standard error	1.0102265
Durbin-Watson	1.3061489

Analysis of Variance:

	DF	Sum of squares	Mean square
Regression	4	632572.39	158143.10
Residuals	8	8.16	1.02

Variables in the equation:

	B	SEB	BETA	T	Sig T
D_2	229.79267	0.58325454	0.500422486	393.98350	0.0000000
D_1	227.90667	0.58325454	0.49631767	390.74992	0.0000000
D_3	230.29667	0.58325454	0.50152243	394.84762	0.0000000
D_4	230.37967	0.58325454	0.50170318	394.98992	0.0000000

Interestingly, the lag variables (Y_{t-1}, Y_{t-2}, Y_{t-3}, Y_{t-4}) were not significant (α not ≤ 0.05). All the four dummy variables (D_1, D_2, D_3, and D_4) were significant. Remember that this situation has occurred with the particular set of data. If the data contain different figures, the estimation and consequently the seasonal variation may be different. The estimated model is

$$Y_t = 227.90667 \times D_1 + 229.79267 \times D_2 + 230.29667 \times D_3 + 230.37967 \times D_4$$

If we want to estimate the CPI in the third quarter, we can do it in the following way:

$$CPI_3 = 227.90667 \times 0 + 229.79267 \times 0 + 230.29667 \times 1 + 230.37967 \times 0$$
$$= 230.29667$$

Note that since there is no lag variable in the estimated model, the estimated CPI for any quarter of any year will remain the same for that particular quarter.

References

Raw Data from Federal Reserve Bank of Dallas: re-arranged by author. Index 2005 = 100
Raw data from Consumer Price Index (CPI-U), Department of Labor, Bureau of Labour Statistics,
    Washington D.C., 20212; rest arranged by author
Stock Exchange of Thailand Index during March 2013

Chapter 17
Experimental Design

Abstract Technical terms such as experimental design, factor, treatment, controlled experiment, and factorial design (full, half, fractional, balanced/unbalanced, contrast/orthogonal) are explained. A systematic procedure is set. A flow chart showing the commonly used designs of experiments is provided. Full illustrations are provided showing development of completely randomized design and a 2^3 full factorial design. It is shown how the concept of ANOVA is utilized in the analysis. A chart is provided. Using SPSS, the techniques of how the concept of ANOVA is utilized to design single factor experiments are demonstrated. Examples are provided.

Keywords Design of experiment · Flow chart · Factorial design · Randomized design · ANOVA · SPSS · Diagram · Factor action

Scientists and engineers conduct experiments to examine certain phenomena and to conclude how good or useful the outcomes of the experiments are.

17.1 Definition of Design of Experiments

As per Sigma (Sigma, isixsigma.com/dictionary/Design_of_Experiments 22 October 2009), three definitions of design of experiments (DOE) may be cited as follows:

(a) DOE is a structured, organized method for determining the relationship between factors (xs) affecting a process and the output of that process (y).
(b) Design is conducting and analyzing controlled tests to evaluate the factors that control the value of a parameter or group of parameters.
(c) DOE refers to the experimental methods used to quantify indeterminate measurements of factors and interactions between factors statistically through observance of forced changes made methodically as directed by mathematically systematic tables.

© Springer Science+Business Media Singapore 2016

A.Q. Miah, *Applied Statistics for Social and Management Sciences*,
DOI 10.1007/978-981-10-0401-8_17

In an experiment, the design is made, the experiment is conducted and the result is analyzed to conclude to test the hypotheses that were perceived beforehand.

17.2   Terms Related to Experimental Design

(i) Experimental Design

The experimenter conducts an experiment to get an answer to his question. He generates data and runs the experiment and analyzes the results. He is to ensure that he collects right type of data, enough data, and analyze the results in such a way that it provides efficient and clear interpretation. The whole process is called experimental design.

(ii) Factor

A factor in an experiment is a controlled independent variable whole levels are set by the experimenter.

(iii) Treatment

A treatment is something that an experimenter administers to the experimental units. For example, an experimenter wants to study the effect of alloy concentration on the product. The factor is alloy concentration. He administers three concentrations 10, 20, and 30 % to study the effect. These 10, 20, and 30 % of concentration are three treatments on the factor. The treatments may also be called levels. Treatments are subdivisions of a factor.

(iv) Controlled Experiment

A controlled experiment is one in which controlled treatments are applied.

(v) Factorial Design

A factorial design is one in which two or more factors are tested. In each factor, two or more levels (treatments) are chosen.

Effect on the response variable (experimental unit) may come from level 1, level 2, or combination of levels 1 and 2 and interaction). This interaction may be studied if the experimenter thinks that there may be significant effect of the interaction on the response variable.

A factorial design helps us in the following ways:

(a) We can identify which factor has a significant effect on the response variable.

(b) We can examine whether the interaction among the factors is significant.

(c) We can identify which factor has the most important effect on the response variable.

(d) Based on the above, we can decide whether we need further investigation.

(vi) Full Factorial Design

Number of runs in a full factorial design depends on the number of factors and the levels. For example, 2^3 is a full factorial design. In this design there are three factors and each factor has two levels. Total number of runs in this design $= 2^3$. If there are k factors and two levels in each factor, then total number of runs $= 2^k$.

(vii) Half Factorial Design

In a half factorial design, number of runs = half of the runs in the full factorial design. For example, in a 2^4 half factorial design, we test 8 experiments instead of 16.

(viii) Fractional Factorial Design

It is a factorial design where the number of tests is less than that in a full factorial design.

(ix) Balanced/Unbalanced Factorial Design

If the number of levels in all the factors is the same, the design is called a balanced design. If the number of levels in all the factors is not same, then the design is called an unbalanced design.

(x) Contrast/Orthogonal

A contrast is a linear combination of two or more factor level means with coefficients that sum to zero.

Two contrasts are orthogonal if the sum of the products of corresponding coefficients (i.e., coefficients for the same means) adds to zero. Mathematically,

$$C = c_1\mu_1 + c_2\mu_2 + c_3\mu_3 + \cdots + c_k\mu_k$$

where

$$\sum_{j=1}^{k} c_j = c_1 + c_2 + c_3 + \cdots + c_k = 0$$

17.3 Procedure for Design of Experiments

(i) Identification of the Problem

Problem identification is the first step in the experimental design. Once the problem is identified, other steps will follow. If the problem can be identified precisely, the volume of works will be reduced and the results can be pinpointed. The design should be as simple as possible so that the results or outcomes of the experiments can be correctly interpreted.

(ii) Response Variable

Two types of variables are involved. One is the response or dependent variable and the other is the independent variable. This is also called factor. A response variable or dependent variable is one which depends on the measurements of other variables called independent variables. If a mistake is made in determining the dependent variable, the whole experiment may go wrong. The dependent variable must be measurable.

Get a clear understanding of the inputs and outputs under investigation.

(iii) Factors or Independent Variables

In the experimental design, the independent variables may also be called "factors." Factors are independent variables. The factors influence the response or dependent variables. For example, compressive strength of concrete depends on the curing method. So compressive strength is a dependent variable and the curing method is an independent variable. The independent variables must also be measurable.

(iv) Factor Levels

Determine the number of levels of the factors. The number of levels also called treatments will depend on the type and the objective of the experiment. For example, if we want to study the effect of curing time on compressive strength of concrete, then the levels of the factor (curing time) may be three (7, 10, and 15 days). The levels should be realistic. Avoid attribute measures such as yes/no, pass/fail. If necessary, categorical measures may be numerically coded, for example, 1 for low and 2 for high.

(v) Possible Interactions

The main effect comes from the factors individually. The effect on the response variable may also come from various combinations of the factors. For example, if there are two independent variables A and B, then the meaningful effect may come from A, B, and AB together.

(vi) Replicates

Set the number of replicates for the experiment. This will however depend on the type of the experiment. The purpose is to obtain statistically significant data for interpretable outcomes.

(vii) Randomization

In the statistical theory of DOE, randomization involves randomly allocating the experimental units across the treatment groups. Randomization provides reliable data and the results.

(viii) Design Matrix

It is important to create a design matrix for the factors under investigation. In the design matrix, the main effects of the factors as well as interactions are shown. It is very useful to develop a specific mathematical model for the specific experiment. This will make the experiment valid and eliminate the anomalies otherwise possible.

(ix) Running the Experiment

The experiment should be supported by actual and factual data. Collect and use the actual and relevant data. Avoid collection of unnecessary data. If unnecessary data are used, the model may produce erroneous results.

17.4   Types of Designs

The type of design depends on the particular experiment and the objective set out by the experimenter before the experiment is conducted. The chart appearing hereafter shows the commonly used designs. An explanation of each of these designs is provided after the design chart.

Flow Chart Showing the Commonly Used Designs of Experiments.

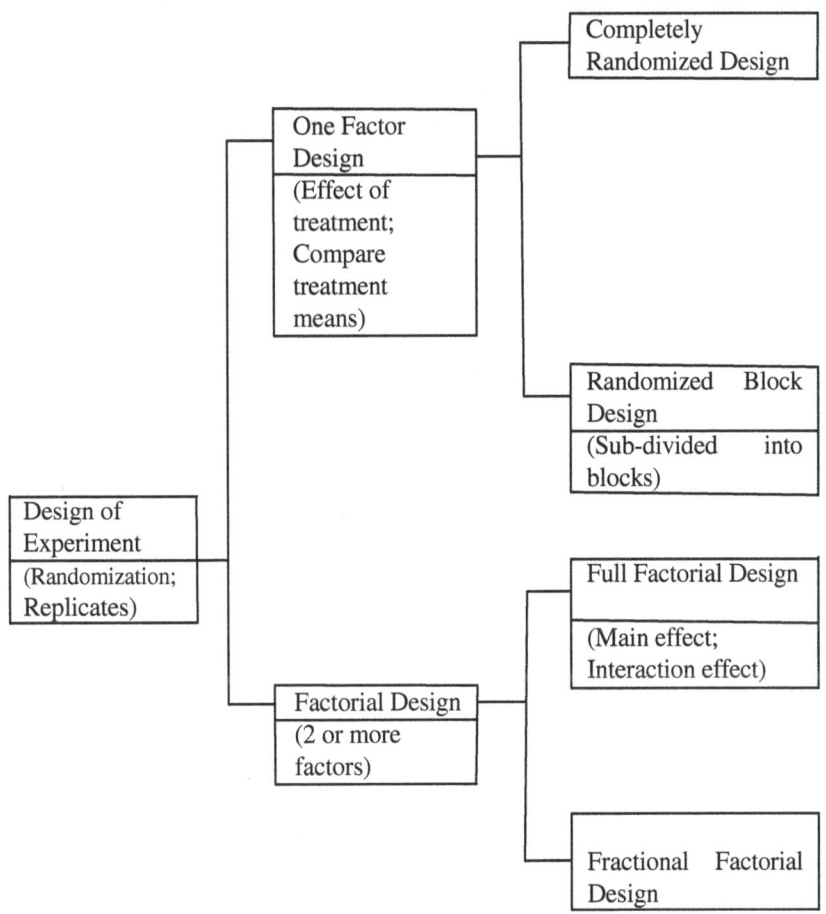

(a) Regression (not shown in the chart)

We can also call it as a traditional design. This is particularly useful if there are several factors perhaps four or more) involved. However, it cannot be controlled for a few factors. Regression may be used for estimating the statistical relationship between the factor and the response variable. But it is not efficient to measure the changes caused to the response variable as a result of certain level change of the factor values.

(b) One Factor design

The design involves one factor (independent variable) that is expected to affect the response variable. The treatment effect can be tested. Also the individual treatment means can be tested.

(c) Completely Randomized design

This is a specific case of one factor design. This is the primary factor. We are primarily interested in studying the effect of this factor on the response variable. There may be other variables which may be nuisance variables. We control those nuisance variables. As the name indicates, the design will also involve randomization. The sequence of the experimental units is determined at random. Randomization provides the validity of the results.

In a completely randomized design, the sample size (number of runs) is given by

$$N = k \times L \times n$$

where

N sample size (number of runs)
K number of factors ($=1$ for completely randomized design)
L number of levels
N number of replications

If the design is a balanced design, the number of replications should be same in all levels of the factor.

17.5 Illustration of a Completely Randomized Design

The following is a layout of a completely randomized design with three levels and five replicates. The X values are responses. In the experiment, replace the X values by the actual responses and the level 1, 2, and 3 by the actual values of the levels.

T. level	Replicate					Replicate total
	1	2	3	4	5	
1	X_{11}	X_{12}	X_{13}	X_{14}	X_{15}	X_1
2	X_{21}	X_{22}	X_{23}	X_{24}	X_{25}	X_2
3	X_{31}	X_{32}	X_{33}	X_{34}	X_{35}	X_3

If this is translated into the SPSS spreadsheet, it will look like the following:

Treatment	X value
1	X_{11}
1	X_{12}
1	X_{13}
1	X_{14}
1	X_{15}
2	X_{21}
2	X_{22}
2	X_{23}
2	X_{24}
2	X_{25}
3	X_{31}
3	X_{32}
3	X_{33}
3	X_{34}
3	X_{35}

The SPSS output for ANOVA will look like the following:

ANOVA

Source of variation	Sum of squares	Degrees of freedom	Mean square	F
Model (treatment)				
Error				
Total				

With the help of F value, results can be interpreted and a conclusion can be drawn whether the treatment effect is significant or not.

(d) Randomized Block Design

Sometimes the experimental units may not be homogeneous. There may be some grouping that may cause some effect on the analysis. However, grouping may be incorporated into the design if it is clear that certain grouping is available in the experimental units. Example of such a situation is the income of the people. When we want to study the income of the people, the grouping in terms of sex (male and female) may be evident. In this case we should introduce the block design involving two blocks (male and female). The analysis will reveal the effect of the treatments as well as blocks on the response variable.

Illustration of a Randomized Block Design.

The following is an illustration layout for a randomized block design. In this illustration, number of treatment (level) is 3 and number of block is 4.

T. level	Block				Replicate total
	1	2	3	4	
1	X_{11}	X_{12}	X_{13}	X_{14}	X_1
2	X_{21}	X_{22}	X_{23}	X_{24}	X_2
3	X_{31}	X_{32}	X_{33}	X_{34}	X_3
Block total					

When the values are translated into SPSS spreadsheet, it will look like the following:

Treatment	Block	X value
1	1	X_{11}
1	2	X_{12}
1	3	X_{13}
1	4	X_{14}
2	1	X_{21}
2	2	X_{22}
2	3	X_{23}
2	4	X_{24}
3	1	X_{31}
3	2	X_{32}
3	3	X_{33}
3	4	X_{34}

The SPSS output for ANOVA will appear like the following:

ANOVA

Source of variation	Sum of squares	Degrees of freedom	Mean square	F
Model (treatment)				
Block				
Error				
Total				

With the help of the calculated F values, we can interpret the result and conclude whether the treatment and block effects are significant or not.

(e) Factorial Design

Factorial Design is one in which there are two or more factors involved. This design can be divided into two sets. One is full factorial design and the other is fractional factorial design. There is no limitation as to how many factors can be. Also, the number of levels can be 2 or more. In a full factorial design, we can test the main effect as well as the interaction effect. Main effect is the effect of a factor that it provides to the response variable. Interaction effect is the effect of two or more factors jointly. This may be explained in the following way.

Let us suppose there are two factors A and B in an experiment. In addition to the direct effect of A and B to the response variable, A may have some effect on B and B may have some effect on A. These are interaction effects. These interaction effects may have some effect on the response variable.

17.6   Illustration of a 2^3 Full Factorial Design

The layout of the 2^3 Full Factorial Design is given in the following table. The design has three factors and two levels as indicated by 2^3. The figures shown inside the table are levels. Each of the three factors has two levels (1 and 2). The interaction levels are automatically set by the computer (SPSS Window). The experimenter does not need to set it. These are shown in the table for completeness. Note the following:

Levels 1 and 2 are not the actual level values. These are only notations. In the experiment actual values are to be used.

Level 1 represents the lower value of the factor.
Level 2 represents the higher value of the factor.
Lower value of factor 1 × lower value of factor 2 = higher value of interaction 1 × 2.
Lower value of factor 1 × higher value of factor 2 = lower value of interaction 1 × 2.
Higher value of factor 1 × higher value of factor 2 = higher value of interaction 1 × 2.

The concept is like (when + represents higher value and—represents lower value):

$$(-) \times (-) = (+)$$
$$(-) \times (+) = (-)$$
$$(+) \times (-) = (-)$$
$$(+) \times (+) = (+)$$

Run	Factor			Interaction				Response
	1	2	3	1 × 2	1 × 3	2 × 3	1 × 2 × 3	
1	1	1	1	2	2	2	1	
2	1	1	2	2	1	1	2	
3	1	2	1	1	2	1	2	
4	1	2	2	1	1	2	1	
5	2	1	1	1	1	2	2	
6	2	1	2	1	2	1	1	
7	2	2	1	2	1	1	1	
8	2	2	2	2	2	2	2	

The interactions act like factors. Whether some or all interactions produce significant effect on the response variable will be evident after testing with the help of ANOVA.

If this design is translated into the SPSS spreadsheet, it will look like the following:

Factor 1	Factor 2	Factor 3	Response
1	1	1	
1	1	2	
1	2	1	
1	2	2	
2	1	1	
2	1	2	
2	2	1	
2	2	2	

If this is run into SPSS with appropriate ANOVA options, then the output for ANOVA will look like the following:

ANOVA

Source of variation	Sum of squares	Degrees of freedom	Mean square	F
Model				
Factor 1				
Factor 2				
Factor 3				
Factor 1 * factor 2				
Factor 1 * factor 3				
Factor 2 * factor 3				
Factor 1 * factor 2 * factor 3				
Error				
Total				

By calculating the respective F values, it can be concluded which factor has significant effect on the response variable.

(f) Fractional Factorial Design

A fractional factorial design is one in which the sample size (number of runs) is less than that in a full factorial design. An example may be given here to explain the concept. The number of runs for a 2^3 full factorial design is 8 ($2^3 = 8$). If we talk of a half factorial design, the number of runs will be 4. If in a design, the experimenter is confident that the interaction effect will not be significant, then the interaction terms may be omitted. In such a case the design will be a fractional factorial design.

17.7   Concept of Anova

17.7.1   Procedure in the Analysis

Main purpose of an analysis in a factorial design is to study the effect of the selected factors on the response variable. In other words, we want to see whether the selected factors have statistical significant effect on the response variable. We may also want to see the change in the effect of the response variable as a result of change in the treatment (level) of a factor.

The effect of factors on the response variable is studied with the help of Analysis of Variance (ANOVA). We have seen in case of regression-based ANOVA that the total sum of squares is divided into sum of squares due to model and sum of squares due to error. But in the case of ANOVA for factorial design, the sums of squares are partitioned into sums of squares due to model, each of the factors and error. From these sums of squares, the respective mean square is calculated. From the mean squares, the corresponding F statistic is calculated. This F statistic is compared with the tabulated critical F values. Then we conclude whether a certain factor has significant effect on the response variable or not. Follow the steps as outlined hereafter.

Run the analysis program such as SPSS.

Prepare the ANOVA

State the null hypothesis.

Partition the sums of squares (variability).

Calculate the mean squares.

Calculate the F ratios.

Decide whether the null hypothesis is rejected or not based on a preset α level.

Interpret the result.

The concept of factor action and the analysis may be summarized as per following diagram.

Concept

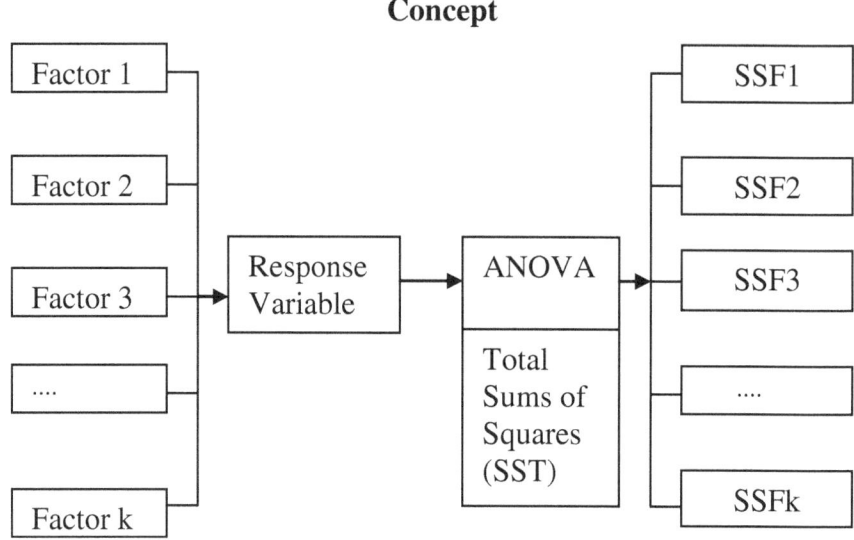

SST = sum of squares total; SSF1 = sum of squares due to factor 1; SSF2 = sum of squares due to factor 2; SSF3 = sum of squares due to factor 3; SSFk = sum of squares due to factor k; SSE = sum of squares due to error.

$$SST = SSF1 + SSF2 + SSF3 + \cdots + SSFk = SSE$$

17.7.2   Test with ANOVA

We have used Z test and t test for comparing the population means. It may be noted that in these tests, we could compare the two population means. We can compare population means of more than two populations by analysis of variance (ANOVA). We shall do it with the help of an example.

Example 1
Randomly selected students of a course were assigned four different teaching methods (assignment, tutorial, group discussion, none) in addition to lecture. Their scores were found as shown in the following table:

1 (None)	2 (Assignment)	3 (Tutorial)	4 (Group discussion)
56	60	70	68
58	65	75	80
45	54	68	75
60	60	74	82
55	66	82	90
Col total = 274	305	369	395

Here,

$$n_1 = 5, \ n_2 = 5, \ n_3 = 5, \ n_4 = 5.$$
$$n = n_1 + n_2 + n_3 + n_4 = 20.$$
$$\bar{x} = 67.15.$$

Our interest is to see whether the four teaching methods produce the same result. In other words, whether the mean scores produced by the four different teaching methods are same. In order to test this, we formulate the following hypotheses:

$$H_0 : \mu_1 = \mu_2 = \mu_3 = \mu_4$$

H_1: All the means are not same.
For the test let us assume $\alpha = 0.05$.
In order to test the null hypothesis, we calculate the sums of squares as follows:

$$SS_{Total} = \sum_{i=1}^{4} \sum_{j=1}^{n_i} x_{ij}^2 - n(\bar{x})^2$$
$$= (\text{sum of squares of all } x \text{ values}) - n(\bar{x})^2$$
$$= (56)^2 + (58)^2 + (45)^2 + \ldots + (90)^2 - 20(67.15)^2$$
$$= 92{,}669.00 - 90{,}182.45$$
$$= 2486.55$$

$$SS_{Treatment} = \sum_{i=1}^{4} \{(col)_i total\}^2 / n_i - n(\bar{x})^2$$
$$= (\text{sum of squares of treatment totals, each square divided by its } n \text{ value}) - n(\bar{x})^2$$
$$= (274)^2/5 + (305)^2/5 + (369)^2/5 + (395)^2/5 - 20(67.15)^2$$
$$= 92{,}057.40 - 90{,}182.45$$
$$= 1874.95$$

$$SS_{error} = SS_{total} - SS_{Treatment}$$
$$= 2{,}486.55 - 1874.95$$
$$= 611.60$$

These may be arranged in the following ANOVA table:

Source of variation	SS	df	MS	F_0 value
Treatment	1874.95	3	624.98	16.36
Error	611.60	16	38.23	
Total	2486.55	19		

Here the calculated value of F_0 is 16.38. The F_0 value $F_{0.05,3,16}$ is 3.239. This is less than the calculated F_0 value. In other words, $F_0 > F_{0.05,3,16}$. The calculated value of F falls in the rejection region of H_0. The null hypothesis is, therefore, rejected. This means that the means are not same. In other words, four different teaching methods produce different results. In this example, it has been demonstrated how analysis of variance helps to conclude whether the treatments have significant effects on the students' performance.

The main concept in the ANOVA concept is to calculate the sums of squares of the items of our interest and/or partition the sums of squares, calculate the mean squares (MS), calculate the F_0 value, and compare this F_0 value with that in the table for the given degrees of freedom and α. Then based on the F values, either reject the null hypothesis or do not reject it. Thereafter, make inference or interpretation. This procedure can be shown in the following sequence:

(i) Formulate the null hypothesis. The null hypothesis is "there is no difference between the means."
(ii) Calculate the sums of squares (for total, treatments, errors).
(iii) Calculate the degrees of freedom for all items.
(iv) Calculate the mean squares.
(v) Calculate the F values.
(vi) Based on the F values, decide whether to reject or not to reject the null hypothesis.
(vii) Interpret the results.

17.8   Single Factor Experiments

A single factor experiment is one in which the number of factor is one. In the following sections, the techniques are explained with the help of examples how to analyze the variance and interpret the test results.

17.8.1   Analysis of Variance

In a manufacturing process, an experimenter wanted to study the effect of content of an element on the hardness of an alloy which uses the element to manufacture the alloy. Four content levels (5, 10, 15, and 20 %) were used. For each concentration, six measurements (replicates) were used. The experiment is shown in the following table:

Example 2

Content of an Element in an Alloy

Cont.	Measurement						Total
	1	2	3	4	5	6	
5	8	10	19	13	11	12	73
10	14	20	16	22	23	18	113
15	17	22	23	20	19	22	123
20	22	30	26	28	22	24	152
Total							461

The calculations proceed in the following way:

$$SST = (\text{sum of squares of observations}) - (\text{sum of squares of observations})^2 / (\text{no. of observations})$$

$$SST = (8)^2 + (14)^2 + (17)^2 + \cdots + (24)^2 - \frac{(451)^2}{24}$$

$$= 9599.00 - 8855.04$$

$$= 743.96$$

$$SS_{\text{Treatment}} = (\text{sum of squares of treatment totals, divided by respective no of observations})$$
$$\quad - (\text{sum of squares of observations})^2 / (\text{no of observations})$$

$$SS_{\text{Treatment}} = \frac{(73)^2}{6} + \frac{(113)^2}{6} + \frac{(1123)^2}{6} + \frac{(152)^2}{6} - \frac{(461)^2}{24}$$

$$= 9388.50 - 8855.04$$

$$= 533.40$$

$$SS_{\text{Error}} = SST - SS_{\text{Treatment}}$$
$$= 743.96 - 533.46$$
$$= 210.50$$

The following is the format of the ANOVA table.

ANOVA

Source of variation	Sums of squares	Degree of freedom	Mean squares	F_0
Between treatment	$SS_{Treatment}$	(No of treatments) $- 1$	$MS_{Treatment} = SS_{Treatment}/$ DF	$MS_{Treatment}/$ MSE
Error (within treatments)	SSE	(No of observations) $-$ (no of treatment)	MSE = SSE/DF	
Total	SST	(No of observation) $- 1$		

The results are summarized in the following ANOVA table:

ANOVA

Source of variation	SS	DF	MS	F_0
Treatment	533.46	3	177.82	16.89
Error	210.50	20	10.53	
Total	743.96	23		

The hypotheses may be stated as follows:

H_0: $\mu_1 = \mu_2 = \mu_3 = \mu_4$. There is no effect of treatment levels on the experimental units.

H_a: The means are not same.

Here treatments are contents of the element.

The F value $F_{0.05,3,20}$ from table = 3.10. The calculated F value (F_0) is larger than $F_{0.05,3,20}$. Therefore, the hull hypothesis (H_0) is rejected. This means that the treatments (content levels of the element in the alloy) have significant effect on the experimental units (hardness).

17.8.2 Tests on Individual Treatment Means

In the previous example, the null hypothesis has been rejected and we concluded that there is a significant difference between treatment means. But we could not identify and conclude on the nature of the difference. To examine the nature of the difference or to conclude which treatment mean is larger or less than which treatment mean, we can use the concept of contrast. Hereafter, the contrast is defined mathematically.

Let

$$X = a_1 x_1 + a_2 x_2 + a_3 x_3 + a_4 x_4$$
$$Y = b_1 y_1 + b_2 y_2 + b_3 y_3 + b_4 y_4$$

Then X is contrast if,

$$a_1 + a_2 + a_3 + a_4 = 0$$

Y is a contrast if,

$$b_1 + b_2 + b_3 + b_4 = 0$$

X and Y are orthogonal if the above two relations are true and plus if,

$$a_1 b_1 + a_2 b_2 + a_3 b_3 + a_4 b_4 = 0$$

If we want to test the means at level 1 and level 2 of a factor, we may formulate the hypothesis as follows:

$$H_0: \mu_1 = \mu_2$$
$$H_a: \mu_1 \neq \mu_2$$

We can test this hypothesis using linear combination of level totals as follows: $y_1 - y_2 = 0$, where y_1 and y_2 are respective level (treatment) means.

Note that in the linear combination of the above sum of all the coefficients $= +1 - 1 = 0$. So this is a contrast.

If we want to test the hypothesis

$$H_0: \mu_1 + \mu_3 = \mu_2 + \mu_4$$
$$H_a: \mu_1 + \mu_3 \neq \mu_2 + \mu_4$$

We can test this hypothesis using linear combination of level totals as follows:

$$y_1 - y_2 + y_3 - y_4 = 0$$

We use the same example as above to test the hypothesis on treatment means. In order to do this, the null hypothesis and their point estimates for the contrasts are shown in the following table.

Null hypothesis	Point estimate
$H_0: \mu_1 = \mu_2$ or, $\mu_1 - \mu_2 = 0$	$y_1 - y_2 = 0$
$H_0: \mu_2 = \mu_3$ or, $\mu_2 - \mu_3 = 0$	$y_2 - y_3 = 0$
$H_0: \mu_1 + \mu_4 = \mu_2 + \mu_3$ or, $\mu_1 + \mu_4 - \mu_2 - \mu_3 = 0$	$y_1 + y_4 - y_2 - y_3 = 0$
$H_0: 2\mu_1 + \mu_3 = 2\mu_2 + \mu_4$ or, $2\mu_1 + \mu_3 - 2\mu_2 - \mu_4 = 0$	$2y_1 + y_3 - 2y_2 - y_4 = 0$

The calculations for the contrasts and their respective sums of squares are shown hereafter.

$$C1: 73-123 = -50 \quad SS_{C1} = \frac{(-50)^2}{6(2)} = 208.33$$

$$C2: 113-123 = -10 \quad SS_{C2} = \frac{(-10)^2}{6(2)} = 8.33$$

$$C3: 73+152-113-123 = 50 \quad SS_{C3} = \frac{(50)^2}{6(4)} = 104.17$$

$$C4: 2 \times 73 + 123 - 2x113 - 152 = -109 \quad SS_{C3} = \frac{(-109)^2}{6(10)} = 198.02$$

These may be summarized in the following ANOVA table:

ANOVA

Source of variation	SS	df	MS	F_0	$F_{0.05}$ from table
Treatment	533.46	3	177.82	16.90	3.10
C1 (1 vs. 3)	208.33	1	208.33	19.79	4.35
C2 (2 vs. 3)	8.33	1	8.33	0.79	4.35
C3 (1,4 vs. 2,3)	104.17	1	104.17	9.90	4.35
C4 (1,3 vs. 2,4)	198.02	1	198.02	18.81	4.35
Error	210.50	20	10.53		
Total	743.96	23			

The sum of squares of the treatment (533.46) has been partitioned into four contrasts ($C1$, $C2$, $C3$, and $C4$). These four contrasts explain 97.26 % of the sums of squares of treatment.

Looking at the F values from table with $\alpha = 0.05$ and appropriate degrees of freedom, we can conclude that there are significant differences between 1 and 3 ($C1$), averages of 1 & 4 and 2 & 3 ($C3$). Their calculated F values are greater than the respective F values from table. There is no significant difference between 2 and 3 ($C2$). Its calculated F value is less than the F value from the table.

17.8.3 Estimation of Treatment for Completely Randomized Design

(a) Single Treatment Mean

Point Estimate: Population treatment mean = sample treatment mean.

$$\mu = \overline{T}$$

Interval Estimate:

$$\overline{T} - t_{\alpha/2} * s/\sqrt{n} \leq \mu \leq \overline{T} + t_{\alpha/2} * s/\sqrt{n}$$

where

$$\overline{T} = \text{treatment mean,}$$
$$s = \sqrt{\text{MSE}}; \ df = n - k$$
$$n = \text{sample size } (n = n_1 + n_2 + n_3 + \cdots n_k)$$

(b) Difference between two treatment means

Point Estimate:

$$\mu_1 - \mu_2 = \overline{T_1} - \overline{T_2}$$

The interval estimate is given by

$$(\overline{T_1} - \overline{T_2}) - t_{\alpha/2} * s/\sqrt{(1/n_1 + 1/n_2)} \leq \mu \leq (\overline{T_1} - \overline{T_2}) + t_{\alpha/2} * s/\sqrt{1/n_1 + 1/n_2}$$

where

$$\overline{T_1} = \text{mean of treatment 1}$$
$$\overline{T_2} = \text{mean of treatment 2}$$
$$df = n - k$$

17.8.4 Multiple Factor Design

A multiple factor experiment is one in which the number of factors is 2 or more.

17.8.5 Example of a 2^3 Factorial Design

An experimenter wanted to test the effect of two factors on a manufacturing process. The levels are as follows:

Factor 1: 2 levels (100 and 150)
Factor 2: 2 levels (1 and 2)
Factor 3: 2 levels (1 and 2)

The design is shown in the following table. The responses are also shown. Our interest is to see whether the factors have significant effect on the response variable ($\alpha = 0.05$).

The design of the experiment is shown in the following table:

No.	Factor			Response
	Factor 1	Factor 2	Factor 3	
1	100	1	1	60
2	100	1	2	65
3	100	2	1	67
4	100	2	2	69
5	150	1	1	68
6	150	1	2	67
7	150	2	1	69
8	150	2	2	74

ANOVA

Source of variation	SS	DF	MS	F_0	$F_{0.05}$ from table
Model	6449.500	4	1612.275	678.85	6.388
Factor 1	6373.250	2	3186.625	1341.74	6.944
Factor 2	55.125	1	55.125	23.211	7.709
Factor 3	21.125	1	21.125	8.895	7.709
Error	9.500	4	2.375		
Total	6459.000	8			

17.8.6 Randomized Block Design

The completely randomized design is appropriate when the experimental units are somewhat homogeneous. Sometimes, this homogeneity is not available in certain data sets. In those situations, there may be some groups within the experimental units. Within the groups the experimental units may be somewhat homogeneous. But there may be significant differences between the groups. In such a situation, we may divide the experimental units into certain groups. These groups are called Blocks.

As an illustration, we may think of three groups of people when studying their income. These may be people working in the (1) farm sector, (2) industries sector, (3) service sector, and (4) IT sector. We may want to see whether incomes of the people of these groups are different. Let us also say that we apply three treatments based on the age groups: 25–35, 36–50, 50+. This may be represented as follows:

Blocks

	1	2	3	4
Treatments 1 (25-35)	y3	y2	y1	y3
2 (36-50)	Y1	y1	y3	y1
3 (50+)	Y2	y3	y2	y2

The treatments are to be assigned randomly. Each treatment will appear only once in each block. Based on responses, we can analyze the variances and use the ANOVA concept to test whether the blocks have significant difference. We shall have to find out the total sums of squares and partition the total sums of squares to treatments and blocks to test the significances of the blocks and treatments.

Treatment	Block				Total
	1	2	3	4	
1	1.00	2.00	3.00	3.50	9.50
2	1.50	2.50	2.90	4.50	11.40
3	2.00	3.00	4.00	5.00	14.00
Total	4.50	7.50	9.90	13.00	34.90

Notes Responses are in US\$(100). Let T represent treatment, B represent block, y represent observation. Also i indicates from left to right and j indicates from top to bottom.

$$SS_{Total} = \sum_{i=1}^{4}\sum_{j=1}^{n_i} y_{ij}^2 - \frac{\sum y}{n}$$

$$= (\text{sum of squares of all observations}) - (\text{sum of all observations})^2/n$$

$$= (1.0)^2 + (1.5)^2 + (2.0)^2 + \cdots + (13.0)^2 - (34.90)^2/12$$

$$= 117.41 - 101.50$$

$$= 15.91$$

$$SS_{Treatment} = \frac{\sum_{i=1}^{4} (T_{i\,total})^2}{n_i} \frac{(\sum y)^2}{n}$$

$$= (\text{sum of squares of treatments, each square divided by its } n \text{ value}) - \frac{(\sum y)^2}{n}$$

$$= (9.50)^2/4 + (11.40)^2/4 + (14.00)^2/4 - (34.90)^2/12$$

$$= 104.05 - 101.50$$

$$= 2.55$$

$$SSB = \frac{\sum_{i=1}^{4} B_i^2}{n_i} - \frac{(\sum y)^2}{n}$$

$$= (4.50)^2/3 + (7.50)^2/3 + (9.90)^2/3 + (13.00)^2 - (34.90)^2/12$$

$$= 114.05 - 101.50$$

$$= 13.00$$

$$SSE = SS_{Total} - SS_{Treatment} - SS_{Block}$$

$$= 15.91 - 2.55 - 13.00$$

$$= 0.35$$

These may be arranged in the following ANOVA table:

Source of variation	SS	DF	MS	F_0
Block	13.00	3	4.33	73.25
Treatment	2.55	2	1.28	21.56
Error	0.35	6	0.06	
Total	15.91	11		

We may state the null hypotheses as follows:

For Blocks H_0: There is difference between block means.

From F table, $F_{0.05,3,6} = 4.757$. Thus, $F_0 > F_{0.05,3,6}$. Therefore, the null hypothesis is rejected and we may conclude that there is significant difference between block means. In other words, blocks (sectors) have significant effect on the experimental units.

For Treatment H_0: There is no difference between treatment means.

From F table, $F_{0.05,2,6} = 5.143$. Here also, $F_0 > F_{0.05,2,6}$. Therefore, the null hypothesis is rejected and we may conclude that there is significant difference between treatment means. In other words, treatments (age groups) have significant effect on the experimental units.

Problems

17.1   An insurance company wants to study how total premiums obtained in the insurance are affected by premiums from different sectors. He collects insurance data for 3 consecutive years (2006, 2007, and 2008) for four factors. The data area summarized in the following table.

Code:

Factor 1 = Fire
Factor 2 = Marine
Factor 3 = Automobile
Factor 4 = Miscellaneous
Response Variable = Insurance Premium (billion $)

Treatment	Factor 1	Factor 2	Factor 3	Factor 4
2006	0.217	0.117	1.720	0.827
2007	0.215	0.116	1.861	0.863
2008	0.227	0.127	1.944	0.920

Prepare the ANOVA Table

(a)   State the hypotheses whether treatment and block have significant effect on the insurance premiums. Use $\alpha = 0.05$.

17.2   An experimenter wants to study the compressive strength of concrete. He believes that compressive strength varies according to grade. He collected the data shown in the following table. Each grade has five replicates.

Grade	Compressive strength (N/mm^2)				
30	30	32	29	32	33
40	39	40	41	40	41
50	48	48	49	50	50

Formulate the null and alternate hypotheses to test whether compressive strength varies by grade.

Prepare the ANOVA and test the null hypothesis and conclude.

17.3   With the data given in the previous problem, test the individual means to see whether the individual treatment means are same or not.

17.4   15 students from a school were selected at random. Again 5 students (from the same 15 students) at random were assigned to each of the three teaching methods (lecture, tutorial, and assignment). Their scores in percentage are summarized in the following table:

Teaching method	Students' score (%)				
1 = Lecture	68	72	64	63	60
2 = Tutorial	75	74	71	75	76
3 = Assignment	69	70	67	66	68

Formulate hypotheses and test whether teaching methods have effect on the students' scores.

Test by linear contrast whether individual treatment means are same or different.

17.5 In an experiment, the effect of feeding frequency on growth of a particular fish species was studied in Asian Institute of Technology. The results are shown in the following table. Unit of weight is gram. Feeding frequency is 4, 5, 6 times a day (day–night). The raw data are from Baouthong, Pornpimon, AIT Thesis no. AE-95-23.

Test whether the feeding frequency has significant effect on the weight gain. Use $\alpha = 0.05$.

Times	Replicate				Total
	1	2	3	4	
4	56.01	49.12	32.36	67.06	204.55
5	64.73	29.34	50.14	33.44	177.65
6	26.47	62.18	19.6	55.26	163.51
Total	147.21	140.64	102.1	155.76	545.71

17.6 In an experiment for the effect of speed on the walking tractor vibration, an experimenter recorded the acceleration amplitudes (m/s^2) at three speeds 1000, 1200, and 1400 as shown in the following table. Raw data are from Sookkumnerd, Chanoknun, AIT Thesis no. AE-00-2.

Prepare the ANOVA and test the following:

(a) Is the treatment effect on the acceleration amplitude significant ($\alpha = 0.05$)?
(b) Is factor effect on the acceleration amplitude significant?

Speed	Factor								Total
	1	2	3	4	5	6	7	8	
1000	1.250	0.698	0.679	0.430	0.420	0.391	0.259	0.156	4.283
1200	0.896	0.623	0.623	0.623	0.536	0.429	0.312	0.292	4.334
1400	1.172	0.762	0.762	0.547	0.449	0.449	0.449	0.254	4.844
Total	3.318	2.083	2.064	1.600	1.405	1.269	1.020	0.702	13.461

17.7   An experimenter conducted an experiment in the Red River Delta of Vietnam to study the rice yields with two varieties of rice and eight levels of nitrogen. The records of rice yields from the experiment are shown in the following table. Raw data are from Trung, Nguyen Manh, AIT Thesis no. AE-91-43.

Rice	N-Level								Total
	0	30	45	60	75	90	120	150	
V18	2.130	5.07	5.800	6.440	6.990	7.45	8.1	8.39	50.370
IR8	2.23	6.84	7.12	7.32	7.44	7.49	7.36	6.92	52.720
Total	4.360	11.91	12.92	13.76	14.430	14.94	15.46	15.310	103.09

Prepare the ANOVA and test the following at $\alpha = 0.05$.

(a)   Does rice variety affect the rice yield significantly?

(b)   Does nitrogen level affect rice yield significantly?

17.8   Shear Strength of soil in six different fields (A, B, C, D, E, and F) was investigated at four treatment levels (depth) 5, 10, 15, and 20 cms. The observed shear strength (kg/cm^2) values are shown in the following table. Raw data are from Ramalingam, Nagarajan, AIT Thesis no. AE-98-1.

Prepare the ANOVA and test at $\alpha = 0.05$ whether,

(a)   Depth has significant effect on shear strength.

(b)   Fields have significant effect on shear strength.

Depth	A	B	C	D	E	F	Total
5	17.5	14.0	14.2	13.5	21.0	22.5	102.7
10	30.0	28.0	19.2	22.5	27.5	36.0	163.2
15	47.5	54.0	33.5	29.5	47.5	51.5	263.5
20	68.5	69.5	50.0	56.0	76.0	73.5	393.5
Total	163.5	165.5	116.9	121.5	172.0	183.5	922.9

17.9   In an experimental design to study the water quality, the experimenter used ammonia concentration in water. Ammonia concentration ($\mu g/l$) in water was recorded at different experimental days and at five treatment levels (0, 25, 50, 75, and 100 open water fish culture ratio). The recorded observations are shown in the following table. Raw data source: Begam, Rowshan Ara, AIT Thesis AE-94-20).

Ratio	Experimental days							Total
	24	38	52	55	66	80	91	
0	0.09	0.10	0.24	3.28	0.130	0.16	0.13	4.13
25	0.07	0.07	0.23	3.44	0.29	0.25	0.21	4.56
50	0.13	0.09	0.27	3.37	0.57	0.52	0.42	5.37
75	0.14	0.15	0.18	3.32	1.97	0.42	0.28	6.46
100	0.15	0.16	0.2	3.51	2.11	0.49	0.40	7.02
Total	0.58	0.57	1.12	16.92	5.07	1.84	1.44	27.54

Prepare the ANOVA and test at $\alpha = 0.05$ whether,

   (a) Ratio (treatment) has significant effect on ammonia concentration.
   (b) Experimental days have significant effect on ammonia concentration.

17.10 In an experimental design for fish culture in AIT, the experimenter used phytoplankton concentration (mg/l) at five treatment levels (0, 25, 50, 75, and 100 open water fish culture ratio, OWFCR) on 7 experimental days of the experiment. The observations are as shown in the following table. Raw data source: Begam, Rowshan Ara, AIT Thesis AE-94-20.

OWFCR	Experimental days							Total
	26	40	54	56	68	82	94	
0	0.72	0.90	0.540	1.090	2.960	1.71	0.76	8.68
25	0.32	1.00	0.67	1.97	3.79	1.58	1.01	10.34
50	0.25	0.68	0.54	1.52	1.64	1.03	0.56	6.22
75	0.47	1.79	1.07	1.34	3.22	1.34	1.45	10.68
100	0.43	1.61	0.81	1.61	1.61	2.44	1.64	10.15
Total	2.19	5.98	3.63	7.53	13.22	8.10	5.42	46.07

Prepare the ANOVA and test at $\alpha = 0.05$ whether,

(a) Ratio (treatment) has significant effect on ammonia concentration.
(b) Experimental days have significant effect on phytoplankton concentration.

Answers

17.1 Treatment effect is not significant; block effect is significant.
17.2 Conclusion: Compressive strength varies with grade.
17.4 1. Teaching methods have effect on students' score.
      2. $\mu_1 \neq \mu_2$.
      3. $\mu_1 = \mu_3$.
      4. $\mu_2 \neq \mu_3$.
17.5 Treatment effect is not significant. Growth does not vary significantly with feeding frequency ($F = 0.356$).

17.6 (a) Treatment effect is not significant ($F = 1.385$).
    (b) Block effect is significant ($F = 25.415$).

17.7 1. Effect of variety on rice yield is not significant ($F = 0.613$).
    2. Effect of nitrogen level on rice yield is significant ($F = 11.864$).

17.8 1. Treatment effect (depth) is significant ($F = 127.9$).
    2. Field effect is significant ($F = 9.129$).

17.9 1. Treatment effect is not significant ($F = 1.735$).
    2. Effect of experimental days on ammonia concentration is significant ($F = 56.871$).

17.10 1. Treatment effect is significant at $\alpha = 0.10$ ($F = 1.735$).
    2. Effect of experimental days on phytoplankton concentration is significant at $\alpha = 0.05$ ($F = 11.207$).

Chapter 18
Statistical Quality Control

Abstract Quality control deals with the subject matter of assurances and testing for failure of the products or services in the manufacturing processes. An important tool is control chart. Several types of control charts and their uses are explained. In each chart, control lines showing Upper Control Limit (UCL), Centerline (CL), and Lower Control Limit (LCL) are developed and drawn. The sequence of construction of control charts and the limits is provided. The technique for development of "Control Zones" and the interpretations are provided. For examining the "Out of Control" Processes, WECO rules are used. The rules are summarized. Examples are provided in each case. Techniques for calculating tolerance limit, process capability, and sample size based on statistical process control are demonstrated with the help of examples.

Keywords Statistical quality control · Manufacturing process · Control charts · Control limits · Control zones · Out of control · WECO rules

18.1 History

Manufacturers are having a tough time in the competitive market environment. Users are getting products of higher quality not only in any domestic market but also in the international market. This situation is creating a lot of pressure on the manufacturers not only in maintaining the quality but also improvement of the products over time.

Quality control is a branch of engineering. It deals with the subject matter of assurance and testing for failure of the products or services in the design and production. The purpose is to see that the products meet or exceed users' requirements or expectation. Quality is to be built into the process of production. Quality controls are tools that help the manufacturers identify the quality problems to the production process. Statistical Quality Control (SQC) is a tool that stems from the area of statistics. Our focus is on the quality control in the manufacturing process. Statistical Process Control (SPC) is an alternative term of SQC.

© Springer Science+Business Media Singapore 2016   353
A.Q. Miah, *Applied Statistics for Social and Management Sciences*,
DOI 10.1007/978-981-10-0401-8_18

Walter A. Shewart did the pioneering works in the statistical process control in the early 1920s. Later during the World War II W. Edwards Deming applied SPC methods in the US. This led to the successfully improving quality in the manufacture of munitions and other strategically important products. Deming was also instrumental in introducing SPC methods to Japanese industry after war had ended (Wikipedia, free encyclopedia, website, 19 November 2009).

18.2   Areas of Statistical Quality Control

The quality professionals view the statistical quality control as divided into three parts (webiste: http://www.wiley.com/college/sc/reid/pdf 19 November 2009). These are discussed hereafter.

(a) Descriptive Statistics: These are used to describe quality characteristics and relationships. These are mean, standard deviation, range, and a measure of the distribution of data.
(b) Statistical Process Control (SPC): This is a statistical tool. It involves a random sample of the output from a process and deciding whether the process is producing products with the characteristics that fall within a predetermined range. Based on the results of the SPC, we can also see whether the process is functioning properly or not.
(c) Accepting Sampling: This is a process of randomly inspecting a sample of goods and deciding whether to accept the entire lot based on the results. Based on the results, we can decide whether a batch of goods produced should be accepted or rejected.

18.3   Variation

Variation occurs in the manufacturing process for various reasons. Due to variation, defective items are produced and the manufacturers are concerned whether a batch is to be accepted or rejected. They are concerned because final acceptance lies with the end users.

Variation may occur due to two broad sets of causes. One cause is called Common Cause Variation. This occurs due to such causes as difference in materials, workers, machines, tools and similar factors. The other set of causes is called Assignable Cause Variation. These are variations due to such causes as poor quality of raw materials, people with no or low training, machines that need repair but not done.

Whatever is the source of variation, it is to be identified and corrective measures must be taken. In practical situations, completely eliminating the variations is impossible. Therefore, reducing the variation is the fundamental purpose of statistical process control. In other words, it is the purpose to maintain the process stability.

18.4 Control Chart

For control chart, considerable coverage has been provided in the website: http://www.balancedscorecard.org/portals/o/PDF/control.pdf, 19 November 2009. A control chart is a chart showing the limits within which the process variation is expected to lie. Control chart is a tool to use for maintaining process stability. Process stability may be defined as "a state in which a process has displayed a certain degree of consistency during the past and is expected to maintain this consistency." This consistency refers to the state that all the data will lie within the predetermined control limits.

We can use a control chart to monitor variation in the process over time, identify the causes of variation, examine the effectiveness of the process, and improve the process. It also helps us to examine how the process performed during a certain period of time.

We shall cover in brief the theoretical part of the control charts, the methods of their construction and their interpretation.

18.4.1 Type of Data for Control Chart

The data that are used in constructing control charts may be divided into the following two sets:

(a) Attribute Data: These are discrete data and may also be called categorical data. These are measured by counting. Example is defect in a process measured by number such as how many number of items are defective. This defect may also be measured by such counts as number of items having defects and number of items having no defect.
(b) Variable Data: Values of these data have continuous measurements. Example is diameter (in cm, etc.) of piston rings produced by a process.

18.4.2 Types of Control Charts

There are a number of control charts. The following are the more commonly used charts:

(a) X-Bar (average) Chart
(b) R (range) Chart
(c) I Chart
(d) p Chart
(e) c Chart
(f) u Chart
(g) np Chart

Different techniques are use to construct these charts. Even in constructing X-Bar charts, there are two techniques namely ordinary averages and moving averages. In this Chapter we shall cover the charts mentioned above. These are more common for application in many processes.

18.4.3   Control Limits in Control Charts

Generally there are three control limits namely, Upper Control Limit (UCL), Centerline, and Lower Control Limit (LCL). See Fig. 18.1.

An important question naturally arises how to establish or construct these control lines. The next section will provide answer to this question.

18.4.4   Theoretical Basis of the Control Limits

In the chapter on estimation, we have shown that in case of a normal population, the interval estimation or in other words the confidence interval is given by

$$\bar{x} - Z_{\alpha/2} * \sigma/\sqrt{n} \leq \mu \leq \bar{x} + Z_{\alpha/2} * \sigma/\sqrt{n}$$

If we use the population parameter μ, then the expression comes out to be

$$\mu - Z_{\alpha/2} * \sigma/\sqrt{n} \leq \mu \leq \mu + Z_{\alpha/2} * \sigma/\sqrt{n}$$

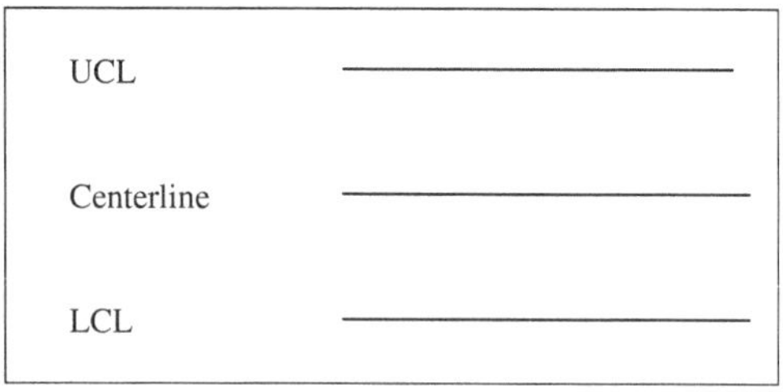

Fig. 18.1 Control limits in a control chart

UCL   ————————————

Centerline   ————————————

LCL   ————————————

The left hand part of this expression $(\mu - Z_{\alpha/2} * \sigma/\sqrt{n})$ represents the lower limit (LCL) and the right side $(\mu + Z_{\alpha/2} * \sigma/\sqrt{n})$ the upper limit (UCL). The value of Z depends on the level of the confidence we want such as 95 % confidence level.

This is the basis of establishing the control limits. Now, the question is how to estimate μ, σ, and set the α level. For α level, for 95 % confidence level, the z value is set to 1.96σ. In constructing the control charts, it is generally set to 3σ. In such a case, the z value becomes

$$z = \frac{x - \mu}{\sigma} = \frac{\mu + 3\sigma - \mu}{\sigma} = \frac{3\sigma}{\sigma} = 3$$

and the UCL and LCL now may assume the form

$$LCL = (\mu - 3 * \sigma/\sqrt{n})$$
$$UCL = (\mu + 3 * \sigma/\sqrt{n})$$

Note that $(\sigma/\sqrt{n})$ is the standard error. So the control limits (UCL and LCL) may be stated as $\mu \pm 3$ * (standard error).

Now two parameters are to be estimated. These are μ and σ. The mean (μ) may be estimated from the grand mean of the samples. The other parameter σ may be estimated by standard deviation method or by range method. We shall show how to estimate these two parameters in different situations.

18.4.5 Control Chart for X-Bar and R

X-Bar stands for average and R stands for range. In construction of a control chart, three control lines are to be drawn. These are (1) Centre line (CL), (2) Upper control limit (UCL) and (3) Lower control limit (LCL).

The sequence of construction of the control chart is as follows:

(a) Examine the data set and identify the sub-groups;
(b) Calculate the average (mean) of each subgroup;
(c) Calculate the average of the subgroup ranges and the mean of the sub-group ranges;
(d) Calculate the locations of the three lines (CL, UCL, and LCL);
(e) Determine the appropriate scale;
(f) Plot the control lines (CL, UCL and LCL) as well as the X-bar lines.

Note that in this section subgroups are involved. Data that can be subdivided into subgroups may be analyzed according to the principle laid down in this section. If the are no subgroups or in other words if there is only one subgroup, then the principle will be different.

Example 1
In a manufacturing process, five subgroups can be identified. Some sample observations have been recorded from the process. The data set is shown in the following table.

	X_1	X_2	X_3	X_4	X_5
	15.0	12.5	14.2	11.8	16.1
	16.0	12.8	14.8	11.1	16.8
	14.5	14.0	15.2	12.3	15.6
	14.0	15.5	16.2	15.1	14.3
	14.8	15.0	16.1	16.0	14.0
X-bar	14.86	13.96	15.30	13.26	15.36
R	2.00	3.00	2.00	4.90	2.80

Grand average is calculated as follows:

$$\bar{\bar{x}} = \frac{\bar{x}_1 + \bar{x}_2 + \bar{x}_3 + \bar{x}_4 + \bar{x}_5}{5}$$
$$= \frac{14.86 + 13.96 + 15.30 + 13.26 + 15.36}{5}$$
$$= 14.55$$

Range R in a sub-group = (highest value in the sub-group) − (smallest value in the sub-group). Average of the R's is calculated as follows:

$$\bar{R} = \frac{2.00 + 3.00 + 2.00 + 4.90 + 2.80}{5} = \frac{14.7}{5} = 2.94$$

Let us examine now how to calculate the control limits.

$$\text{Centreline} = \bar{\bar{x}} = 14.55$$

We know that the UCL is given as

$$\mu \pm 3 * (\text{standard error})$$

In calculating the standard error, the range method is more commonly used. According to this method, the standard error is given as $\dfrac{\bar{R}}{d_2\sqrt{n}}$. Therefore, the UCL may be written as

$$\text{UCL} = \bar{\bar{X}} + \frac{3\bar{R}}{d_2\sqrt{n}}$$

Here, d_2 is a function of n and are usually found in the statistical tables. Setting $\dfrac{3}{d_2\sqrt{n}} = A_2$, we get

$$\text{UCL} = \bar{\bar{X}} + A_2\bar{R}$$

Since d_2 is a function of n, A_2 is also a function of n and values of A_2 for various n values can also be found in a statistical table. Thus, the UCL is calculated as follows:

$$\begin{aligned}
\text{UCL} &= \bar{\bar{X}} + A_2\bar{R} \\
&= 14.55 + 0.577 \times 2.94 \\
&= 14.55 + 1.70 \\
&= 16.25
\end{aligned}$$

Value of A_2 is 0.577 for $n = 5$ from the table.
Similarly, the LCL may be calculated as follows:

$$\begin{aligned}
\text{LCL} &= \bar{\bar{X}} - A_2\bar{R} \\
&= 14.55 - 0.577 \times 2.94 \\
&= 14.55 - 1.70 \\
&= 12.85
\end{aligned}$$

Using the X-bars, the CL, UCL and LCL, the appropriate control chart is drawn as is provided in Fig. 18.2.

For Ranges

The control chart limits are given by

$$\begin{aligned}
\text{UCL} &= D_4 * \bar{R} \\
\text{CL} &= \bar{R} \\
\text{LCL} &= D_3 * \bar{R}
\end{aligned}$$

D_3 and D_4 are functions of n and can be found from the statistical table for various values of n. For our particular example, the control limits are calculated as follows:

$$\begin{aligned}
\text{UCL} &= D_4 * \bar{R} = 2.115 * 2.94 = 6.22 \\
\text{CL} &= \bar{R} = 2.94 \\
\text{LCL} &= D_3 * \bar{R} = 0 * 2.94 = 0.
\end{aligned}$$

The control chart showing the control limits for the ranges is shown in Fig. 18.3.

Fig. 18.2 X-bar control chart

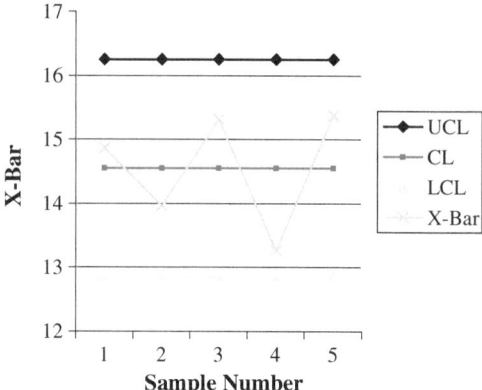

Fig. 18.3 R control chart

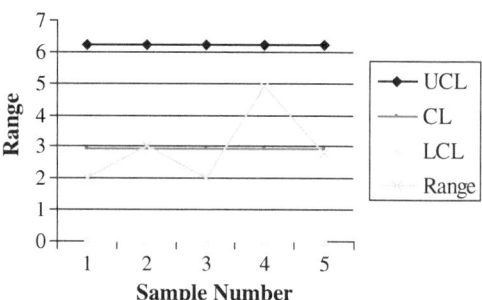

18.4.6   I Chart (Individual Observation Chart)

In some processes, subgroups cannot be identified. In those situations "I control charts" are used using individual observations. Dates are to be in actual sequence starting from the start date until the finished date. In the table that follows, 1–10 indicates the observation numbers in correct sequence.

Date	1	2	3	4	5	6	7	8	9	10
Individual X	30	29	25	28	26	32	30	31	33	35
Moving R		1	4	3	2	6	2	1	2	2

$$\bar{x} = \frac{30 + 29 + 25 + 28 + 26 + 32 + 30 + 31 + 33 + 35}{10} = \frac{299}{10} = 29.9$$

Ranges for this I chart are calculated as follows:
In this case, it is moving range.

Range in a cell = (observation in that cell) − (observation in the immediate previous cell). See the table above. The R-bar (moving range bar) is calculated as follows:

$$\bar{R} = \frac{1+4+3+2+6+2+1+2+2}{(10-1)} = \frac{23}{9} = 2.56$$

For X-Bar

In the processes where subgroups can be identified, the UCL is given by

$$UCL = \bar{\bar{X}} + A_2\bar{R}$$

For individual observation chart (I chart), the control limits for average are given by

$$UCL = \bar{x} + E * \bar{R}$$
$$CL = \bar{x}$$
$$LCL = \bar{x} - E * \bar{R}$$

In these formulas, E is a control chart factor and can be found from a statistical table. For two-point average, its value is always is 2.66. For a particular value of n, E can also be calculated from $E = 3/d_2$. The control limits are calculated as follows:

$$UCL = \bar{x} + E * \bar{R} = 29.9 + 2.66 * 2.56 = 36.71$$
$$CL = \bar{x} = 29.9$$
$$LCL = \bar{x} - E * \bar{R} = 29.9 - 2.66 * 2.56 = 23.09$$

For R (moving rage)

The control limits are calculated from the following formulas:

$$UCL = D_4 * \bar{R}$$
$$CL = \bar{R}$$
$$LCL = D_3 * \bar{R}$$

The values of D_3 and D_4 can be obtained from the statistical table.
Therefore,
UCL = 3.267 * 2.56 = 8.364 (from table D_4 = 3.267 for n = 2. Note that for two consecutive point range n = 2)

$$CL = \bar{R} = 2.56$$
$$LCL = 0.0 * 2.56 = 0 \text{ (from table } D_3 = 0 \text{ for } n = 2)$$

The UCL, CL, LCL, and R-bar are plotted as shown in Fig. 18.4.

Fig. 18.4 I chart average

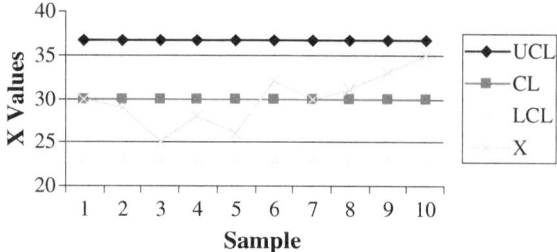

Fig. 18.5 I chart (moving range)

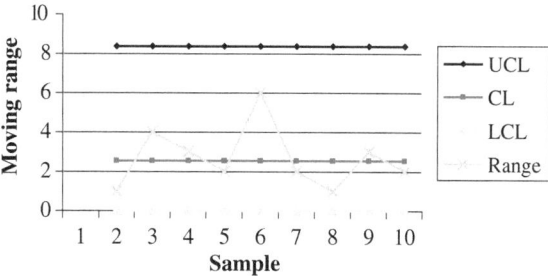

The moving range control limits in the control chart is plotted as shown in Fig. 18.5.

18.4.7   P Chart (Proportion Defective)

A P chart is one that monitors the proportion of nonconforming units in a sample. The appropriate data for P chart are attribute (discrete) data. Example of such data are good/bad, defective/non-defective, conforming/nonconforming, etc. The conforming/nonconforming should be clearly defined. It is better if the sample size is large, say about 100. Also the observation should be recorded according to the days of a month sequentially.

Example 2

In a manufacturing process, observations were made for checking the nonconforming units during a month. The data are shown in the following table.

Day of month	No. observed	No. defective	Proportion defective
1	26	7	0.269
2	28	8	0.286
3	25	9	0.360
4	26	9	0.346
5	27	10	0.370

(continued)

(continued)

Day of month	No. observed	No. defective	Proportion defective
6	28	10	0.357
7	30	11	0.367
8	31	11	0.355
9	32	12	0.375
10	31	13	0.419
11	30	14	0.467
12	35	10	0.286
13	33	9	0.273
14	32	10	0.313
15	28	8	0.286
16	29	7	0.241
17	30	9	0.300
18	29	10	0.345
19	31	8	0.258
20	25	7	0.280
21	26	8	0.308
22	28	7	0.250
23	29	7	0.241
24	31	6	0.194
25	31	5	0.161
26	27	7	0.259
27	29	8	0.276
28	30	9	0.300
29	32	10	0.313
30	30	8	0.267
Average proportion			0.304

$\bar{P} = 0.304$

If the process proportion (population proportion) P is known, then the following formulas may be used for determining the control limits:

$$UCL = P + 3\sqrt{\frac{P(1-P)}{n}}$$

$$CL = P$$

$$LCL = P - 3\sqrt{\frac{P(1-P)}{n}}$$

If the process proportion (population proportion) P is not known, then it has to be estimated from the sample proportion. It is identified as follows:

$\bar{p}$ = average of the individual sample proportions

Thus the control limits are given by

$$\text{UCL} = \bar{p} + 3\sqrt{\frac{\bar{p}(1-\bar{p})}{n}}$$

$$\text{CL} = \bar{p}$$

$$\text{LCL} = \bar{p} - 3\sqrt{\frac{\bar{p}(1-\bar{p})}{n}}$$

In the example that we are working on, the population proportion P is not known. So, we have to use the second set of formulas as follows:

$$\text{UCL} = \bar{p} + 3\sqrt{\frac{\bar{p}(1-\bar{p})}{n}}$$

$$= 0.304 + 3\sqrt{\frac{0.304(1-0.304)}{30}}$$

$$= 0.304 + 3\sqrt{\frac{0.304 * 0.896}{30}}$$

$$= 0.304 + 3 * 0.084$$

$$= 0.304 + 0.252$$

$$= 0.556$$

$$\text{CL} = \bar{p}$$

$$= 0.304$$

$$\text{LCL} = \bar{p} - 3\sqrt{\frac{\bar{p}(1-\bar{p})}{n}}$$

$$= 0.304 - 0.252$$

$$= 0.052$$

The control limits are shown in the following Fig. 18.6.

18.4.8   c Chart (Control Chart for Defects)

A c chart is a technique by which the number of defects in a process is intended to be controlled. In a chart, c stands for "counts." In some situations, it is necessary to control the number of defects in the manufacturing process rather than the percentage of defects. In c charts, it is assumed that the sample sizes of the subgroups are equal. However, no assumption is made regarding the size of the sample.

Fig. 18.6 Control chart for proportion

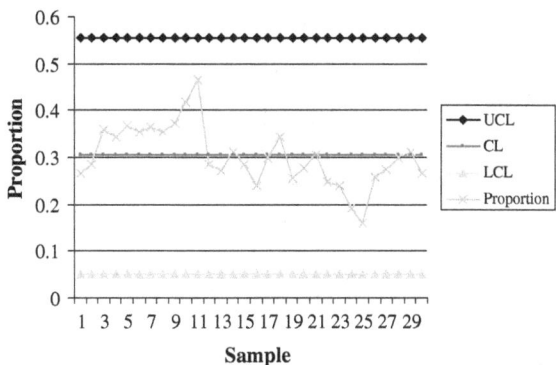

Let

c_i the number of defects in a manufacturing unit

k number of units

Then, the average of the number of defects $\bar{c} = \dfrac{1}{k}\sum_{i=1}^{k} c_i$

The control limits are given by

$$\text{UCL} = \bar{c} + 3\sqrt{\bar{c}}$$
$$\text{CL} = \bar{c}$$
$$\text{LCL} = \bar{c} - 3\sqrt{\bar{c}}$$

Example 3

In a process, 20 observations are made and the corresponding number of defects (nonconforming) is recorded. This is shown in the following Table. We shall calculate the control limits and draw the control chart. This is demonstrated hereafter.

Sample	No. of defects	Sample	No. of defects
1	5	11	12
2	6	12	10
3	4	13	9
4	8	14	9
5	7	15	7
6	9	16	9
7	12	17	6
8	12	18	8
9	7	19	8
10	5	20	7

Average number of defects = 8

The control limits are calculated as follows:

$$\text{UCL} = \bar{c} + 3\sqrt{\bar{c}} = 8 + 3\sqrt{8} = 8 + 3 * 2.828 = 8 + 8.484 = 16.828$$
$$\text{CL} = \bar{c} = 8$$
$$\text{LCL} = \bar{c} - 3\sqrt{\bar{c}} = 8 - 3\sqrt{8} = 8 - 3 * 2.828 = 8 - 8.484 = 0$$

If the value of LCL is zero or minus, it is set to zero (Fig. 18.7).

18.4.9 u Chart (Defects Per Unit)

In c chart, we have worked with the total number of defects. But in some manu-facturing processes, it may be necessary to work with the number of defects per unit rather than the total number of defects. In such a case u chart is used. A u chart is for number of defects per unit. The u stands for "units." The distribution of number of units is assumed to be a Poisson distribution.

Let,

n sample size, and
c total number of defects in the sample
k number of samples

Then, the average number of units per unit (sample) is given by

$$u = \frac{c}{n}$$

The average of the k number of samples is given by

$$\bar{u} = \frac{1}{k}\sum_{i=1}^{k} u_i$$

Fig. 18.7 C control chart

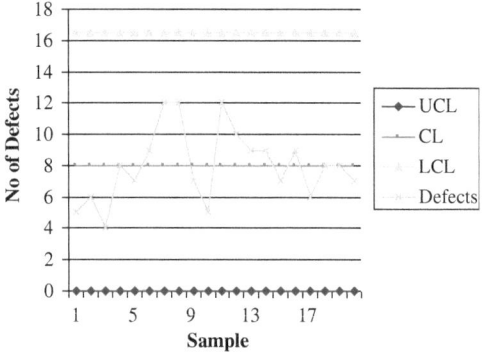

The control limits are given by

$$UCL = \bar{u} + 3\sqrt{\frac{\bar{u}}{n}}$$

$$CL = \bar{u}$$

$$UCL = \bar{u} - 3\sqrt{\frac{\bar{u}}{n}}$$

Similar to a c chart, the distribution of number of defects is assumed to be a Poisson distribution. In a u chart, it is not necessary that the sample sizes in each unit be equal. Also there is no assumption regarding the size of the sample.

Example 4

Sample	Sample size n	No of defects	Defects/Unit
1	5	8	1.60
2	6	8	1.33
3	8	10	1.25
4	8	10	1.25
5	7	8	1.14
6	9	10	1.11
7	10	15	1.50
8	10	14	1.40
9	11	12	1.09
10	12	16	1.33
11	8	10	1.25
12	8	14	1.75
13	9	10	1.11
14	10	12	1.20
15	11	16	1.45
16	10	12	1.20
17	9	10	1.11
18	8	11	1.38
19	11	15	1.36
20	12	18	1.50
21	12	15	1.25
22	12	16	1.33
23	10	8	0.80
24	10	15	1.50
25	11	9	0.82
Sum			32.03
Average n	9.48		1.28

Control limits are calculated as follows. Since the sample sizes are not constant, we have used average sample size. This is simple, provides a straight line and easy to interpret. However, if this is not done, UCL and LCL for each observation (sample) may be calculated and plotted. In this case, the UCL and LCL will not be straight lines.

$$\sqrt{\bar{u}} = \frac{32.03}{25} = 1.28$$

$$UCL = \bar{u} + 3\sqrt{\frac{\bar{u}}{n}}$$

$$= 1.28 + 3 * \sqrt{\frac{1.28}{9.48}}$$

$$= 1.28 + 3 * 0.367 = 1.28 + 1.101 = 2.381$$

$$CL = \bar{u} = 1.28$$

$$UCL = \bar{u} - 3\sqrt{\frac{\bar{u}}{n}}$$

$$= 1.28 - 1.101$$

$$= 0.179$$

With these, the control chart is plotted as shown in the following figure (Fig. 18.8).

18.4.10 np Control Chart

In industrial statistics, the np control chart is a type of control chart in which the total number of nonconforming items is of importance. There are some similarities between an np chart and a p chart. In a p chart the proportions of nonconforming units are of interest. In an np chart, the total number of defective units

Fig. 18.8 u control chart

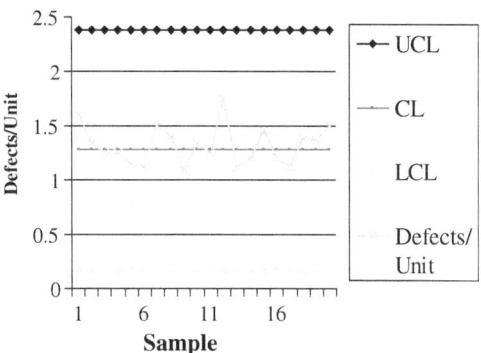

(nonconforming) is of interest. Essentially, in np chart construction and interpretation, the subgroup sample sizes must be equal. Otherwise, there is no point in comparing. In "np", n stands for number and p stands for proportion. So the product of n and p np equals the average number of nonconforming units.

The control limits are set as follows:

$$UCL = \overline{np} + 3 * \sqrt{\overline{np}(1 - \bar{p})}$$

$$CL = \overline{np}$$

$$LCL = \overline{np} - 3 * \sqrt{\overline{np}(1 - \bar{p})}; \quad \text{if this value is 0 or negative, use 0.}$$

Consider the following example.

Example 5

From a process, 20 subgroups were identified. From each subgroup, number of defective units was recorded. Sample size in each subgroup was 25. The data recorded and calculated are as follows:

Sample	No. observed	No. defective	Proportion defective
1	25	7	0.280
2	25	8	0.320
3	25	9	0.360
4	25	9	0.360
5	25	10	0.400
6	25	10	0.400
7	25	11	0.440
8	25	11	0.440
9	25	12	0.480
10	25	13	0.520
11	25	14	0.560
12	25	10	0.400
13	25	9	0.360
14	25	10	0.400
15	25	8	0.320
16	25	7	0.280
17	25	9	0.360
18	25	10	0.400
19	25	8	0.320
20	25	7	0.280
Average	25	9.6	0.384

Fig. 18.9 np control chart

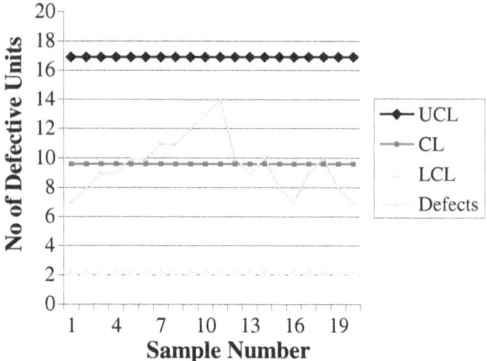

The control limits are calculated as follows (see Fig. 18.9):

$$UCL = 25 * 0.384 + 3 * \sqrt{25 * 0.384(1 - 0.384)}$$
$$= 9.6 + 3 * 2.43$$
$$= 9.6 + 7.29$$
$$= 16.89$$
$$CL = \overline{np} = 9.6$$
$$LCL = \overline{np} - 3 * \sqrt{\overline{np}(1 - \bar{p})}$$
$$= 9.6 - 7.29$$
$$= 2.31$$

The control limits are plotted and shown in Fig. 18.10.

Interpretation: All the observations fall within the control limits (UCL and LCL). So, the process is running within control.

Fig. 18.10 Control chart zones

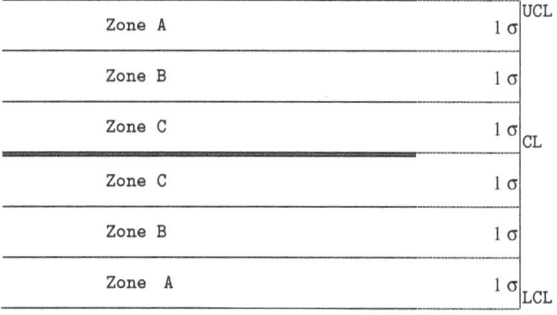

18.4.11 Control Chart Zones

Control zones may be divided into three zones for interpreting purposes. The zones are as follows. See Fig. 18.10.

Zone C: It is the next to centre line zone. There are two C zones, one each on either side of the centre line. Its width is one σ.

Zone B: This is next to C. Its width is also one σ. Like zone C, Zone B also lies on both sides of the centre line.

Zone A: It is the farthest zone. There are 2 zones A, one each on either side of the centre line. Its width is also one σ.

Note the following:

(a) Zone C lies between centre line and one σ limit of the CL.
(b) Zone B lies between 1σ and 2σ limit from the CL.
(c) Zone A lies between 2σ and 3σ limit from the CL.

Generally, if any point lies outside the control limits (UCL and LCL), the process is said to be out of control. If more observations fall close to the centre line, this is an indication of less variability in the design.

For examining the "Out of Control" processes, WECO rules may be applied. WECO stands for Western Electric Company. The rules are summarized in the following table.

Zone	Observation	Remark
A	A single point falls outside 3σ (Zone A)	Process out of control
A	Two out of three successive points fall in Zone A or beyond; the odd point may be anywhere	Process out of control
B	Four out of five successive points fall in Zone B; the odd point may be anywhere	Process out of control
C	Eight successive points fall in Zone C or beyond	Process out of control

Further, to WECO rules, note also the interpretations summarized hereafter. These relate to the comments regarding the process (ref.: website: www.stasoft. com/textbook/stqacon.html dated 11 December 2009).

Sl. no.	Observation	Remark
1	Nine points fall in Zone C or beyond (on one side of centre line)	The process average has probably changed
2	Eight points in a row steadily increasing or decreasing	Perhaps drift in process average

(continued)

(continued)

Sl. no.	Observation	Remark
3	Fourteen points in a row alternatively up and down	Two systematically alternating causes are producing different results
4	Two out of three points in a row in Zone or beyond	Early warning of a process shift
5	Four out of five points in a row in Zone B or beyond	Early warning of potential process shift
6	Fifteen points in a row in Zone C (above and below the centre line	Smaller variability than expected
7	Eight points in a row in Zone B, A or beyond, on either side of the centre line (without points in Zone C)	An indication that different samples are affected by different factors

18.4.12   Guide for Using Type of Control Chart

The situations in which case which chart may be used are summarized in the following table. This is for guide only. Before deciding the type of chart to be constructed more pros and cons need to be considered.

Sl. No.	Chart type	Description	Use	Type of data
1	X-bar and R	Average and range chart	If 2 or more sub-groups are available	Measurement (continuous measurement)
2	I chart	Average and moving range	If one group/sub-group	Measurement (continuous measurement)
3	P chart	Proportion	If proportion is of interest	Attribute/discrete
4	C chart	Number of defects	If number of defects is of interest	Attribute/discrete
5	U chart	Defects per unit	If number of defects per unit is of interest	Attribute/discrete
6	Np chart	Total no. of defective units	If total no. of defective units is of interest	Attribute/discrete

18.4.13   Special Topics

(a) Tolerance Limits

Sometimes it is necessary to know the capability of the process. In this effort, the tolerance chart may be useful. Tolerance chart, histogram and regression analysis may be used to assess the process capability. In the tolerance chart, the upper specification limit (USL) and lower specification limit (LSL) are set. These USL and LSL are not the same as UCL and LCL. The UCL and LCL are the limits showing the proportion of samples would lie in. The proportion such as about 99.99 % (for 6σ) is fixed by the control limits. In case of USL and LSL, the limits are determined to indicate to include a specified proportion of the samples.

The USL and LSL for the normally distributed population may be set by the following:

$$USL = \bar{X} + k_2 s$$
$$LSL = \bar{X} - k_2 s$$

where, $k_2 = a$ factor and $s =$ sample standard deviation.

The value of k_2 may be determined by the following (NIST/SEMATECH e-Handbook of Statistical Methods, http://www.itl.nist.gov/div898/handbook/, 14 December 2009)

$$k_2 = \sqrt{\frac{(N-1)\left(1 + \frac{1}{N}\right)\left(z^2_{(1-p)/2}\right)}{\chi^2_{\gamma,N-1}}}$$

where,

N sample size,
P proportion of the parameter will lie (this is pre-set),
γ probability
$N - 1$ degree of freedom for χ^2
K_2 is for two-sided tolerance limit

Example 6

See the data in the example of Sect. 17.4.6 (I chart). We need to calculate the USL and LSL.

In the said example, we have the following:

$$N = 10$$
$$\bar{X} = 29.9$$
$$s = 3.0714 \text{ (calculated now)}$$

Let us assume that we want to set the limit that 90 % of the observations will fall within the limits (USL and LSL) This means that we setting $p = 0.90$.

Probability that the observations will fall within 90 is 99 %. This means that $\gamma = 0.99$.

Now,

$P = 0.90$

Therefore,

$(1 - p)/2 = (1 - 0.90)/2 = 0.10/2 = 0.05.$

Therefore, $z = 1.645$ (from the table)

Again,

$\gamma = 0.99$

Therefore,

$\chi^2_{(0.99,9)} = 2.0879$ (from the table)

$$k_2 = \sqrt{\frac{(N-1)\left(1+\frac{1}{N}\right)\left(z^2_{(1-p)/2}\right)}{\chi^2_{\gamma,N-1}}}$$

$$= \sqrt{\frac{(10-1)\left(1+\frac{1}{10}\right)(1.645)^2}{2.0897}}$$

$$= \sqrt{\frac{9\left(\frac{11}{10}\right)(1.645)^2}{2.0897}} = 3.582$$

Therefore,

$$\begin{aligned} \text{USL} &= \bar{X} + k_2 s \\ &= 29.9 + 3.582 \times 3.0714 \\ &= 29.9 + 11.00 \\ &= 40.90 \\ \text{LSL} &= \bar{X} - k_2 s \\ &= 29.9 - 3.582 \times 3.0714 \\ &= 29.90 - 11.00 \\ &= 18.90 \end{aligned}$$

Note that standard deviation was not calculated in the example of Sect. 17.4.6. It is now calculated and used in this example.

From the example of Sect. 17.4.6, we know

$$\text{UCL} = 36.71$$
$$\text{LCL} = 23.09$$

Therefore, USL, LSL, UCL, and LCL are shown here for comparison

Fig. 18.11 Tolerance limits

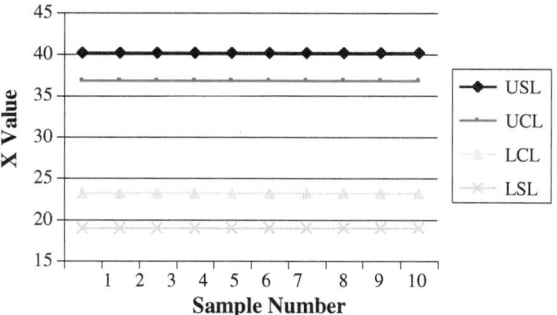

$$\text{USL} = 40.90$$
$$\text{UCL} = 36.71$$
$$\text{LCL} = 23.09$$
$$\text{LSL} = 18.90$$

UCL and LCL are natural control limits.

The limits are plotted and shown in Fig. 18.11.

Note that in this particular example, the USL and LSL fall outside UCL and LCL

(b) Process Capability

In a manufacturing process, quite often it becomes necessary to check the capability of the process when the process is operating in control. Tolerance chart, histogram and process capability ration (PCR) are the tools that may be used for the purpose. We shall discuss here only the PCR.

Process capability Ratio (PCR) is given by

$$\text{PCR} = \frac{\text{USL} - \text{LSC}}{6\sigma}$$

The USL and LSL will be known from the specific process set by the management or the process design engineer, the spread 6σ will need to be calculated. The sigma (σ) may be estimated by

$$\sigma = \frac{\bar{R}}{d_2}$$

For the previous example, we calculate the PCR as follows:

$$\sigma = \frac{\bar{R}}{d_2}$$
$$= \frac{2.56}{2.97} = 0.862$$

Therefore,

$$PCR = \frac{USL - LSC}{6\sigma}$$
$$= \frac{40.90 - 18.90}{6 * 0.862}$$
$$= \frac{22.00}{5.172}$$
$$= 4.254$$

Interpretation:
The interpretation of the PCR may be made in the following way:

1. If the PCR = 1.00, there will be a good number of nonconforming units.
2. If PCR < 1.00, a large number of nonconforming units will be produce.
3. If PCR > 1, very few nonconforming units will be produced.
4. Percentage of tolerance band used by the process = (1/PCR) * 100.

In the previous example,

(a) The PCR is greater than one. Therefore, very few nonconforming units will be produced.
(b) Percentage of tolerance band used by the process = (1/PCR)*100 = (1/4.254) * 100 = 23.51 % only.

The previous calculations were made based on the assumption that the process centered (normal population). For the non-centered process, the PCR should be calculated in the following way (Hines and Montgomery 1990).

Two PCRk need to be calculated, one based on the USL and the other based on LSL. Note that in this case we are using the notation PCR$_k$. See the following calculations.

First (based on USL)

$$PCR_k = \frac{USL - \bar{\bar{X}}}{3\sigma}$$
$$= \frac{40.90 - 29.90}{3 * 3.0714}$$
$$= \frac{40.90 - 29.90}{3 * 3.0714} = 1.194$$

Second (based on LCL)

$$\text{PCR}_k = \frac{\bar{\bar{X}} - \text{LSL}}{3\sigma}$$
$$= \frac{29.90 - 18.90}{3 * 3.0714}$$
$$= \frac{29.90 - 18.90}{3 * 3.0714} = 1.194$$

The minimum of the two is to be taken as PCR_k. In this case both the PC_ks are equal. So accepted the $\text{PCR}_k = 1.194$.

If $\text{PCR} = \text{PCR}_k$, the process is centered. In this particular example, $\text{PCR} \neq \text{PCR}_k$. Therefore, the process is off centered. The percentage band may now be recalculated as follows:

Percentage band $= (1/\text{PCR}_k) * 100 = (1/1.194) * 100 = 83.75\%$. This means that the process is using 83.75 % of the tolerance band.

(c) Sampling based on Statistical Process Control

The principle for determining the sample size is for nonparametric tolerance limits (Himes and Montgomery 1990). This is not the exact figure but is an approximate one and alright for practical purposes.

The formula for determining the required sample size is

$$n = \frac{1}{2} + \frac{1+p}{1-p} * \frac{\chi^2_{\alpha,4}}{4}$$

Example 7
In our previous example, the following calculation is made for the sample size.

$$n = \frac{1}{2} + \frac{1+p}{1-p} * \frac{\chi^2_{\alpha,4}}{4}$$
$$n = \frac{1}{2} + \frac{1+0.90}{1-0.90} * \frac{13.28}{4}$$
$$n = 0.50 + \frac{1.90}{0.10} * \frac{13.28}{4}$$
$$= 0.50 + 63.08$$
$$= 63.58, \text{say } 64.$$

This means that if we want to see that 90 % of the process units will be within the specified tolerance limits with 99 % probability, we need to select the sample size of 64 to examine the validity.

18.4.14   Summary of Control Chart Formulas

For easy reference, the control chart formulas are summarized here. However, the user should study the details first before using these summary formulas. This is a summary only and it does not cover many important details.

No.	Chart type	Mean/R	UCL	CL	LCL
1	X-bar chart	Mean	$\bar{\bar{X}} + A_2\bar{R}$	$\bar{\bar{X}}$	$X - A_2\bar{R}$
		Range	$D_4\bar{R}$	$\bar{R}$	$D_3\bar{R}$
2	I chart	Mean	$\bar{X} + E*\bar{R}$	$\bar{X}$	$\bar{X} - E*\bar{R}$
		Range	$D_4\bar{R}$	$\bar{R}$	$D_3\bar{R}$
3	p chart	Mean	$\bar{p} + 3\sqrt{\frac{p(1-\bar{p})}{n}}$	$\bar{p}$	$\bar{p} - 3\sqrt{\frac{p(1-\bar{p})}{n}}$
4	c chart	Mean	$\bar{c} + 3\sqrt{\bar{c}}$	$\bar{c}$	$\bar{c} - 3\sqrt{\bar{c}}$
5	u chart	Mean	$\bar{u} + 3\sqrt{\frac{\bar{u}}{n}}$	$\bar{u}$	$\bar{u} - 3\sqrt{\frac{\bar{u}}{n}}$
6	np chart	Mean	$\overline{np} + 3*\sqrt{\overline{np}(1-\bar{p})}$	$\overline{np}$	$\overline{np} - 3*\sqrt{\overline{np}(1-\bar{p})}$
7	Tolerance limits	Mean	$USL = \bar{X} + k_2 s$		$LSL = \bar{X} - k_2 s$

The interrelationships of the control chart constants are given here.

$$A = \frac{3}{\sqrt{n}}$$

$$A_2 = \frac{3}{d_2\sqrt{n}}$$

$$D_1 = d_2 - 3d_3; \quad \text{if negative}, D_1 = 0$$

$$D_2 = d_2 + 3d_3$$

$$D_3 = 1 - 3\frac{d_3}{d_2}; \quad \text{if negative}, D_3 = 0$$

$$D_4 = 1 + 3\frac{d_3}{d_2}$$

$$E = \frac{3}{d_2}$$

$$D_3 + D_4 = 2$$

Problems

18.1 In a manufacturing process observations of a particular product were recorded five times a day for 10 days. The observations are shown in the following table.

Sample	x_1	x_2	x_3	x_4	x_5
1	6.00	7.00	6.00	5.50	5.90
2	6.30	6.90	7.00	8.00	8.50
3	5.50	7.20	6.90	8.10	7.80
4	5.90	6.50	6.90	6.30	7.00
5	7.80	7.00	5.80	8.00	6.50
6	8.00	9.00	7.00	6.20	7.10
7	8.00	9.00	6.90	7.50	8.10
8	6.70	6.80	7.50	7.70	8.20
9	7.20	7.10	7.80	6.80	8.10
10	6.90	7.50	7.20	6.50	6.80

(a) Calculate $\bar{\bar{X}}, \bar{R}$, and UCL, CL, and LCL for mean.
(b) Calculate UCL, CL. and LCL for Range.
(c) Plot the control chart for average and Range.

18.2 Is the process in 18.1 running within control?

18.3 Based on 20 days observations, each day five observations, in a manufacturing process, the process design engineer calculated the following:

$$\bar{\bar{X}} = 5.56$$
$$\bar{R} = 1.80$$

(a) Calculate the UCL, CL, and LCL for mean.
(b) Calculate UCL, CL, and LCL for Range.

18.4 In a fish culture pond, pH values were observed during a period of 20 days in a month. The observations are shown hereafter.

Day	1	2	3	4	5	6	7	8	9	10
x value	6.2	6.0	5.9	5.5	6.0	5.8	6.1	6.5	7.0	7.1

Day	11	12	13	14	15	16	17	18	19	20
x value	6.5	6.1	7.0	7.4	7.0	5.5	6.6	6.8	7.1	7.4

(a) Calculate the UCL, CL, and LCL for the average.
(b) Calculate the UCL, CL, and LCL for the moving range.
(c) Plot the control charts for average and moving range.

18.5 In a manufacturing process, the designer has calculated the following for the average:

$$UCL = 23.55$$
$$LCL = 18.85$$

(a) What is CL?
(b) What is the control limit bandwidth?
(c) What is the value of $\bar{\bar{X}}$?

18.6 How an $\bar{X}$ chart differs from an I chart?
18.7 In a process the number of nonconforming units was recorded during a month of operation. The number of items checked and the corresponding nonconforming units were recorded. The data are shown in the following table.

Day	No. observed	No. nonconforming	Day	No. observed	No. nonconforming
1	25	6	16	29	4
2	30	8	17	30	5
3	26	9	18	31	6
4	25	7	19	25	6
5	29	6	20	27	9
6	19	6	21	19	7
7	22	5	22	22	6
8	27	9	23	28	5
9	28	6	24	25	6
10	23	5	25	22	8
11	26	7	26	26	9
12	23	8	27	23	7
13	28	7	28	24	5
14	22	6	29	25	4
15	25	5	30	22	4

(a) Calculate the p-bar.
(b) Calculate the UCL, CL, and LCL.
(c) Draw the control chart.

18.8   In a process the process design engineer wanted to check whether the number of defect items were within control. During one month, he counted the number of defective units by using equal number of sample units every day. The number of defective units was counted as shown in the following table.

Sample no.	No. of defective units	Sample no.	No. of defective units	Sample no.	No. of defective units
1	8	11	7	21	9
2	7	12	8	22	10
3	9	13	9	23	11
4	10	14	5	24	8
5	12	15	6	25	7
6	12	16	7	26	11
7	14	17	3	27	8
8	11	18	4	28	9
9	10	19	7	29	6
10	9	20	6	30	5

(a) Calculate the average number of defective units counted.
(b) Calculate the control limits.
(c) Draw the control chart.
(d) Interpret the result.

18.9   In a process, the number of defects per unit was of interest to the management. Thirty samples were drawn in sequence of time during the process was running in a month. The sample sizes and the number of defects found are shown in the following table.

(a) Calculate the average number of defects per unit.
(b) Calculate the control limits.
(c) Draw the control chart.
(d) Interpret the result.

18.10 The management of a manufacturing process wants to study the total number of nonconforming units being produced. A team was assigned to study the situation. They took a sample of size 20 in each case every day for 15 days. The number of nonconforming units was observed as shown in the following table.

(a) Calculate the control limits.
(b) Draw the control chart.

Sample	No. nonconforming	Sample	No. nonconforming	Sample	No. nonconforming
1	4	6	5	11	3
2	3	7	3	12	4
3	3	8	4	13	2
4	4	9	2	14	4
5	2	10	1	15	3

18.11   What is the main difference between control limits and tolerance limits?

18.12   In a manufacturing process the management wants to set the tolerance limits in such a way that 90 % of the observations will fall within the tolerance limits and the probability of this happening will be 99.00 %.
In an exercise, 15 samples were drawn sequentially from the process as per days of months. The individual measurements as observed are shown in the following table.

(a) Calculate the tolerance limits.
(b) Draw the tolerance limits chart.

Date	x value	Date	x value	Date	x value
1	41	6	38	11	43
2	39	7	44	12	41
3	42	8	46	13	39
4	44	9	39	14	37
5	38	10	38	15	38

18.13   Use the data for problem 18.12 and assume that the process centers on the normal population. Then,

(a) Calculate the PCR.
(b) Interpret the result.
(c) Find out the percentage of tolerance band used by the process.

18.14   In the problem 18.13, check the assumption made by calculating the PCR_k. And find out the new percentage band of the tolerance band used by the process.

18.15   In a process, it is intended that 90 % of the observations should fall within the tolerance limits with a probability of 95 % (see data in problem 18.12). Calculate the sample size required.

Answers

18.2   no
18.3   Mean
       UCL = 8.80
       CL = 5.56
       LCL = 2.32
       Range
       UCL = 2.855
       CL = 1.80
       LCL = 0.745
18.4   Average
       UCL = 7.90
       CL = 6.52
       LCL = 5.14

       (a)   Range
             UCL = 0.825
             CL = 0.52
             LCL = 0.215

18.5   (a) 21.2
       (b) 4.70
       (c) 21.2
18.6   (a) 0.256
       (b) UCL = 0.495
       CL = 0.256
       LCL = 0.017
18.8   (a) c-bar = 8.267
       (b) UCL = 16.893, CL = 8.267; LCL = 0
       (d) The process is running within control.
18.9   (a) 0.903
       (b) UCL = 1.783; CL = 0.903; LCL = 0.023
       (d) The process is not running within control.
18.10 (a) UCL = 8.021; CL = 3.14; LCL = 0
18.12 (a) USL = 47.360; LSL = 33.574
18.13 (a) 2.672
       (b) PCR > 1, so very few nonconforming units will be produced.
       (c) 37.43 %.
18.14 PCRk = 2.672. So assumption ok; Process centers around the normal population.
       % of t band = 37.43 %
18.15 n = 46.

References

webiste: http://www.wiley.com/college/sc/reid/pdf 19 November 2009)
http://www.balancedscorecard.org/portals/o/PDF/control.pdf, 19 November 2009
www.stasoft.com/textbook/stqacon.html dated 11 December 2009)
NIST/SEMATECH e-Handbook of Statistical Methods, http://www.itl.nist.gov/div898/handbook/, 14 December 2009)

Chapter 19
Summary for Hypothesis Testing

Abstract Some topics in brief considered to be understood before testing of hypotheses are explained. A summary table is attached. This table shows the relevant type of data, tests, application for tests, and the hypotheses to serve as examples.

Keywords Variable · Hypothesis · Testing · Types of tests · Measurement of data · Distributions · Probability · Data collection · Sample size · Parameter and statistics · SPSS

19.1 Prerequisite

When we make an analysis of data, the researchers/users sometimes become confused or do not know which type of analysis is to be done with the data available on their hand. This chapter has been prepared to help them in this respect. But it should be kept in mind that this chapter does not show everything necessary. This chapter should be used only as guides. For detailed procedures the respective chapters should be read.

For analysis of data and working on the hypothesis some concepts are required to be known. Read these concepts first unless you have already in the knowledge of these concepts. The required concepts are in short explained in the following sections.

19.1.1 Variable

Variable in simple language means the one that varies (not fixed). Actually, the value of a variable is not the variable itself. In a formal language it may be said that a variable is any factor, trait, or condition that can exist in differing amounts or types.

On the basis of vales a variable is of two types: a continuous variable and a categorical variable. On the basis of an experiment a variable may be of three kinds: an independent variable, a dependent variable and a controlled variable. The independent variable is the one that is changed by the scientist.

19.1.2   Hypothesis

A hypothesis may be termed in different ways based on its application. Our concern is a statistical hypothesis. A statistical hypothesis is an assumption about a population parameter. This assumption may or may not be true.

19.1.3   Hypothesis Testing

Testing refers to the formal procedures used by statisticians to accept or reject statistical hypotheses. It is a method how statistical inference is drawn.

Two hypotheses are formulated—a null hypothesis and an alternative hypothesis. Testing is concentrated on the null hypothesis. Based on the test result, we either say the null hypothesis is rejected or we say the null hypothesis is not rejected. Based on this we make the inference (conclusion). Remember, interpretation of the result is a crucial point.

Testing of hypothesis is sometimes called confirmatory data analysis because the tests confirm or disconfirm the preliminary findings.

19.1.4   Measurement of Data

For analysis purpose data are to be measured. How to measure data depends on the purpose and the objectives. The level of measurement is of importance. Statisticians categorize measurements according to levels. Each level corresponds to how this measurement can be treated mathematically.

There are four levels of measurements: nominal, ordinal, interval, and ratio. Each level is explained hereafter.

Nominal: Nominal level indicates namely of categories only, such as names of cities, categories of industries, etc. There is no order.

Ordinal: Ordinal data have order. The interval between measurements is not meaningful. Example is first, second, third, etc. There is no distance property.

Interval: In interval data there is interval property; distance between two successive intervals is meaningful. There is no zero in the measurement.

Ratio: Ratio level of measurement is the highest level. This measurement has the property of interval measurement and in addition there is defined zero.

19.1.5   Objective

Data should be accepted (secondary data) or collected. Only such data should be collected as are useful to match the objective. The collection of data should be done in such a way that the objectives can be met by analyzing. A lot depends on the possibility of analyzing the data and a lot of analysis depends on the type measurements of the data.

19.1.6 Distributions

A good idea is required regarding the distributions commonly used in statistical analysis. The following distributions have been explained in this book:

Discrete Probability Distribution

Binomial distribution
Multinomial distribution
Hypergeometric distribution
Poisson distribution.

Continuous Distribution

Normal distribution
Student t-distribution
F-distribution.

Other Distribution

Chi-square distribution

Test of hypotheses using chi-square distribution is very simple. It is easy to understand and calculate. As such it is very popular in testing hypothesis. The chi-square test makes very few assumptions about the underlying population. For this reason, it is sometimes called a nonparametric test.

19.1.7 Sample Size

Data are collected based on samples. It would be best approach to collect from the entire population not using samples. But due to time, budget and practicability it is not feasible. Hence, we have to go for sample. Therefore, it is imperative to see that the conclusions based on sample data are valid for the entire population.

Two items are important. One is sample size. Generally, larger the sample size, better are the conclusions. The other is the method of sampling such as random sampling. Usually, random sampling is adopted.

19.1.8 Probability

In hypothesis testing the test is always associated with probability. The error acceptable is also called significance level. Although it is not fixed, usually the acceptable significance level (α) is taken to be 0.05 (5 %).

19.1.9 Data Collection

Data collection is the process of collecting and measuring the information on targeted variables in an established systematic way, which then enables us to answer relevant questions and evaluate outcomes. Data can be collected from secondary sources and direct collection from the field.

Whatever method is used in collection of data some items must be kept in mind when collection. The following is a guide:

1. Make a systematic design and planning of data collection. Data are to be collected according to the need only. No information that is not necessary to achieve the objectives will be collected. Otherwise, the purpose will be defeated.
2. Collect continuous measurable data whenever possible. For example, income data can be collected using the actual amount. This is a continuous data. Income can also be collected using group such as amount between $1000 and 5000. This becomes categorical. If you collect categorical data, in analysis some very much necessary information (such as mean, standard deviation) is lost and analysis is restricted. It is important to note that continuous data, if wanted, can be converted to categories later. The reverse is not true.
3. Before collection of data think what type of analysis you are going to use in data analysis to match your objective.

19.1.10 Graph for Initial Idea

During analysis if you do not have a god of what type of analysis is to be done, it is better for you to plot the data in graph such as scatter plot. Looking at the graph you may have a good idea of what type of relationship seems to be between two or more variables.

19.1.11 Type of Test

In one way, there are two types of tests—parametric test and nonparametric test. In parametric test, distribution is associated such as normal distribution test (z test). In nonparametric test, no assumption is necessary for distribution of population.

Another type when some distribution is used, the tests are divided into two types—one tail test and two tail test. If one tail of the distribution is involved, it is a one tail test. If two tails of a distribution is involved, it is a two tail test.

Tests using chi-square distribution are called chi-square test. Tests using normal distribution is called z test. Tests using t-distribution are called t test. Tests using F-distribution are called F test. Tests using ANOVA are called ANOVA tests.

Yet in another type, hypotheses are null hypothesis and alternative hypothesis.

19.1.12 Parameter and Statistics

Parameter is a characteristic of population. Examples are population mean μ, population standard deviation σ. A statistic is a characteristic of a sample. Examples of statistics are sample mean $\bar{x}$, sample standard deviation s.

19.1.13 Model Fitting

In Regression model (statistical and econometric model), fitting of the model is important. If the model fitting is not good, it may hardly be possible to use it for prediction/forecasting.

As a guide, R^2 value should be as high as possible, 0.90 or 90 % for example; F-value should be as small as possible, say close to zero; significance level (α) should be as small as possible, for example, $\alpha < 0.05$.

19.1.14 Dummy Variable

Categorical data have limitations. One limitation is that these cannot generally be used in the analyses that are applicable for continuous data. Regression is one important analysis for continuous data. Fortunately, categorical data can be used in the regression analysis by converting these into dummy variables. For this technique, see the chapter on multiple regression.

19.1.15 SPSS

SPSS means Statistical Package for Social Sciences. There are other few packages. But SPSS is versatile and it is very useful. In examples in this book SPSS has been used. So the analyses that have been used are according to SPSS.

19.2   Summary

Data	Types of test	What is tested	Example of hypothesis
Cat-georgical	Chi-square		
	Goodness-of-fit test	Comparing a set of observed frequencies to a set of expected frequencies	Null: two distributions are Similar; alternative: two distributions are different
	Test of independence	Testing the difference between the observed frequencies of several classifications of two variables	Null: the two variables are independent (no relationship). Alternative: the two variables are not independence (related)
Continuous	Parametric test		
	(1) Z test; t test for small sample ($n < 30$)	Tests one population mean	Null: population mean has a specific value; alternative: population mean does not have the same value
		Tests equality of two population means	Null: two population means are same/equal; alternative: the two population means are not same.
		Tests one population proportion	Null: the population proportion has a specific value; alternative: the population proportion does not have the same value
		Tests equality of two population proportions	Null: two population proportions are the same; alternative: the two population proportions are not same
	Power of test	Calculation of type I error (α) and type II error (β). Cannot do for open ended hypothesis	Calculation of power
	Nonparametric test—the sign test	Test the medians of a continuous variable	Null: the median is a specified value; alternative: the median is not equal to that specified value (greater or less)
	Nonparametric test—the rank test		

(continued)

(continued)

Data	Types of test	What is tested	Example of hypothesis
	(1) The wilcoxon rank-sum test	Compare two population means	Null: incomes of two population groups are equal; alternative: incomes of the same two population groups are not equal (greater or less)
	(2) Spearman rank correlation	Measure the degree to which two numerical variables are monotonically related or associated	Null: there is no spearman correlation between the populations; alternative: spearman correlation coefficient between the two populations is not zero (greater or less)
	Kruskal–Wallis test	Analysis of variance (ANOVA) is carried out by this method. It is nonparametric method and does not assume normal population. It uses ranks	Null: there is no difference between groups (all group rank means are same); alternative: there is difference between groups (all group rank means are not same)
	Simple regression	It uses the dependent variable and one independent variable	Used in statistical estimation and model building. Can be used for prediction also. Check the model fitting
	Multiple regression	It uses the dependent variable and two or more independent variables. Can use categorical data also as dummy variables	Used in statistical estimation. Can be used for prediction/forecasting also. Check the model fitting (R^2, α value, F-value)

Statistical Tables

© Springer Science+Business Media Singapore 2016
A.Q. Miah, *Applied Statistics for Social and Management Sciences*,
DOI 10.1007/978-981-10-0401-8

Table A1 Areas under the normal curve

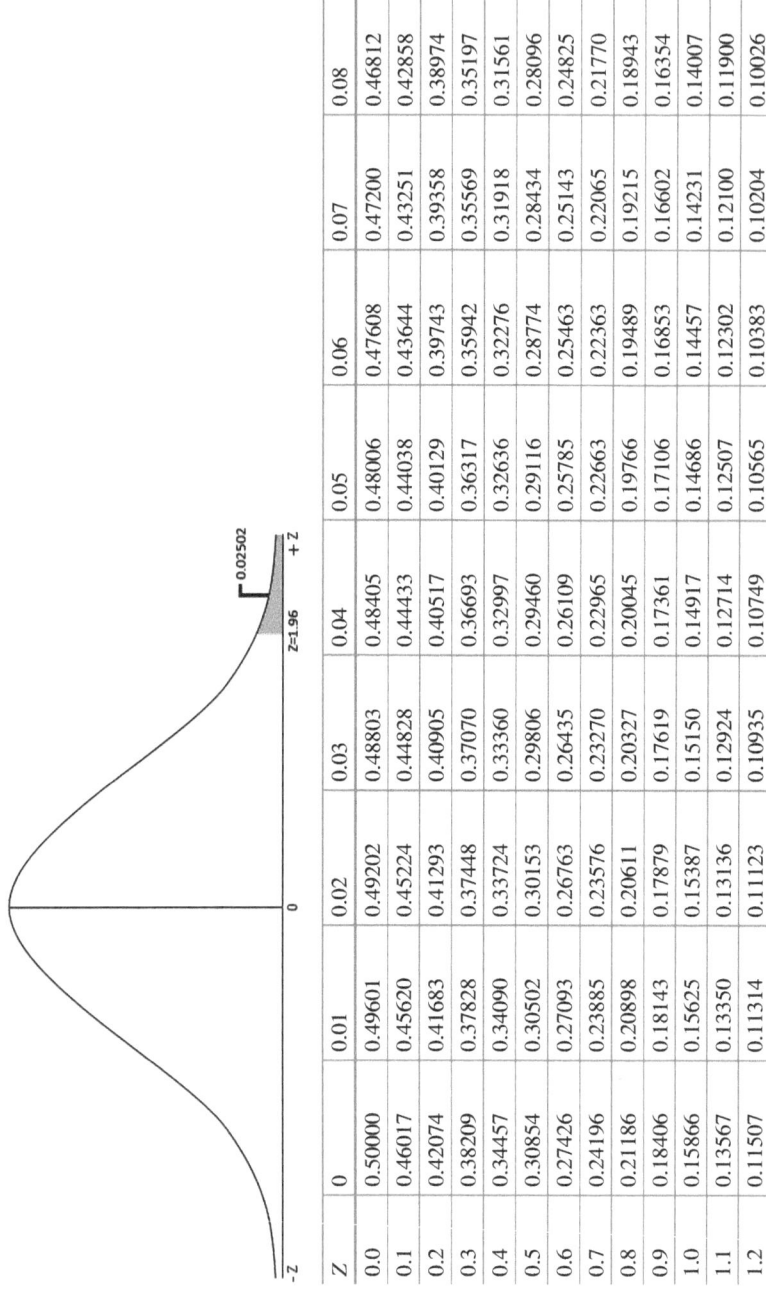

Z	0	0.01	0.02	0.03	0.04	0.05	0.06	0.07	0.08	0.09
0.0	0.50000	0.49601	0.49202	0.48803	0.48405	0.48006	0.47608	0.47200	0.46812	0.46414
0.1	0.46017	0.45620	0.45224	0.44828	0.44433	0.44038	0.43644	0.43251	0.42858	0.42465
0.2	0.42074	0.41683	0.41293	0.40905	0.40517	0.40129	0.39743	0.39358	0.38974	0.38591
0.3	0.38209	0.37828	0.37448	0.37070	0.36693	0.36317	0.35942	0.35569	0.35197	0.34827
0.4	0.34457	0.34090	0.33724	0.33360	0.32997	0.32636	0.32276	0.31918	0.31561	0.31207
0.5	0.30854	0.30502	0.30153	0.29806	0.29460	0.29116	0.28774	0.28434	0.28096	0.27759
0.6	0.27426	0.27093	0.26763	0.26435	0.26109	0.25785	0.25463	0.25143	0.24825	0.24510
0.7	0.24196	0.23885	0.23576	0.23270	0.22965	0.22663	0.22363	0.22065	0.21770	0.21476
0.8	0.21186	0.20898	0.20611	0.20327	0.20045	0.19766	0.19489	0.19215	0.18943	0.18673
0.9	0.18406	0.18143	0.17879	0.17619	0.17361	0.17106	0.16853	0.16602	0.16354	0.16109
1.0	0.15866	0.15625	0.15387	0.15150	0.14917	0.14686	0.14457	0.14231	0.14007	0.13786
1.1	0.13567	0.13350	0.13136	0.12924	0.12714	0.12507	0.12302	0.12100	0.11900	0.11702
1.2	0.11507	0.11314	0.11123	0.10935	0.10749	0.10565	0.10383	0.10204	0.10026	0.09850

(continued)

Table A1 (continued)

Z	0	0.01	0.02	0.03	0.04	0.05	0.06	0.07	0.08	0.09
1.3	0.09779	0.09511	0.09340	0.09179	0.09013	0.08849	0.08690	0.08534	0.08378	0.08226
1.4	0.08075	0.07928	0.07780	0.07636	0.07493	0.07353	0.07213	0.07078	0.06942	0.06811
1.5	0.06679	0.06551	0.06424	0.06300	0.06176	0.06057	0.05937	0.05821	0.05705	0.05590
1.6	0.05478	0.05370	0.05263	0.05155	0.05051	0.04947	0.04846	0.04746	0.04648	0.04551
1.7	0.04456	0.04363	0.04272	0.04181	0.04094	0.04006	0.03922	0.03834	0.03755	0.03671
1.8	0.03595	0.03515	0.03439	0.03364	0.03288	0.03216	0.03144	0.03072	0.03005	0.02936
1.9	0.02873	0.02805	0.02741	0.02681	0.02618	0.02558	0.02502	0.02442	0.02385	0.02330
2.0	0.02275	0.02221	0.02169	0.02118	0.02067	0.02018	0.01970	0.01923	0.01876	0.01831
2.1	0.01787	0.01743	0.01700	0.01659	0.01618	0.01578	0.01540	0.01501	0.01465	0.01425
2.2	0.01389	0.01357	0.01321	0.01289	0.01253	0.01221	0.01189	0.01161	0.01130	0.01102
2.3	0.01074	0.01046	0.01018	0.00990	0.00966	0.00938	0.00914	0.00890	0.00866	0.00842
2.4	0.00818	0.00798	0.00775	0.00755	0.00735	0.00715	0.00695	0.00675	0.00659	0.00639
2.5	0.00623	0.00603	0.00587	0.00571	0.00555	0.00539	0.00523	0.00507	0.00495	0.00479
2.6	0.00467	0.00451	0.00439	0.00427	0.00415	0.00403	0.00392	0.00380	0.00368	0.00356
2.7	0.00348	0.00336	0.00328	0.00316	0.00308	0.00300	0.00288	0.00280	0.00272	0.00264
2.8	0.00256	0.00248	0.00240	0.00232	0.00224	0.00220	0.00212	0.00204	0.00200	0.00192
2.9	0.00188	0.00180	0.00176	0.00168	0.00164	0.00160	0.00152	0.00145	0.00144	0.00140
3.0	0.00136	0.00132	0.00128	0.00124	0.00120	0.00116	0.00112	0.00108	0.00104	0.00100
3.1	0.00096	0.00092	0.00092	0.00088	0.00084	0.00082	0.00080	0.00076	0.00072	0.00072
3.2	0.00068	0.00068	0.00064	0.00062	0.00060	0.00056	0.00054	0.00052	0.00050	0.00048
3.3	0.00480	0.00046	0.00044	0.00042	0.00040	0.00040	0.00038	0.00036	0.00036	0.00034
3.4	0.00032	0.00032	0.00030	0.00028	0.00028	0.00027	0.00026	0.00025	0.00024	0.00025

(continued)

Table A1 (continued)

Z	0	0.01	0.02	0.03	0.04	0.05	0.06	0.07	0.08	0.09
3.5	0.00022	0.00021	0.00021	0.00020	0.00020	0.00019	0.00018	0.00017	0.00017	0.00016
3.6	0.00016	0.00015	0.00015	0.00014	0.00014	0.00013	0.00013	0.00012	0.00012	0.00011
3.7	0.00011	0.00011	0.00010	0.00009	0.00009	0.00009	0.00009	0.00008	0.00008	0.00008
3.8	0.00007	0.00007	0.00007	0.00007	0.00006	0.00006	0.00006	0.00006	0.00006	0.00006
3.9	0.00005	0.00005	0.00005	0.00005	0.00004	0.00004	0.00004	0.00004	0.00004	0.00003

Table A2 Percentage points of the t distribution

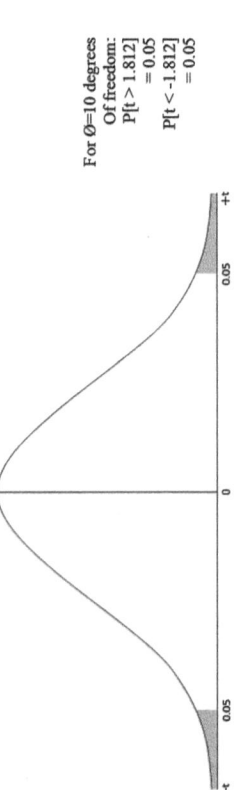

For $\emptyset=10$ degrees
Of freedom:
$P[t > 1.812]$
$= 0.05$
$P[t < -1.812]$
$= 0.05$

df	α								
	0.25	0.20	0.15	0.10	0.05	0.025	0.010	0.005	0.001
1	1.000	1.376	1.963	3.078	6.314	12.706	31.821	63.657	318.309
2	0.816	1.061	1.386	1.886	2.92	4.303	6.965	9.93	22.327
3	0.765	0.978	1.250	1.638	2.353	3.182	4.541	5.841	10.215
4	0.741	0.941	1.190	1.533	2.132	2.776	3.747	4.604	7.173
5	0.727	0.920	1.156	1.476	2.015	2.571	3.365	4.032	5.893
6	0.718	0.906	1.134	1.440	1.943	2.447	3.143	3.707	5.208
7	0.711	0.896	1.119	1.415	1.895	2.365	2.998	3.499	4.785
8	0.706	0.889	1.108	1.397	1.860	2.306	2.896	3.355	4.501
9	0.703	0.883	1.100	1.383	1.833	2.262	2.821	3.25	4.297
10	0.700	0.879	1.093	1.372	1.812	2.228	2.764	3.169	4.144
11	0.697	0.876	1.088	1.363	1.796	2.201	2.718	3.106	4.025
12	0.695	0.873	1.083	1.356	1.782	2.179	2.681	3.055	3.930
13	0.694	0.870	1.079	1.350	1.771	2.160	2.650	3.012	3.852
14	0.692	0.868	1.076	1.345	1.761	2.145	2.624	2.977	3.787

(continued)

Table A2 (continued)

df	α									
	0.25	0.20	0.15	0.10	0.05	0.025	0.010	0.005	0.001	
15	0.691	0.866	1.074	1.341	1.753	2.131	2.602	2.947	3.733	
16	0.690	0.865	1.071	1.337	1.746	2.120	2.583	2.921	3.686	
17	0.689	0.863	1.069	1.333	1.74	2.110	2.567	2.898	3.646	
18	0.688	0.862	1.067	1.33	1.734	2.101	2.552	2.878	3.610	
19	0.688	0.861	1.066	1.328	1.729	2.093	2.539	2.861	3.579	
20	0.687	0.860	1.064	1.325	1.725	2.086	2.528	2.845	3.552	
21	0.686	0.859	1.063	1.323	1.721	2.080	2.518	2.831	3.527	
22	0.686	0.858	1.061	1.321	1.717	2.074	2.508	2.819	3.505	
23	0.685	0.858	1.060	1.319	1.714	2.069	2.500	2.807	3.485	
24	0.685	0.857	1.059	1.318	1.711	2.064	2.492	2.797	3.467	
25	0.684	0.856	1.058	1.316	1.708	2.060	2.485	2.787	3.450	
26	0.684	0.856	1.058	1.315	1.706	2.056	2.479	2.779	3.435	
27	0.684	0.855	1.057	1.314	1.703	2.052	2.473	2.771	3.421	
28	0.683	0.855	1.056	1.313	1.701	2.048	2.467	2.763	3.408	
29	0.683	0.854	1.055	1.311	1.699	2.045	2.462	2.756	3.396	
30	0.683	0.854	1.055	1.310	1.697	2.042	2.457	2.750	3.385	
40	0.681	0.851	1.050	1.303	1.684	2.021	2.423	2.704	3.307	
50	0.679	0.849	1.047	1.299	1.676	2.009	2.403	2.678	3.261	
60	0.679	0.848	1.046	1.296	1.671	2.000	2.390	2.660	3.232	
70	0.678	0.847	1.044	1.294	1.667	1.994	2.381	2.648	3.211	
80	0.678	0.846	1.043	1.292	1.664	1.990	2.374	2.639	3.195	
90	0.677	0.846	1.042	1.291	1.662	1.987	2.368	2.532	3.183	
100	0.677	0.845	1.042	1.290	1.660	1.984	2.364	2.626	3.174	
120	0.677	0.845	1.041	1.289	1.658	1.980	2.358	2.617	3.160	

Source Calculated with the help of StaTable, Software from Cytel Software, Cambridge, MA (08 October 2009)

Table A3(1) Chi-square (χ^2) critical points

df	\(\alpha\) 0.9999	0.99	0.98	0.975	0.95	0.925	0.90	0.85	0.80
1	0.000	0.0002	0.0006	0.001	0.004	0.009	0.016	0.036	0.064
2	0.0002	0.020	0.040	0.051	0.103	0.156	0.211	0.325	0.446
3	0.005	0.115	0.185	0.216	0.352	0.472	0.584	0.800	1.005
4	0.028	0.297	0.429	0.484	0.711	0.897	1.064	1.366	1.649
5	0.082	0.554	0.752	0.831	1.145	1.394	1.610	1.994	2.343
6	0.172	0.872	1.134	1.237	1.635	1.941	2.204	2.661	3.070
7	0.300	1.239	1.564	1.690	2.167	2.528	2.833	3.358	3.822
8	0.463	1.646	2.032	2.180	2.733	3.144	3.490	4.078	4.594
9	0.661	2.088	2.532	2.700	3.325	3.785	4.168	4.816	5.380
10	0.889	2.558	3.059	3.247	3.940	4.446	4.865	5.570	6.179
11	1.145	3.053	3.609	3.816	4.575	5.124	5.578	6.336	6.988
12	1.427	3.571	4.178	4.404	5.226	5.818	6.304	7.114	7.807
13	1.734	4.107	4.765	5.009	5.892	6.524	7.042	7.901	8.634
14	2.061	4.660	5.368	5.629	6.571	7.242	7.790	8.696	9.467
15	2.408	5.229	5.985	6.262	7.261	7.369	8.547	9.499	10.307

(continued)

Table A3(1) (continued)

df	α								
	0.9999	0.99	0.98	0.975	0.95	0.925	0.90	0.85	0.80
16	2.774	5.812	6.614	6.908	7.962	8.707	9.312	10.309	11.152
17	3.156	6.408	7.255	7.564	8.672	9.452	10.085	11.125	12.002
18	3.555	7.015	7.506	8.231	9.390	10.205	10.865	11.946	12.857
19	3.967	7.633	8.567	8.906	10.117	10.965	11.651	12.773	13.716
20	4.395	8.260	9.237	9.591	10.851	11.732	12.443	13.604	14.578
25	6.707	11.524	12.697	13.120	14.611	15.645	16.473	17.818	18.940
30	9.258	14.953	16.306	16.791	18.493	19.664	20.599	22.110	23.364
35	11.996	18.509	20.027	20.569	22.465	23.763	24.800	26.460	27.836
40	14.883	22.164	23.838	24.433	26.509	27.926	29.051	30.856	32.345
45	17.894	25.901	27.720	28.366	30.612	32.140	33.350	35.290	36.884
50	21.009	29.707	31.664	32.357	34.764	36.397	37.689	39.754	41.449
55	24.214	33.570	35.659	36.398	38.958	40.691	42.060	44.245	46.036
60	27.497	37.485	39.699	40.482	43.188	45.016	46.459	48.759	50.641
65	30.848	41.444	43.779	44.603	47.450	49.370	50.883	53.293	55.262
70	34.261	45.442	47.893	48.757	51.739	53.748	55.329	57.844	59.898
75	37.728	49.475	52.039	52.942	56.054	58.148	59.795	62.412	64.547
80	41.244	53.640	56.213	57.153	60.391	62.567	64.278	66.994	69.207
85	44.806	57.634	60.412	61.389	64.749	67.005	68.777	71.589	73.878
90	48.408	61.754	64.635	65.647	69.126	71.460	73.291	76.195	78.558
95	52.049	65.898	68.879	69.925	73.520	75.929	77.818	80.813	83.248
100	55.725	70.065	73.142	74.222	77.929	80.412	82.358	85.441	87.945

Source Calculated with help of online calculator: http://www.danielsoper.Com/statcalc/calc12.aspx, (29 December 2009)

Table A3(2) Chi-square (χ^2) critical points

df	α									
	0.25	0.20	0.15	0.10	0.05	0.025	0.010	0.005	0.001	
1	1.32	1.64	2.07	2.71	3.84	5.02	6.63	7.88	10.83
2	2.77	3.22	3.79	4.61	5.99	7.38	9.21	10.60	13.82
3	4.11	4.64	5.32	6.25	7.81	9.35	11.34	12.84	16.27
4	5.39	5.99	6.74	7.78	9.48	11.14	13.28	14.86	18.47
5	6.63	7.29	8.12	9.24	11.07	12.83	15.09	16.75	20.52
6	7.84	8.56	9.45	10.64	12.59	14.45	16.81	18.55	22.46
7	9.04	9.80	10.75	12.02	14.07	16.01	18.48	20.28	24.32
8	10.22	11.03	12.00	13.36	15.51	17.53	20.09	21.96	26.12
9	11.39	12.24	13.29	14.68	16.92	19.02	21.67	23.59	27.88
10	12.55	13.44	14.53	15.99	18.31	20.48	23.21	25.19	29.59
11	13.70	14.63	15.77	17.28	19.68	21.92	24.72	26.76	31.26
12	14.85	15.81	16.99	18.55	21.03	23.34	26.22	28.30	32.91
13	15.98	16.98	18.20	19.81	22.36	24.74	27.69	29.82	34.53
14	17.12	18.15	19.41	21.06	23.68	26.12	29.14	31.32	36.12
15	18.25	19.31	20.60	22.31	25.00	27.49	30.58	32.80	37.70
16	19.37	20.47	21.79	23.54	26.30	28.85	32.00	34.27	39.25

(continued)

Table A3(2) (continued)

df	α								
	0.25	0.20	0.15	0.10	0.05	0.025	0.010	0.005	0.001
17	20.49	21.61	22.98	24.77	27.59	30.19	33.41	35.72	40.79
18	21.60	22.76	24.16	25.99	28.87	31.53	34.81	37.16	42.31
19	22.72	23.90	25.33	27.20	30.14	32.85	36.19	38.58	43.82
20	23.83	25.04	26.50	28.41	31.41	34.17	37.57	40.00	45.32
25	29.34	30.68	32.28	34.38	37.65	40.65	44.31	46.93	52.62
30	34.80	36.25	37.99	40.26	43.77	46.98	50.89	53.67	59.70
35	40.22	41.78	43.64	46.06	49.80	53.20	57.34	60.27	66.62
40	45.62	47.27	49.24	51.80	55.76	59.34	63.69	66.77	73.40
45	50.98	52.73	54.81	57.51	61.66	65.41	69.96	73.17	80.08
50	56.33	58.16	60.35	63.17	67.50	71.42	76.15	79.49	86.66
55	61.67	63.58	65.86	68.80	73.31	77.38	82.29	85.75	93.17
60	66.98	68.97	71.34	74.40	79.08	83.30	88.38	91.95	99.61
65	72.29	74.35	76.81	79.97	84.82	89.18	94.42	98.11	105.99
70	77.58	79.72	82.26	85.53	90.53	95.02	100.40	104.20	112.30
75	82.86	85.07	87.69	91.06	96.22	100.84	106.39	110.29	118.60
80	88.13	90.41	93.11	96.58	101.80	106.60	112.30	116.30	124.80
85	93.39	95.73	98.51	102.08	107.52	112.39	118.24	122.33	131.04
90	98.65	101.05	103.90	107.60	113.10	118.10	124.10	128.30	137.20
95	103.90	106.36	109.29	113.04	118.75	123.86	129.97	134.25	143.34
100	109.10	111.67	114.66	118.50	124.30	129.60	135.80	140.20	149.50

Source Calculated with help of *StaTable*, Software from Cytel Software, Cambridge, MA (08 October 2009)

Table A4(1) F distribution (F values for $\alpha = 0.10$)

d_2 \ d_1	1	2	3	4	5	6	7	8	9	10
1	39.863	49.500	53.593	55.833	57.240	58.204	58.906	59.439	59.858	60.195
2	8.526	9.000	9.162	9.243	9.293	9.326	9.349	9.367	9.381	9.392
3	5.538	5.462	5.391	5.343	5.309	5.285	5.266	5.252	5.240	5.230
4	4.545	4.325	4.191	4.107	4.051	4.010	3.979	3.955	3.936	3.920
5	4.060	3.780	3.619	3.520	3.453	3.405	3.368	3.339	3.316	3.297
6	3.776	3.463	3.289	3.181	3.108	3.055	3.014	2.983	2.958	2.937
7	3.589	3.257	3.074	2.961	2.883	2.827	2.785	2.752	2.725	2.703
8	3.458	3.113	2.924	2.806	2.726	2.668	2.624	2.589	2.561	2.538
9	3.360	3.006	2.813	2.693	2.611	2.551	2.505	2.469	2.440	2.416
10	3.285	2.924	2.728	2.605	2.522	2.461	2.414	2.377	2.347	2.323
11	3.225	2.860	2.660	2.536	2.451	2.389	2.342	2.304	2.274	2.248
12	3.177	2.807	2.606	2.480	2.394	2.331	2.283	2.245	2.214	2.188
13	3.136	2.763	2.560	2.434	2.347	2.283	2.234	2.195	2.164	2.138
14	3.102	2.726	2.522	2.395	2.307	2.243	2.193	2.154	2.122	2.095
15	3.073	2.695	2.490	2.361	2.273	2.208	2.158	2.119	2.086	2.059
16	3.048	2.668	2.462	2.333	2.244	2.178	2.128	2.088	2.055	2.028
17	3.026	2.645	2.437	2.308	2.218	2.152	2.102	2.061	2.028	2.001
18	3.007	2.624	2.416	2.286	2.196	2.130	2.079	2.038	2.005	1.977
19	2.990	2.606	2.397	2.266	2.176	2.109	2.058	2.017	1.984	1.956
20	2.975	2.589	2.380	2.249	2.158	2.091	2.040	1.999	1.965	1.937
21	2.961	2.575	2.365	2.233	2.142	2.075	2.023	1.982	1.948	1.920
22	2.949	2.561	2.351	2.219	2.128	2.060	2.008	1.967	1.933	1.904
23	2.937	2.549	2.339	2.207	2.115	2.047	1.995	1.953	1.919	1.890

(continued)

Table A4(1) (continued)

d_2 \ d_1	1	2	3	4	5	6	7	8	9	10
24	2.927	2.538	2.327	2.195	2.103	2.035	1.983	1.941	1.906	1.877
25	2.918	2.528	2.317	2.184	2.092	2.024	1.971	1.929	1.895	1.866
26	2.909	2.519	2.307	2.174	2.082	2.014	1.961	1.919	1.884	1.855
27	2.901	2.511	2.299	2.165	2.073	2.005	1.952	1.909	1.874	1.845
28	2.894	2.503	2.291	2.157	2.064	1.996	1.943	1.900	1.865	1.836
29	2.887	2.495	2.283	2.149	2.057	1.988	1.935	1.892	1.857	1.827
30	2.881	2.489	2.276	2.142	2.049	1.980	1.927	1.884	1.849	1.819
40	2.835	2.440	2.226	2.091	1.997	1.927	1.873	1.829	1.793	1.763
50	2.809	2.412	2.197	2.061	1.966	1.895	1.840	1.796	1.760	1.729
60	2.791	2.393	2.177	2.041	1.946	1.875	1.819	1.775	1.738	1.707
70	2.779	2.380	2.164	2.027	1.931	1.860	1.804	1.760	1.723	1.691
80	2.769	2.370	2.154	2.016	1.921	1.849	1.793	1.748	1.711	1.680
90	2.762	2.363	2.146	2.008	1.912	1.841	1.785	1.739	1.702	1.670
100	2.756	2.356	2.139	2.002	1.906	1.834	1.778	1.732	1.695	1.663

d_2 \ d_1	12	15	20	25	30	40	60	80	100	120
1	60.705	61.220	61.740	62.055	62.265	62.529	62.794	62.927	63.007	63.061
2	9.408	9.425	9.441	9.451	9.458	9.466	9.475	9.479	9.482	9.483
3	5.216	5.200	5.184	5.175	5.168	5.160	5.151	5.147	5.144	5.143
4	3.896	3.870	3.844	3.828	3.817	3.804	3.790	3.782	3.778	3.775
5	3.268	3.238	3.207	3.187	3.174	3.157	3.140	3.132	3.126	3.123
6	2.905	2.871	2.836	2.815	2.800	2.782	2.762	2.752	2.746	2.742
7	2.668	2.632	2.595	2.571	2.556	2.535	2.514	2.504	2.497	2.493

(continued)

Table A4(1) (continued)

d_2	d_1									
	12	15	20	25	30	40	60	80	100	120
8	2.502	2.464	2.425	2.400	2.383	2.361	2.339	2.328	2.321	2.316
9	2.379	2.340	2.298	2.273	2.255	2.232	2.209	2.197	2.189	2.184
10	2.284	2.244	2.201	2.174	2.155	2.132	2.107	2.095	2.087	2.082
11	2.209	2.167	2.123	2.095	2.076	2.052	2.026	2.013	2.005	2.000
12	2.147	2.105	2.060	2.031	2.012	1.986	1.960	1.946	1.938	1.932
13	2.097	2.053	2.007	1.978	1.958	1.932	1.904	1.890	1.882	1.876
14	2.054	2.010	1.962	1.933	1.912	1.885	1.857	1.843	1.834	1.828
15	2.017	1.972	1.924	1.894	1.873	1.845	1.817	1.802	1.793	1.787
16	1.985	1.940	1.891	1.860	1.839	1.811	1.782	1.766	1.757	1.751
17	1.958	1.912	1.862	1.831	1.809	1.781	1.751	1.735	1.726	1.719
18	1.933	1.887	1.837	1.805	1.783	1.754	1.723	1.707	1.698	1.691
19	1.912	1.865	1.814	1.782	1.759	1.730	1.699	1.683	1.673	1.666
20	1.892	1.845	1.794	1.761	1.738	1.708	1.677	1.660	1.650	1.643
21	1.875	1.827	1.776	1.742	1.719	1.689	1.657	1.640	1.630	1.623
22	1.859	1.811	1.759	1.726	1.702	1.671	1.639	1.622	1.611	1.604
23	1.845	1.796	1.744	1.710	1.686	1.655	1.622	1.605	1.594	1.587
24	1.832	1.783	1.730	1.696	1.672	1.641	1.607	1.590	1.579	1.572
25	1.820	1.771	1.718	1.683	1.659	1.627	1.593	1.576	1.565	1.557
26	1.809	1.760	1.706	1.671	1.647	1.615	1.581	1.563	1.551	1.544
27	1.799	1.749	1.695	1.660	1.636	1.603	1.569	1.550	1.539	1.531
28	1.790	1.740	1.685	1.650	1.625	1.593	1.558	1.539	1.528	1.520
29	1.781	1.731	1.676	1.640	1.616	1.583	1.547	1.529	1.517	1.509
30	1.773	1.722	1.667	1.632	1.607	1.573	1.538	1.519	1.507	1.499

(continued)

Table A4(1) (continued)

d_2	d_1										
	12	15	20	25	30	40	60	80	100	120
40	1.715	1.662	1.605	1.568	1.541	1.506	1.467	1.447	1.434	1.425
50	1.680	1.627	1.568	1.529	1.502	1.465	1.424	1.402	1.390	1.379
60	1.657	1.603	1.543	1.504	1.476	1.437	1.395	1.372	1.358	1.348
70	1.641	1.587	1.526	1.486	1.457	1.418	1.374	1.350	1.335	1.325
80	1.629	1.574	1.513	1.472	1.443	1.403	1.358	1.334	1.318	1.307
90	1.620	1.564	1.503	1.461	1.432	1.391	1.346	1.321	1.304	1.293
100	1.612	1.557	1.494	1.453	1.423	1.382	1.336	1.310	1.293	1.282

d_1 degrees of freedom, numerator; d_2 degrees of freedom, denominator

Source Calculated using distribution calculator (dostat.stat.sc.edu/prototype/calculators/index.php3?dist=F; 12 October 2009)

Table A4(2) F distribution (F values for $\alpha = 0.05$)

d_2	d_1									
	1	2	3	4	5	6	7	8	9	10
1	161.45	199.5	215.71	224.58	230.16	233.99	236.77	238.88	240.54	241.88
2	18.513	19.00	19.164	19.247	19.296	19.33	19.353	19.371	19.385	19.396
3	10.128	9.552	9.277	9.117	9.013	8.941	8.887	8.845	8.812	8.786
4	7.709	6.944	6.591	6.388	6.256	6.163	6.094	6.041	5.999	5.964
5	6.608	5.786	5.409	5.192	5.050	4.950	4.876	4.818	4.772	4.735
6	5.987	5.143	4.757	4.534	4.387	4.284	4.207	4.147	4.099	4.060
7	5.591	4.737	4.347	4.120	3.972	3.866	3.787	3.726	3.677	3.637
8	5.318	4.459	4.066	3.838	3.687	3.581	3.500	3.438	3.388	3.347
9	5.117	4.256	3.863	3.633	3.482	3.374	3.293	3.230	3.179	3.137
10	4.965	4.103	3.708	3.478	3.326	3.217	3.135	3.072	3.020	2.978
11	4.844	3.982	3.587	3.357	3.204	3.095	3.012	2.948	2.896	2.854
12	4.747	3.885	3.490	3.259	3.106	2.996	2.913	2.849	2.796	2.753
13	4.667	3.806	3.411	3.179	3.025	2.915	2.832	2.767	2.714	2.671
14	4.600	3.739	3.344	3.112	2.958	2.848	2.764	2.699	2.646	2.602
15	4.543	3.682	3.287	3.056	2.901	2.790	2.707	2.641	2.588	2.544
16	4.494	3.634	3.239	3.007	2.852	2.741	2.657	2.591	2.538	2.494
17	4.451	3.592	3.197	2.965	2.810	2.699	2.614	2.548	2.494	2.450
18	4.414	3.555	3.160	2.928	2.773	2.661	2.577	2.510	2.456	2.412
19	4.381	3.522	3.127	2.895	2.740	2.628	2.544	2.477	2.423	2.378
20	4.351	3.493	3.098	2.866	2.711	2.599	2.514	2.447	2.393	2.348
21	4.325	3.467	3.072	2.840	2.685	2.573	2.488	2.420	2.366	2.321
22	4.301	3.443	3.049	2.817	2.661	2.549	2.464	2.397	2.342	2.297
23	4.279	3.422	3.028	2.796	2.640	2.528	2.442	2.375	2.320	2.275

(continued)

Table A4(2) (continued)

d_2	d_1									
	1	2	3	4	5	6	7	8	9	10
24	4.260	3.403	3.009	2.776	2.621	2.508	2.423	2.355	2.300	2.255
25	4.242	3.385	2.991	2.759	2.603	2.490	2.405	2.337	2.282	2.236
26	4.225	3.369	2.975	2.743	2.587	2.474	2.388	2.321	2.265	2.220
27	4.210	3.354	2.960	2.728	2.572	2.459	2.373	2.305	2.250	2.204
28	4.196	3.340	2.947	2.714	2.558	2.445	2.359	2.291	2.236	2.190
29	4.183	3.328	2.934	2.701	2.545	2.432	2.346	2.278	2.223	2.177
30	4.171	3.316	2.922	2.690	2.534	2.421	2.334	2.266	2.211	2.165
40	4.085	3.232	2.839	2.606	2.449	2.336	2.249	2.180	2.124	2.077
50	4.034	3.183	2.790	2.557	2.400	2.286	2.199	2.130	2.073	2.026
60	4.001	3.150	2.758	2.525	2.368	2.254	2.167	2.097	2.040	1.993
70	3.978	3.128	2.736	2.503	2.346	2.231	2.143	2.074	2.017	1.969
80	3.960	3.111	2.719	2.486	2.329	2.214	2.126	2.056	1.999	1.951
90	3.947	3.098	2.706	2.473	2.316	2.201	2.113	2.043	1.986	1.938
100	3.936	3.087	2.696	2.463	2.305	2.191	2.103	2.032	1.975	1.927

d_2	d_1									
	12	15	20	25	30	40	60	80	100	120
1	243.9	246.0	248.0	249.3	250.1	251.1	252.2	252.7	253.0	253.3
2	19.413	19.429	19.446	19.46	19.462	19.471	19.479	19.48	19.49	19.487
3	8.745	8.703	8.660	8.634	8.617	8.594	8.572	8.561	8.554	8.549
4	5.912	5.858	5.803	5.769	5.746	5.717	5.688	5.673	5.664	5.658
5	4.678	4.619	4.558	4.521	4.496	4.464	4.431	4.415	4.405	4.399
6	4.000	3.938	3.874	3.835	3.808	3.774	3.740	3.722	3.712	3.705
7	3.575	3.511	3.445	3.404	3.376	3.340	3.304	3.286	3.275	3.268

(continued)

Table A4(2) (continued)

d_2 \ d_1	12	15	20	25	30	40	60	80	100	120
8	3.284	3.218	3.150	3.108	3.079	3.043	3.005	2.986	2.975	2.967
9	3.073	3.006	2.936	2.893	2.864	2.826	2.787	2.768	2.756	2.748
10	2.913	2.845	2.774	2.730	2.700	2.661	2.621	2.601	2.588	2.580
11	2.788	2.719	2.646	2.601	2.571	2.531	2.490	2.469	2.457	2.448
12	2.687	2.617	2.544	2.498	2.466	2.426	2.384	2.363	2.350	2.341
13	2.604	2.533	2.459	2.412	2.380	2.339	2.297	2.275	2.261	2.252
14	2.534	2.463	2.388	2.341	2.308	2.266	2.223	2.201	2.187	2.178
15	2.475	2.403	2.328	2.280	2.247	2.204	2.160	2.137	2.123	2.114
16	2.425	2.352	2.276	2.227	2.194	2.151	2.106	2.083	2.068	2.069
17	2.381	2.308	2.230	2.181	2.147	2.104	2.058	2.035	2.020	2.011
18	2.342	2.269	2.191	2.141	2.107	2.063	2.017	1.993	1.978	1.968
19	2.308	2.234	2.155	2.106	2.071	2.026	1.980	1.955	1.940	1.930
20	2.278	2.203	2.124	2.074	2.039	1.994	1.946	1.922	1.907	1.896
21	2.250	2.176	2.096	2.045	2.010	1.965	1.917	1.891	1.876	1.866
22	2.226	2.151	2.071	2.020	1.984	1.938	1.889	1.864	1.849	1.838
23	2.204	2.128	2.048	1.996	1.961	1.914	1.865	1.839	1.823	1.813
24	2.183	2.108	2.027	1.975	1.939	1.892	1.842	1.816	1.800	1.790
25	2.165	2.089	2.007	1.955	1.919	1.872	1.822	1.796	1.779	1.768
26	2.148	2.072	1.990	1.938	1.901	1.853	1.803	1.776	1.760	1.749
27	2.132	2.056	1.974	1.921	1.884	1.836	1.785	1.758	1.742	1.731
28	2.118	2.041	1.959	1.906	1.869	1.820	1.769	1.742	1.725	1.714
29	2.104	2.027	1.945	1.891	1.854	1.806	1.753	1.726	1.710	1.698
30	2.092	2.015	1.932	1.878	1.841	1.792	1.740	1.712	1.695	1.684

(continued)

Table A4(2) (continued)

d_2	d_1									
	12	15	20	25	30	40	60	80	100	120
40	2.003	1.924	1.839	1.783	1.744	1.693	1.637	1.608	1.589	1.577
50	1.952	1.871	1.784	1.727	1.649	1.634	1.576	1.544	1.525	1.511
60	1.917	1.836	1.748	1.690	1.649	1.594	1.534	1.502	1.481	1.467
70	1.893	1.812	1.722	1.664	1.622	1.566	1.505	1.471	1.450	1.435
80	1.875	1.793	1.703	1.644	1.602	1.545	1.482	1.448	1.426	1.411
90	1.861	1.779	1.688	1.629	1.586	1.528	1.465	1.429	1.407	1.391
100	1.850	1.768	1.676	1.616	1.573	1.515	1.450	1.415	1.392	1.376

d_1 degrees of freedom, numerator; d_2 degrees of freedom, denominator

Source Calculated using distribution calculator (dostat.stat.sc.edu/prototype/calculators/index.php3?dist=F; 12 October 2009)

Table A4(3) *F* Distribution (*F* values for $\alpha = 0.01$)

d_2 \ d_1	1	2	3	4	5	6	7	8	9	10
1	4052.2	4999.5	5403.3	5624.6	5763.7	5859.0	5928.3	5981.1	6022.5	6055.9
2	98.502	99.000	99.166	99.249	99.300	99.333	99.356	99.374	99.388	99.399
3	34.116	30.816	29.457	28.710	28.237	27.911	27.672	27.489	27.345	27.229
4	21.198	18.000	16.694	15.977	15.522	15.207	14.976	14.799	14.659	14.546
5	16.258	13.274	12.060	11.392	10.967	10.672	10.456	10.289	10.158	10.051
6	13.745	10.925	9.780	9.148	8.746	8.466	8.260	8.102	7.976	7.874
7	12.246	9.547	8.451	7.847	7.460	7.191	6.993	6.840	6.719	6.620
8	11.259	8.649	7.591	7.006	6.632	6.371	6.178	6.029	5.911	5.814
9	10.561	8.022	6.992	6.422	6.057	5.802	5.613	5.467	5.351	5.257
10	10.044	7.559	6.552	5.994	5.636	5.386	5.200	5.057	4.942	4.849
11	9.646	7.206	6.217	5.668	5.316	5.069	4.886	4.744	4.632	4.539
12	9.330	6.927	5.953	5.412	5.064	4.821	4.640	4.499	4.388	4.296
13	9.074	6.701	5.739	5.205	4.862	4.620	4.441	4.302	4.191	4.100
14	8.862	6.515	5.564	5.035	4.695	4.456	4.278	4.140	4.030	3.939
15	8.683	6.359	5.417	4.893	4.556	4.318	4.142	4.004	3.895	3.805
16	8.531	6.226	5.292	4.773	4.437	4.202	4.026	3.890	3.780	3.691
17	8.400	6.112	5.185	4.669	4.336	4.102	3.927	3.791	3.682	3.593
18	8.285	6.013	5.092	4.579	4.248	4.015	3.841	3.705	3.597	3.508
19	8.185	5.926	5.010	4.500	4.171	3.939	3.765	3.631	3.523	3.434
20	8.096	5.849	4.938	4.431	4.103	3.871	3.699	3.564	3.457	3.368
21	8.017	5.780	4.874	4.369	4.042	3.812	3.640	3.506	3.398	3.310
22	7.945	5.719	4.817	4.313	3.988	3.758	3.587	3.453	3.346	3.258
23	7.881	5.664	4.765	4.264	3.939	3.710	3.539	3.406	3.299	3.211

(continued)

Table A4(3) (continued)

d_2	d_1									
	1	2	3	4	5	6	7	8	9	10
24	7.823	5.614	4.718	4.218	3.895	3.667	3.496	3.363	3.256	3.168
25	7.770	5.568	4.675	4.177	3.855	3.627	3.457	3.324	3.217	3.129
26	7.721	5.526	4.637	4.140	3.818	3.591	3.421	3.288	3.182	3.094
27	7.677	5.488	4.601	4.106	3.785	3.558	3.388	3.256	3.149	3.062
28	7.636	5.453	4.568	4.074	3.754	3.528	3.358	3.226	3.120	3.032
29	7.598	5.420	4.538	4.045	3.725	3.499	3.330	3.198	3.092	3.005
30	7.562	5.390	4.510	4.018	3.699	3.473	3.305	3.173	3.067	2.979
40	7.314	5.179	4.313	3.828	3.514	3.291	3.124	2.993	2.888	2.801
50	7.171	5.057	4.199	3.720	3.408	3.186	3.020	2.890	2.785	2.698
60	7.077	4.977	4.126	3.649	3.339	3.119	2.953	2.823	2.718	2.632
70	7.011	4.922	4.074	3.600	3.291	3.071	2.906	2.777	2.672	2.585
80	6.963	4.881	4.036	3.563	3.255	3.036	2.871	2.742	2.637	2.551
90	6.925	4.849	4.007	3.535	3.228	3.009	2.845	2.715	2.611	2.524
100	6.895	4.824	3.984	3.513	3.206	2.988	2.823	2.694	2.590	2.503

d_2	d_1									
	12	15	20	25	30	40	60	80	100	120
1	6106.4	6157.3	6208.7	6239.8	6260.6	6286.8	6313.0	6326.2	6334.1	6339.4
2	99.416	99.432	99.449	99.460	99.466	99.474	99.482	99.490	99.490	99.491
3	27.052	26.872	26.690	26.580	26.504	26.411	26.316	26.270	26.240	26.221
4	14.374	14.198	14.020	13.910	13.838	13.745	13.652	13.610	13.580	13.558
5	9.888	9.722	9.553	9.449	9.379	9.291	9.202	9.157	9.130	9.111
6	7.718	7.559	7.396	7.296	7.229	7.143	7.057	7.013	6.987	6.969
7	6.469	6.314	6.155	6.058	5.992	5.908	5.824	5.781	5.755	5.737

(continued)

Table A4(3) (continued)

d_2	d_1									
	12	15	20	25	30	40	60	80	100	120
8	5.667	5.515	5.359	5.263	5.198	5.112	5.032	4.989	4.963	4.946
9	5.111	4.962	4.808	4.713	4.649	4.567	4.483	4.441	4.415	4.398
10	4.706	4.558	4.405	4.311	4.247	4.165	4.082	4.039	4.014	3.996
11	4.397	4.251	4.099	4.005	3.941	3.860	3.776	3.734	3.708	3.690
12	4.155	4.010	3.858	3.765	3.701	3.619	3.535	3.493	3.467	3.449
13	3.960	3.815	3.665	3.571	3.507	3.425	3.341	3.298	3.272	3.255
14	3.800	3.656	3.505	3.412	3.348	3.266	3.181	3.138	3.112	3.094
15	3.666	3.522	3.372	3.278	3.214	3.132	3.047	3.004	2.977	2.959
16	3.553	3.409	3.259	3.165	3.101	3.018	2.933	2.889	2.863	2.845
17	3.455	3.312	3.162	3.068	3.003	2.920	2.835	2.791	2.764	2.746
18	3.371	3.227	3.077	2.983	2.919	2.835	2.749	2.705	2.678	2.660
19	3.297	3.153	3.003	2.909	2.844	2.761	2.674	2.630	2.602	2.584
20	3.231	3.088	2.938	2.843	2.779	2.695	2.608	2.563	2.535	2.517
21	3.173	3.030	2.880	2.785	2.720	2.636	2.548	2.503	2.475	2.457
22	3.121	2.978	2.827	2.733	2.667	2.583	2.495	2.450	2.422	2.403
23	3.074	2.931	2.781	2.686	2.620	2.535	2.447	2.401	2.373	2.354
24	3.032	2.889	2.738	2.643	2.577	2.492	2.403	2.357	2.329	2.310
25	2.993	2.850	2.699	2.604	2.538	2.453	2.364	2.317	2.289	2.270
26	2.958	2.815	2.664	2.569	2.503	2.417	2.327	2.281	2.252	2.233
27	2.926	2.783	2.632	2.536	2.470	2.384	2.294	2.247	2.218	2.198
28	2.896	2.753	2.602	2.506	2.440	2.354	2.263	2.216	2.187	2.167
29	2.868	2.726	2.574	2.478	2.412	2.325	2.234	2.187	2.158	2.138
30	2.843	2.700	2.549	2.453	2.386	2.299	2.208	2.160	2.131	2.111

(continued)

Table A4(3) (continued)

d_2	d_1									
	12	15	20	25	30	40	60	80	100	120
40	2.665	2.522	2.369	2.271	2.203	2.114	2.019	1.969	1.938	1.917
50	2.562	2.419	2.265	2.167	2.098	2.007	1.909	1.857	1.825	1.803
60	2.496	2.352	2.198	2.098	2.028	1.936	1.836	1.783	1.749	1.726
70	2.450	2.306	2.150	2.050	1.980	1.886	1.785	1.730	1.695	1.672
80	2.415	2.271	2.115	2.015	1.944	1.849	1.746	1.690	1.655	1.630
90	2.389	2.244	2.088	1.987	1.916	1.820	1.716	1.659	1.623	1.598
100	2.368	2.223	2.067	1.965	1.893	1.797	1.692	1.634	1.598	1.572

d_1 degrees of freedom, numerator; d_2 degrees of freedom, denominator
Source Calculated using distribution calculator (dostat.stat.sc.edu/prototype/calculators/index.php3?dist=F; 12 October 2009)

Table A5 Wilcoxon rank-sum table

n_1	n_2	Lower tail probability						Upper tail probability					
		0.01	0.01	0.03	0.05	0.10	0.20	0.20	0.10	0.05	0.025	0.01	0.005
4	4			10	11	13	14	22	23	25	26		
	5		10	11	12	14	15	25	26	28	29	30	
	6	10	11	12	13	15	17	27	29	31	32	33	34
	7	10	11	13	14	16	18	30	32	34	35	37	38
	8	11	12	14	15	17	20	32	35	37	38	40	41
	9	11	13	14	16	19	21	35	37	40	42	43	45
	10	12	13	15	17	20	23	37	40	43	45	47	48
	11	12	14	16	18	21	24	40	43	46	48	50	52
	12	13	15	17	19	22	26	42	46	49	51	53	55
5	5	15	16	17	19	20	22	33	35	36	38	39	40
	6	16	17	18	20	22	24	36	38	40	42	43	44
	7	16	18	20	21	23	26	39	42	44	45	47	49
	8	17	19	21	23	25	28	42	45	47	49	51	53
	9	18	20	22	24	27	30	45	48	51	53	55	57
	10	19	21	23	26	28	32	48	52	54	57	59	61
	11	20	22	24	27	30	34	51	55	58	61	63	65
	12	21	23	26	28	32	36	54	58	62	64	67	69
6	6	23	24	26	28	30	33	45	48	50	52	54	55
	7	24	25	27	29	32	35	49	52	55	57	59	60
	8	25	27	29	31	34	37	53	56	59	61	63	65
	9	26	28	31	33	36	40	56	60	63	65	68	70
	10	27	29	32	35	38	42	60	64	67	70	73	75
	11	28	30	34	37	40	44	64	68	71	71	78	80
	12	30	32	35	38	42	47	67	72	76	79	82	84

(continued)

Table A5 (continued)

n_1	n_2	Lower tail probability						Upper tail probability					
		0.01	0.01	0.03	0.05	0.10	0.20	0.20	0.10	0.05	0.025	0.01	0.005
7	7	32	34	36	39	41	45	60	64	66	69	71	73
	8	34	35	38	41	44	48	64	68	71	74	77	78
	9	35	37	40	43	46	50	69	73	76	79	82	84
	10	37	39	42	45	49	53	73	77	81	84	87	89
	11	38	40	44	47	51	56	77	82	86	89	93	95
	12	40	42	46	49	54	59	81	86	91	94	98	100
8	8	43	45	49	51	55	59	77	81	85	87	91	93
	9	45	47	51	54	58	62	82	86	90	93	97	99
	10	47	49	53	56	60	65	87	92	96	99	103	105
	11	49	51	55	59	63	69	91	97	101	105	109	111
	12	51	53	58	62	66	72	96	102	106	110	115	117
9	9	56	59	62	66	70	75	96	101	105	109	112	115
	10	58	61	65	69	73	78	102	107	111	115	119	122
	11	61	63	68	72	76	82	107	113	117	121	126	128
	12	63	66	71	75	80	86	112	118	123	127	132	135
10	10	71	74	78	82	87	93	117	123	128	132	136	139
	11	73	77	81	86	91	97	123	129	134	139	143	147
	12	76	79	84	89	94	101	129	136	141	146	151	154
11	11	87	91	96	100	106	112	141	147	153	157	162	166
	12	90	94	99	104	110	117	147	154	160	165	170	174
12	12	105	109	115	120	127	134	166	173	180	185	191	195

Source www.stat.aukland.ac.nz/~wildEnc/Ch10.wilcoxon.pdf (30 September 2009)

Table A6 Spearman rank correlation

Use t test for test of hypothesis related to Spearman Rank Correlation. This is simpler and is valid. Percentage Points of the t distribution are provided in Table A2.

Table A7 Sample size based on CV. (α = 5 %; precision = 5 %)

N	CV									
	0.10	0.20	0.30	0.40	0.50	0.60	0.70	0.80	0.90	1.00
100	13	38	58	71	79	85	88	91	93	94
200	14	47	82	110	132	147	158	166	172	177
300	15	51	95	135	168	195	215	230	242	251
400	15	53	103	152	196	232	261	284	303	317
500	15	55	108	165	217	263	300	331	357	377
600	15	56	112	174	234	288	334	373	405	432
700	15	57	115	182	248	309	363	409	448	481
800	15	57	118	188	260	327	388	441	487	526
900	15	58	120	193	269	343	410	470	522	568
1000	15	58	121	197	278	356	430	496	555	606
1200	15	58	124	204	291	379	463	540	611	674
1400	15	59	126	209	301	397	490	578	659	733
1600	15	59	127	213	310	411	512	609	700	784
1800	15	59	128	216	317	423	531	636	736	829
2000	15	60	129	219	322	433	547	659	767	869
2250	15	60	130	222	328	444	564	684	801	913
2500	15	60	131	224	333	453	579	706	831	952
2750	15	60	132	226	337	461	591	724	857	986
3000	15	60	132	227	341	467	602	741	880	1016
3500	15	60	133	230	346	478	620	768	918	1068
4000	15	61	134	232	350	486	634	789	949	1110
4500	15	61	134	233	354	493	645	807	875	1145
5000	15	61	135	234	357	498	654	822	997	1175

(continued)

Table A7 (continued)

N	CV									
	0.10	0.20	0.30	0.40	0.50	0.60	0.70	0.80	0.90	1.00
5500	15	61	135	235	359	503	662	834	1015	1201
6000	15	61	135	236	361	506	669	845	1031	1223
6500	15	61	135	237	363	510	675	854	1045	1243
7000	15	61	136	238	364	513	680	862	1057	1260
7500	15	61	136	238	365	515	684	869	1068	1275
8000	15	61	136	239	367	517	688	876	1077	1289
8500	15	61	136	239	368	519	692	881	1086	1301
9000	15	61	136	239	368	521	695	887	1093	1313
9500	15	61	136	240	369	523	698	891	1100	1323
10,000	15	61	136	240	370	524	700	895	1107	1332
15,000	15	61	137	242	375	534	717	923	1149	1394
20,000	15	61	137	243	377	538	726	937	1172	1427
25,000	15	61	138	243	378	541	731	946	1186	1448

CV coefficient of variation
Source Author

Table A8 Sample size based on CV. ($\alpha = 5$ %; precision = 3 %)

N	CV									
	0.10	0.20	0.30	0.40	0.50	0.60	0.70	0.80	0.90	1.00
100	30	63	79	87	91	94	95	96	97	98
200	35	92	132	155	168	177	183	186	189	191
300	37	109	168	208	234	251	262	270	276	280
400	39	120	196	252	291	317	336	349	359	366
500	39	127	217	289	340	377	404	423	437	448
600	40	133	234	319	384	432	466	492	511	526
700	40	137	248	346	423	481	524	557	582	601
800	41	141	260	368	457	526	579	619	650	674
900	41	144	269	388	488	568	629	677	714	743
1000	41	146	278	406	516	606	677	732	776	810
1200	41	149	291	435	565	674	763	834	891	937
1400	41	152	301	459	606	733	839	926	996	1054
1600	42	154	310	479	640	784	907	1009	1094	1164
1800	42	156	317	495	670	829	967	1085	1184	1266
2000	42	157	322	509	696	869	1022	1155	1267	1362
2250	42	159	328	524	724	913	1084	1234	1363	1473
2500	42	160	333	536	748	952	1139	1305	1451	1577
2750	42	161	337	547	769	986	1188	1370	1532	1672
3000	42	162	341	556	787	1016	1232	1430	1606	1762
3500	42	163	346	571	818	1068	1309	1534	1739	1923
4000	42	164	350	583	842	1110	1373	1623	1854	2065
4500	42	164	354	593	863	1145	1428	1700	1955	2191
5000	42	165	357	601	879	1175	1475	1767	2044	2303

(continued)

Table A8 (continued)

N	CV									
	0.10	0.20	0.30	0.40	0.50	0.60	0.70	0.80	0.90	1.00
5500	42	166	359	608	894	1201	1515	1825	2123	2403
6000	42	166	361	613	906	1223	1551	1877	2193	2494
6500	42	166	363	618	917	1243	1582	1923	2257	2576
7000	42	167	364	622	926	1260	1610	1965	2314	2652
7500	42	167	365	626	934	1275	1635	2002	2367	2720
8000	42	167	367	629	942	1289	1658	2036	2414	2783
8500	42	167	368	632	948	1301	1679	2067	2458	2842
9000	42	168	368	635	954	1313	1697	2096	2498	2895
9500	42	168	369	637	959	1323	1714	2122	2535	2945
10,000	43	168	370	639	964	1332	1730	2146	2569	2992
15,000	43	169	375	653	996	1394	1836	2311	2810	3323
20,000	43	169	377	660	1013	1427	1894	2404	2948	3518
25,000	43	170	378	665	1023	1448	1930	2463	3037	3646

CV coefficient of variation
Source Author

Table A9 Sample size based on CV. ($\alpha = 5\%$; precision $= 2\%$)

N	CV									
	0.10	0.20	0.30	0.40	0.50	0.60	0.70	0.80	0.90	1.00
100	49	79	90	94	96	97	98	98	99	99
200	65	132	162	177	185	189	192	194	195	196
300	73	168	223	251	267	276	282	286	289	291
400	77	196	273	317	343	359	369	376	380	384
500	81	217	317	377	414	437	452	462	470	475
600	83	234	354	432	480	511	532	547	557	565
700	84	248	387	481	542	582	609	628	642	652
800	86	260	415	526	600	650	684	708	725	738
900	87	269	441	568	655	714	756	785	807	823
1000	88	278	464	606	706	776	825	860	886	906
1200	89	291	502	674	800	891	956	1004	1040	1067
1400	90	301	534	733	884	996	1079	1140	1186	1222
1600	91	310	561	784	960	1094	1194	1270	1327	1372
1800	91	317	584	829	1029	1184	1302	1392	1462	1516
2000	92	322	604	869	1091	1267	1404	1509	1591	1655
2250	92	328	624	913	1162	1363	1522	1647	1745	1823
2500	92	333	642	952	1225	1451	1633	1777	1892	1984
2750	93	337	658	986	1282	1532	1736	1900	2032	2138
3000	93	341	671	1016	1334	1606	1832	2016	2165	2286
3500	93	346	693	1068	1424	1739	2007	2230	2414	2565
4000	94	350	711	1110	1500	1854	2162	2423	2642	2824
4500	94	354	725	1145	1566	1955	2300	2598	2851	3064
5000	94	357	737	1175	1622	2044	2424	2757	3044	3288

(continued)

Table A9 (continued)

N	CV									
	0.10	0.20	0.30	0.40	0.50	0.60	0.70	0.80	0.90	1.00
5500	94	359	747	1201	1671	2123	2536	2903	3222	3497
6000	95	361	756	1223	1715	2193	2637	3036	3387	3693
6500	95	363	763	1243	1753	2257	2730	3159	3541	3876
7000	95	364	769	1260	1788	2314	2814	3273	3685	4049
7500	95	365	775	1275	1819	2367	2892	3378	3819	4211
8000	95	367	780	1289	1847	2414	2963	3476	3944	4364
8500	95	368	785	1301	1872	2458	3029	3567	4062	4509
9000	95	368	789	1313	1895	2498	3090	3652	4173	4646
9500	95	369	792	1323	1917	2535	3147	3732	4277	4776
10,000	95	370	796	1332	1936	2569	3200	3807	4375	4899
15,000	95	375	817	1394	2070	2810	3582	4360	5123	5855
20,000	96	377	829	1427	2144	2948	3810	4702	5601	6488
25,000	96	378	835	1448	2191	3037	3960	4934	5933	6939

CV coefficient of variation
Source Author

Table A10 Sample size based on proportion ($\alpha = 5\%$; precision $= 5\%$)

N	P: 0.05	0.10	0.15	0.20	0.25	0.30	0.35	0.40	0.50
	PQ: 0.05	0.09	0.13	0.16	0.19	0.21	0.23	0.24	0.25
100	43	58	67	71	74	76	78	79	79
200	56	82	100	110	119	123	128	130	132
300	61	95	120	135	148	155	162	165	168
400	64	103	133	152	169	179	188	192	196
500	67	108	143	165	184	196	207	212	217
600	68	112	150	174	196	210	222	228	234
700	69	115	155	182	206	221	235	242	248
800	70	118	160	188	214	230	245	252	260
900	71	120	163	193	220	238	254	262	269
1000	71	121	167	197	226	244	261	269	278
1200	72	124	171	204	235	254	273	282	291
1400	73	126	175	209	242	262	282	292	301
1600	73	127	178	213	247	269	289	300	310
1800	74	128	180	216	251	274	295	306	317
2000	74	129	182	219	255	278	300	311	322
2250	74	130	183	222	258	282	305	317	328
2500	75	131	185	224	261	286	310	321	333
2750	75	132	186	226	264	289	313	325	337
3500	75	133	189	230	269	295	321	334	346
4000	75	134	190	232	272	299	325	338	350
4500	76	134	191	233	274	301	328	341	354
5000	76	135	192	234	276	303	330	343	357
5500	76	135	193	235	277	305	332	346	359

(continued)

Table A10 (continued)

N	P: 0.05 PQ: 0.05	0.10 0.09	0.15 0.13	0.20 0.16	0.25 0.19	0.30 0.21	0.35 0.23	0.40 0.24	0.50 0.25
6000	76	135	193	236	278	306	334	347	361
6500	76	135	194	237	279	307	335	349	363
7000	76	136	194	238	280	308	336	350	364
7500	76	136	195	238	281	309	338	352	365
8000	76	136	195	239	282	310	338	353	367
8500	76	136	195	239	282	311	339	353	368
9000	76	136	195	239	283	312	340	354	368
9500	76	136	196	240	283	312	341	355	369
10,000	76	136	196	240	284	313	341	356	370
15,000	76	137	197	242	286	316	345	360	375
20,000	77	137	198	243	288	318	347	362	377
25,000	77	138	198	243	289	319	349	363	378

Source Author

Table A11 Sample size based on proportion (α = 5 %; precision = 3 %)

N	P: 0.05	0.10	0.15	0.20	0.25	0.30	0.35	0.40	0.50
	PQ: 0.05	0.09	0.13	0.16	0.19	0.21	0.23	0.24	0.25
100	68	79	85	87	89	90	91	91	91
200	103	132	147	155	160	164	166	167	168
300	125	168	195	208	219	225	230	232	234
400	139	196	232	252	268	277	284	288	291
500	150	217	263	289	309	321	331	336	340
600	157	234	288	319	345	359	372	378	384
700	164	248	310	346	376	393	409	416	423
800	168	260	328	368	403	423	441	449	457
900	173	269	343	388	427	449	470	479	488
1000	176	278	357	406	448	473	495	506	516
1200	181	291	379	435	484	513	540	553	565
1400	185	301	397	459	514	546	577	592	606
1600	188	310	412	479	538	575	608	625	640
1800	191	317	424	495	559	598	635	653	670
2000	193	322	434	509	577	619	659	677	696
2250	195	328	445	524	596	641	684	704	724
2500	197	333	454	536	612	660	705	727	748
2750	198	337	462	547	626	676	723	746	769
3000	199	341	468	556	638	690	740	764	787
3500	201	346	479	571	658	714	767	792	818
4000	203	350	487	583	674	732	788	816	842
4500	204	354	494	593	687	747	806	834	863
5000	205	357	499	601	698	760	821	850	879

(continued)

Table A11 (continued)

N	P: 0.05	0.10	0.15	0.20	0.25	0.30	0.35	0.40	0.50
	PQ: 0.05	0.09	0.13	0.16	0.19	0.21	0.23	0.24	0.25
5500	205	359	504	608	707	771	833	864	894
6000	206	361	508	613	714	780	844	875	906
6500	207	363	511	618	721	788	853	885	917
7000	207	364	514	622	727	795	861	894	926
7500	208	365	517	626	732	801	868	901	934
8000	208	367	519	629	736	806	874	908	942
8500	208	368	521	632	740	811	880	914	948
9000	208	368	523	635	744	815	885	920	954
9500	209	369	524	637	747	819	890	925	959
10,000	209	370	526	639	750	823	894	929	964
15,000	210	375	535	653	769	846	921	959	996
20,000	211	377	540	660	779	858	936	975	1013
25,000	212	378	543	665	786	865	945	984	1023

Source Author

Table A12 Sample size based on proportion ($\alpha = 5\%$; precision = 2 %)

N	P: 0.05	0.10	0.15	0.20	0.25	0.30	0.35	0.40	0.50
	PQ: 0.05	0.09	0.13	0.16	0.19	0.21	0.23	0.24	0.25
100	83	90	93	94	95	95	96	96	96
200	141	162	172	177	180	182	183	184	185
300	185	223	242	251	258	261	264	265	267
400	218	273	303	317	328	334	339	341	343
500	245	317	357	377	392	401	408	411	414
600	267	354	405	432	452	462	472	476	480
700	285	387	449	481	506	520	532	537	542
800	300	415	488	526	556	573	587	594	600
900	313	441	523	568	603	622	639	647	655
1000	324	464	555	606	646	669	688	697	706
1200	343	502	612	674	724	752	778	789	800
1400	358	534	660	733	792	826	857	871	884
1600	369	561	701	784	853	892	928	944	960
1800	379	584	737	829	906	951	992	1011	1029
2000	387	604	769	869	954	1004	1050	1071	1091
2250	396	624	803	913	1008	1064	1115	1139	1162
2500	403	642	833	952	1055	1116	1173	1199	1225
2750	409	658	859	986	1097	1164	1225	1254	1282
3000	414	671	882	1016	1135	1206	1272	1303	1334
3500	422	693	920	1068	1199	1280	1354	1390	1424
4000	429	711	952	1110	1253	1341	1423	1462	1500
4500	434	725	977	1145	1298	1393	1482	1524	1566
5000	438	737	999	1175	1337	1437	1532	1578	1622

(continued)

Table A12 (continued)

N	P: 0.05	0.10	0.15	0.20	0.25	0.30	0.35	0.40	0.50
	PQ: 0.05	0.09	0.13	0.16	0.19	0.21	0.23	0.24	0.25
5500	442	747	1018	1201	1370	1476	1576	1624	1671
6000	445	756	1033	1223	1399	1509	1615	1665	1715
6500	447	763	1047	1243	1425	1539	1649	1702	1753
7000	449	769	1060	1260	1447	1566	1679	1734	1788
7500	451	775	1070	1275	1468	1589	1706	1763	1819
8000	453	780	1080	1289	1486	1611	1731	1789	1847
8500	455	785	1089	1301	1502	1630	1753	1813	1872
9000	456	789	1096	1313	1517	1648	1774	1835	1895
9500	457	792	1103	1323	1531	1664	1792	1855	1917
10,000	458	796	1110	1332	1543	1678	1809	1873	1936
15,000	465	817	1153	1394	1627	1778	1925	1998	2070
20,000	469	829	1175	1427	1672	1832	1989	2067	2144
25,000	471	835	1189	1448	1701	1866	2030	2110	2191

Source Author

Table A13 Random numbers

	1	2	3	4	5	6	7	8	9	10
First thousand										
1	1956	1643	0815	7760	1609	636	1830	0502	0086	8638
2	7117	8231	0374	2021	7491	8695	5447	4756	6276	4906
3	6696	4659	1906	5260	4241	5548	0332	6300	5091	3008
4	8621	2412	8982	9932	1753	4922	8562	8802	0096	2344
5	6290	4221	0100	0028	1221	9900	9648	2882	9380	4097
6	3241	1112	1494	3831	7988	7234	4657	4670	0637	3321
7	2335	1461	4926	1097	5165	2819	3234	6963	9854	5361
8	5778	8424	9231	4350	3255	3284	1240	3955	1873	2272
9	0542	4534	5178	2171	2063	3622	9200	3212	8713	9002
10	4196	6120	0702	8744	6626	1747	4915	0304	0337	8422
11	2280	7671	5195	8171	8936	9163	2460	8579	1760	7585
12	8510	7633	2407	3691	5507	2049	3302	0820	7454	0484
13	4662	8834	5469	3861	2157	4675	0147	1064	6919	2693
14	2053	2096	1614	7011	5451	3962	9108	9703	7985	3060
15	0317	9790	0835	3289	6104	6240	3734	9595	8002	2875
16	1188	8204	1174	9817	2228	8900	0408	9127	8838	4845
17	1973	8422	3459	9025	5979	1943	6718	4502	2788	4544
18	6791	9352	2940	5178	2592	9281	5952	1613	3278	5283
19	6847	2239	5470	4025	5055	9589	8932	2450	8744	3121
20	2984	9069	6233	5147	6800	4056	4761	4950	9955	1446
21	5604	8703	5777	2194	7405	2868	2282	9211	6090	8455
22	4728	5550	7327	7565	9724	0836	6903	2236	5523	3372
23	0033	3279	9788	9896	4594	5279	2696	4738	7410	4715

(continued)

Table A13 (continued)

	1	2	3	4	5	6	7	8	9	10
24	0278	9403	3609	8102	2986	0717	1068	8669	3048	3484
25	9392	0177	9215	4003	0366	6968	0901	5564	6538	9371
Second thousand										
26	8628	6011	3344	5198	3389	1497	9596	9364	8933	9045
27	3554	3842	1412	9306	4361	0652	6400	1706	5317	1826
28	3875	4976	1927	2772	8412	6958	9235	4766	3363	7006
29	7571	3514	2647	5736	7329	5622	4847	9306	9145	8250
30	3843	1038	6255	3304	3415	9499	2466	1791	1777	7428
31	3293	3809	9589	2286	5121	8082	8993	1156	9560	4735
32	9000	5822	6203	9170	1918	8940	4673	7980	2156	3401
33	8796	0915	0754	5195	2010	5822	2784	5734	7408	7864
34	7096	5630	1083	6150	4119	9734	1427	6083	9256	0503
35	9311	2398	3862	4542	0145	1750	3009	7496	9808	3594
36	1317	8493	3923	1096	0164	8706	4121	8036	8972	5630
37	6808	6119	4741	5084	7609	4900	1687	5200	829	5962
38	8044	2702	9894	7191	4779	4190	7524	4193	0258	9846
39	5663	1556	3093	8394	7550	6207	2031	7016	8241	3870
40	7877	9758	8704	6874	4863	1130	4744	1710	8989	6738
41	3247	1669	4741	9251	5066	4761	2740	2116	3522	5885
42	6859	0997	6910	4390	1799	2635	6978	8421	2394	4906
43	7916	4540	1722	0561	7100	1354	0927	0128	5022	4920
44	9650	1502	0593	5285	3729	9902	9844	1292	1804	3076
45	9278	7116	9321	5895	6717	1311	3136	9845	4873	3584
46	2294	9867	4581	6263	1342	3427	7505	7225	4170	9466

(continued)

Table A13 (continued)

	1	2	3	4	5	6	7	8	9	10
47	6099	6303	4221	7909	2428	3652	2105	8202	3769	8088
48	2356	9240	2985	5705	6380	511	0000	1590	0821	8834
49	4920	6759	9940	8731	0936	0375	7175	8988	2479	4146
50	4683	6471	0753	4318	5721	7933	5962	8368	9794	0406

Third thousand

	1	2	3	4	5	6	7	8	9	10
51	7642	5917	7854	5728	7711	5918	1170	8467	4624	4561
52	8865	3052	1712	2228	3000	1036	4201	8752	9755	3000
53	0542	3763	0330	6118	6878	3027	8559	8184	9311	3625
54	0519	3194	0161	8370	2541	2901	4659	8466	9051	2241
55	5746	5159	4842	0584	3031	7943	0768	0118	8402	3937
56	3175	0910	0638	3080	2286	3326	7493	1011	4068	0998
57	2700	9454	6371	3785	7909	6335	8185	0698	9912	0506
58	9787	6415	1474	9465	1076	9696	5396	2163	1993	3106
59	8830	2386	1623	2207	4300	2176	5040	9576	5194	6909
60	6052	2259	1711	8481	5533	0909	9825	9066	8427	5383
61	6599	6724	9824	6602	7203	6992	4535	7756	4062	1276
62	7054	5386	5296	4472	7805	2704	3484	7576	5312	1835
63	8780	1592	4158	4665	9582	3630	3352	4916	2736	3114
64	8891	5088	3815	7169	0008	9434	6825	5026	2487	2307
65	4100	6430	2415	0594	8872	5686	9666	1167	2736	1167
66	5301	8089	5546	9361	1590	7839	3825	1501	4632	5500
67	4732	4481	7074	5356	2159	4963	4495	2570	9350	9827
68	9129	5531	8236	1363	9535	1702	4068	7607	4852	2683
69	5939	7904	2810	2896	8991	1208	1848	3753	688	2368

(continued)

Table A13 (continued)

	1	2	3	4	5	6	7	8	9	10
70	9179	0604	3968	9834	2455	4680	8194	3044	8588	9147
71	9996	3259	1124	9793	7926	6741	0423	4397	7741	2474
72	5163	5296	5109	3651	5581	1834	7734	7096	1992	2392
73	5152	6806	9511	6423	3235	5301	6384	7878	1078	2968
74	8614	8873	8014	7244	2067	1012	4450	4736	3163	2616
75	9097	6442	1015	4456	8225	1801	3586	0422	9975	9222

Fourth thousand

	1	2	3	4	5	6	7	8	9	10
76	9377	9461	0936	3524	2059	2951	5384	4295	7740	2722
77	4841	1651	7587	6935	7776	4851	2479	1912	2277	7821
78	6388	5176	9695	2956	3606	1941	8147	5561	1193	1830
79	9030	4298	4918	9767	1577	5989	4882	3974	4532	3061
80	0646	8849	0567	7995	1834	0631	9861	9208	9808	3231
81	4612	3230	4814	3606	8980	3834	8291	8020	1409	1219
82	9458	9167	8882	3554	5371	3836	2719	1344	6580	1533
83	7145	1762	8917	8849	5301	7352	5451	7193	3340	9046
84	8431	5888	8939	0783	8155	2144	4294	3726	3500	4832
85	3880	7043	3835	6569	7547	4907	4419	1289	0913	0403
86	6297	5920	8665	6762	4990	1638	3138	2601	6934	4884
87	3488	7614	4111	0981	1476	0604	3296	5451	9781	4569
88	3983	0841	4024	6667	2005	4733	3706	6335	3757	0847
89	9270	9827	9260	1225	5799	8345	2081	3539	3050	6440
90	2552	4073	9618	4848	0620	8704	7173	6493	5494	4272
91	8567	2219	8651	8545	0723	1255	0506	4300	6699	5878
92	6467	2491	1814	4411	0155	9787	3146	9624	2021	3831

(continued)

Table A13 (continued)

	1	2	3	4	5	6	7	8	9	10
93	6602	5726	0797	9523	2791	9657	3118	4979	4178	7006
94	7174	5317	9494	9817	3943	5686	9907	8414	4730	8016
95	5450	7174	7016	5311	4669	7076	6376	3288	3195	6543
96	0798	9198	6863	1592	0106	3655	2558	7642	8150	4817
97	0454	3293	2391	2462	4844	7798	3710	8497	3474	2200
98	1217	5199	2915	9466	6797	3018	3781	1050	4778	6116
99	0786	7877	4097	1570	7327	2699	0058	8821	9136	4460
100	5816	2331	2700	7884	3569	7837	0330	6051	3787	0624

Table A14 Control chart factors

No. of Observation n	A	A_2	d_2	d_3	D_3	D_4
2	2.121	1.880	1.128	0.853	0	3.276
3	1.732	1.023	1.693	0.888	0	2.575
4	1.501	0.729	2.059	0.880	0	2.282
5	1.342	0.577	2.326	0.864	0	2.115
6	1.225	0.483	2.534	0.848	0	2.004
7	1.134	0.419	2.704	0.833	0.076	1.924
8	1.061	0.373	2.847	0.820	0.136	1.864
9	1.000	0.337	2.970	0.808	0.184	1.816
10	0.949	0.308	3.078	0.797	0.223	1.777
11	0.905	0.285	3.173	0.787	0.256	1.744
12	0.866	0.266	3.258	0.778	0.284	1.719
13	0.832	0.249	3.336	0.770	0.308	1.692
14	0.802	0.235	3.407	0.762	0.329	1.671
15	0.775	0.223	3.472	0.755	0.348	1.652
16	0.750	0.212	3.532	0.749	0.364	1.636
17	0.728	0.203	3.588	0.743	0.379	1.621
18	0.707	0.194	3.640	0.738	0.392	1.608
19	0.688	0.187	3.689	0.733	0.404	1.596
20	0.671	0.180	3.735	0.729	0.414	1.586
21	0.655	0.173	3.778	0.724	0.425	1.575
22	0.640	0.167	3.819	0.720	0.434	1.566
23	0.626	0.162	3.858	0.716	0.443	1.557
24	0.612	0.157	3.895	0.712	0.452	1.548
25	0.600	0.153	3.931	0.709	0.459	1.541
Over 25	$\frac{3}{\sqrt{n}}$					

Source Website www.ct-yankee.com/spc/factors.xls, 03 December 2009

Table A15 Sample size based on tolerance limits

p	α					
	0.10	0.05	0.025	0.01	0.005	
0.80	19	22	26	31	34	
0.81	20	24	28	33	36	
0.82	21	25	29	35	39	
0.83	22	27	31	37	41	
0.84	23	28	33	39	44	
0.85	25	30	35	42	47	
0.86	27	32	38	45	50	
0.87	28	33	41	49	54	
0.88	31	38	45	53	55	
0.89	34	42	49	58	65	
0.90	38	46	54	64	72	
0.91	42	51	60	71	80	
0.92	48	58	68	81	90	
0.93	55	66	78	93	103	
0.94	64	78	91	108	121	
0.95	77	93	110	130	146	
0.96	96	117	137	164	183	
0.97	129	157	184	219	245	
0.98	194	236	277	330	369	
0.99	388	473	555	662	740	

p proportion of the process units to be within the tolerance limits
α probability of p

Table A16 Some common symbols

Symbol	Pronounced as	Meaning
$\sum$	summation	Addition of several values
α	alpha	Significance level; type I error in hypothesis testing
β	beta	Parameter (coefficient of independent variable in regression); type II error
χ^2	Chi-square	Chi-square distribution statistic
π	pai	Its value is 3.1412857
e	–	Base of natural logarithm (2.71828)
e_i	–	Error term
∞	infinity	Indicating infinite value
$\int$	integration	Integrate a continuous variable
Δ	delta	Amount of change
M_0	–	Mode
M_d	–	Median
$\geq$	greater–equal	Example: $x \geq y \Rightarrow x$ is greater than or equal to y
$\leq$	less–equal	Example: $x \leq y \Rightarrow x$ is less than or equal to y
n	–	Sample size
$\bar{x}$	–	Sample mean
s	–	Sample standard deviation
t	tee	t distribution statistic
N	–	Population size
μ	miew	Population mean
σ	sigma	Population standard deviation

Index

© Springer Science+Business Media Singapore 2016
A.Q. Miah, *Applied Statistics for Social and Management Sciences*,
DOI 10.1007/978-981-10-0401-8

MIX

Papier | Fördert
gute Waldnutzung

FSC® C083411

Zeitfracht Medien GmbH
Ferdinand-Jühlke-Straße 7
99095 Erfurt, Deutschland
produktsicherheit@kolibri360.de